Highway Planning, Survey, and Design

Solved Practical Problems in Transportation Engineering

Volume I: Traffic and Pavement Engineering

Volume II: Highway Planning, Survey, and Design

Highway Planning, Survey, and Design

Ghazi G. Al-Khateeb

University of Sharjah, UAE and Jordan University of Science and Technology, Jordan

CRC Press
Taylor & Francis Group
Boca Raton London New York

CRC Press is an imprint of the
Taylor & Francis Group, an **informa** business

CRC Press
Taylor & Francis Group
6000 Broken Sound Parkway NW, Suite 300
Boca Raton, FL 33487-2742

© 2021 by Taylor & Francis Group, LLC

CRC Press is an imprint of Taylor & Francis Group, an Informa business

No claim to original U.S. Government works

Printed on acid-free paper

International Standard Book Number-13: 978-0-367-14986-4 (Hardback)
International Standard Book Number-13: 978-0-367-50012-2 (Paperback)

Visit the Taylor & Francis Web site at
http://www.taylorandfrancis.com

and the CRC Press Web site at
http://www.crcpress.com

This book is dedicated to my parents;

to the soul of my mother, and

my father, Gaseem;

to my beloved family;

my wife, Nuha; my sons, Khalid, Amr, and Ayham; and my daughter, Dana;

to

my sisters, Shareefah and Suhad;

and my brothers, Ja'fer, Mohammed, Faisal, Hashem, and Tareq;

and to

*all my teachers, supervisors, colleagues, friends, and students
in every place I studied, taught, or researched …*

*Without the science, knowledge, support, and encouragement I received
over the years and the tremendous time and persistence devoted
in writing this book, this huge work would not be possible*

Contents

Preface

This book, comprising of two volumes and encompassing a total of five parts, is the outcome of a great deal of effort and time spent over the years in studying and teaching at outstanding academic institutes like Jordan University of Science and Technology, Applied Science University, University of Illinois at Urbana-Champaign, the American University of Sharjah, and the University of Sharjah, and in conducting advanced scientific research at some of the highly recognized research centers worldwide like the Advanced Transportation Research and Engineering Laboratory (ATREL) at the University of Illinois in Urbana-Champaign, the Federal Highway Administration's Turner-Fairbank Highway Research Center (TFHRC), advanced research labs at Jordan University of Science and Technology and the Advanced Pavement Research Lab at the University of Sharjah.

The book implements a unique kind of approach and categorizes transportation engineering topics into five major key areas, as shown below:

- Volume I: Traffic and Pavement Engineering
  - Part I "Traffic Engineering" deals with the functional part of transportation systems and introduces engineering techniques, practices, and models that are applied to design traffic systems, control traffic flow and movement, and construct proper roads and highways to achieve safe and efficient movement of people and traffic on roadways.
  - Part II "Pavement Materials, Analysis, and Design" deals with both the structural and functional parts of transportation facilities and introduces engineering techniques and principles of the uses of high-quality and sustainable materials that are employed to design, maintain, and construct asphalt-surfaced road pavements and concrete rigid pavements. The ultimate goal of pavement engineering is to provide a pavement structure that is safe, durable, sustainable, and capable of carrying the predicted traffic loads under prevailing climatic conditions. Proper structural design of pavements is one that takes into consideration the mechanistic analysis of pavements for stresses and strains that can predict the performance of the pavement with time. This section fulfills this goal by presenting the subject in a unique manner.

- Volume II: Highway Planning, Survey, and Design
  - Part I "Urban Transportation Planning" presents a process that involves a multi-modal approach and comprehensive planning steps and models to design and evaluate a variety of alternatives for transportation systems and facilities, predict travel demand and future needs, and manage the facilities and services for the different modes of transportation to finally achieve a safe, efficient, and sustainable system for the movement of people and goods.
  - Part II "Highway Survey" presents the basic concepts and standard procedures necessary to make precise and accurate distance, angle, and level measurements for highway alignment, cross-sections, and earth quantities used in the design of highways.
  - Part III "Geometric Design of Highways" deals with engineering design techniques, standards, and models that control the three main elements of highway geometric design: horizontal and vertical alignments, profile, and cross-section to achieve the primary objectives of geometric design: safety, efficiency, and sustainability.

The book is designed to benefit students in engineering programs at academic institutes where courses in pavement engineering, highway engineering, transportation engineering, traffic engineering, urban transportation planning, and survey are offered. The book is intended to be used as a state-of-the-art textbook for engineering students at the undergraduate and graduate level as well as professionals and technologists in the civil engineering field.

The main goals of this book are:

(1) To serve as a textbook in traffic and pavement engineering as well as highway and transportation engineering at the undergraduate level.
(2) To serve as a reference book in advanced courses or special topics that deal with contemporary subjects at the graduate level.
(3) To serve as a reliable professional reference for academic professors, practitioners, professional engineers, professional/licenser exams, site engineers, researchers, lab managers, quality control/quality assurance (QC/QA) engineers, and technologists in the field of civil engineering.

The distinctiveness of this book emanates from the plentiful number of problems on each topic and the broad range of ideas and practical problems that are included in all areas of the book. Furthermore, the problems cover theory, concept, practice, and application. The solution of each problem in the book follows a step-by-step procedure that includes the theory and the derivation of the formulas, in some cases, and the computations. Besides this, almost all problems in the five parts of the book include detailed calculations that are solved using MS Excel worksheets where mathematical, trigonometric, statistical, and logical formulas are used by inserting the correct function in the worksheet to perform the computations more rapidly and efficiently. The MS Excel Solver tool is at times used for solving complex equations in several problems in the book. Additionally, numerical methods, linear algebraic methods, and least squares regression techniques are utilized in some problems to assist in solving the problem and make the solution much easier. The advantage of these MS Excel worksheets and computations is that each one can be used to solve other practical problems with similar type of inputs by just changing the input values to obtain the outputs or the results. The book is supplemented by a CD that includes all the MS Excel worksheets for the computational problems of the book.

In summary, the book is designed to be informative and filled with an abundance of solutions to problems in the engineering science of transportation. It is hoped that this book will enrich the knowledge and science in transportation engineering, thereby elevating the civil engineering profession in general and the transportation engineering practice in particular, as well as advancing the transportation engineering field to the best levels possible. It is also hoped that the targeted domain, including students, academic professors, and professionals, will benefit considerably from this book.

Author's Bio

Dr. Ghazi G. Al-Khateeb received a Bachelor of Science (B.S.) and Master of Science (M.S.) in Civil Engineering/ Transportation from Jordan University of Science and Technology (JUST) in 1991 and 1994, respectively, and the doctoral (Ph.D.) degree in Civil Engineering/ Transportation from the University of Illinois at Urbana-Champaign, USA, in 2001.

Dr. Al-Khateeb is currently a professor at the University of Sharjah (UOS) in the United Arab Emirates (September 1, 2015 to present). He also served as a visiting professor at the American University of Sharjah, UAE (September 1, 2014 to August 31, 2015). He is currently on leave from Jordan University of Science and Technology. He has been on the academic staff of JUST since September of 2006. During his work at JUST, Dr. Al-Khateeb held the position of Vice Dean of Engineering for two years (September 2012 to September 2014) and the Vice Director for the Consultative Center for Science and Technology (September 2009 to September 2010).

Previously he worked as a senior research scientist at the Turner-Fairbank Highway Research Center (TFHRC) of the Federal Highway Administration (FHWA) in Virginia, USA, for six years (November 2000 to September 2006). Dr. Khateeb's research is in the area of pavement and transportation engineering. He has published more than 90 papers in international scientific refereed journals and conferences as well as book chapters. Dr. Al-Khateeb teaches undergraduate as well as graduate courses in civil engineering, transportation engineering, and pavement engineering, and has supervised many senior design projects for undergraduate students and thesis work for graduate students.

Dr. Al-Khateeb served as a member of the American Society of Civil Engineers (ASCE), the Association of Asphalt Paving Technologies (AAPT), Jordan Engineers Association (JEA), Jordan Society for Scientific Research (JSSR), and Jordan Road Accidents Prevention (JRAP).

Dr. Al-Khateeb is listed in *Who's Who* in Engineering Academia. In addition, he serves as an editorial / advisory board member for several international journals and publishers such as the *Materials Analysis and Characterization* at Cambridge Scholars Publishing, *Science Progress* in the Materials Science and Engineering Section with Sage Publishing, the *International Journal of Recent Development in Civil and Environmental Engineering.* He also served as a lead guest editor for a special issue on "Innovative Materials, New Design Methods, and Advanced Characterization Techniques for Sustainable Asphalt Pavements," for the *International Journal of Advances in Materials Science and Engineering.* In addition, he is an active reviewer for more than twenty international indexed and refereed scientific journals.

Dr. Al-Khateeb has been the principal investigator and co-investigator for many funded research projects in his field. He has also received several honor awards for his study, teaching, and research accomplishments during his career.

Dr. Al-Khateeb has served on several technical national and international committees such as the Technical Committee for the Road Master Plan of Jordan, the Asphalt and Roads Technical Committee and Superpave Technical Committee for the Jordan's Ministry of Public Works and Housing, and the Technical Traffic Committee for Irbid City. At the universities where he worked, he has served on many technical committees at the university level, college level, and department level.

Acknowledgments

Special thanks to the publisher of this book, Taylor & Francis Group/CRC Press, particularly to Joseph Clements, senior publisher; Lisa Wilford, editorial assistant; Joette Lynch, production editor; and Bryan Moloney, project manager at Deanta Global, for their support and help.

Part I

Urban Transportation Planning

1 Terminology

Chapter 1 of the book consists of questions that cover the terminology and concepts related to urban transportation planning. Knowledge of the four-step travel demand forecasting process is essential. Before that, an understanding of the terminology and concepts of urban transportation planning is also necessary. The terminology involves factors affecting travel demand and the main steps in the travel demand forecasting process, including population and economic analysis, land use analysis, trip generation, trip distribution, transport mode choice, and traffic assignment. Therefore, the questions presented in this chapter offer a review of the basic terminology and concepts used in urban transportation planning. The questions are in a multiple-choice format and the answers are available at the end of Part I.

2 Travel Demand Forecasting

Chapter 2 includes practical problems related to trip generation, the first process in travel demand forecasting. Travel demand forecasting is considered the main element in urban transportation planning. It aims to describe the travel decisions of trip makers, including the necessity to make the trip, the destination of the trip, the mode of transportation to use, and the route that should be taken. There are several factors that affect travel demand; these factors include the location and intensity of land use, the socioeconomic characteristics of people living in the area, the extent, cost, and quality of available transportation services, and the purpose of the trip. Trip generation is the first step of the travel demand forecasting process. Trip generation is the process of determining the number of trips that will begin or end in each traffic analysis zone (TAZ) within a study area. Trips are either produced by a traffic zone or attracted to a traffic zone. They are called trip ends. Trip generation, in general, has two functions: (1) to develop a relationship between trip end production or attraction and land use, and (2) to use the relationship to estimate the number of trips generated in the future under new land-use conditions. The number of productions, attractions, origins, and destinations will be determined for traffic zones. Three major methods are used in trip generation: (1) the cross-classification method, (2) rates based on an activity units-model, and (3) the regression models. Combinations of the above models are also used. In addition, the trip generation models are classified into two types: (1) household-based (disaggregate) models, and (2) zonal-based (aggregate) models. The factors that impact which model to use include the accuracy and preference of the model, the reliability of the model, the availability of the data, and the availability of computer tools and software.

TRIP GENERATION

2.1 Residential zone A, shown in Figure 2.1, has 400 trips coming in and 600 trips going out.

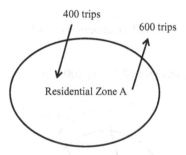

FIGURE 2.1 Residential zone A for Problem 2.1.

Determine the following:

(1) The number of productions for zone A.
(2) The number of attractions for zone A.
(3) The number of trip ends for zone A.
(4) The number of origins for zone A.
(5) The number of destinations for zone A.

Solution:

(1) The number of productions is equal to the total number of trips. In other words, it is equal to:

$$\text{No. of Productions} = \text{Trips In} + \text{Trips Out} \tag{2.1}$$

$\Rightarrow$

$$\text{No. of Productions} = 400 + 600 = 1000$$

(2) The number of attractions for residentials zones is determined as below:

$$\text{No. of Attractions} = 0$$

(3) The number of trip ends is determined using the following formula:

$$\text{No. of Trip Ends} = \text{No. of Productions} + \text{No. of Attractions} \tag{2.2}$$

$\Rightarrow$

$$\text{No. of Trip Ends} = 1000 + 0 = 1000$$

(4) The number of origins is determined, as shown below:

$$\text{No. of Origins} = \text{No. of Trips Going Out} \tag{2.3}$$

$\Rightarrow$

$$\text{No. of Origins} = 600$$

(5) The number of destinations is determined using the formula below:

$$\text{No. of Destinations} = \text{No. of Trips Coming In} \tag{2.4}$$

$\Rightarrow$

$$\text{No. of Destinations} = 400$$

2.2   Business zone B, shown in Figure 2.2, has 400 trips coming in and 600 trips going out.

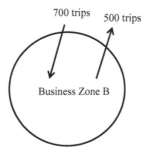

700 trips   500 trips

Business Zone B

FIGURE 2.2   Business zone B for Problem 2.2.

Determine the following:

(1) The number of productions for zone B.
(2) The number of attractions for zone B.
(3) The number of trip ends for zone B.
(4) The number of origins for zone B.
(5) The number of destinations for zone B.

Solution:

(1)The number of productions for business zones is determined as below:

$$\text{No. of Productions} = 0$$

(2) The number of attractions for business zones is determined as below:

$$\text{No. of Attractions} = \text{Trips In} + \text{Trip Out} \qquad (2.5)$$

$\Rightarrow$

$$\text{No. of Attractions} = 700 + 500 = 1200$$

(3) The number of trip ends is determined using the following formula:

$$\text{No. of Trip Ends} = \text{No. of Productions} + \text{No. of Attractions}$$

$\Rightarrow$

$$\text{No. of Trip Ends} = 0 + 1200 = 1200$$

(4) The number of origins is determined, as shown below:

$$\text{No. of Origins} = \text{No. of Trips Going Out}$$

$\Rightarrow$

$$\text{No. of Origins} = 500$$

(5) The number of destinations is determined using the formula below:

$$\text{No. of Destinations} = \text{No. of Trips Coming In}$$

$\Rightarrow$

$$\text{No. of Destinations} = 700$$

2.3   Given the zone below with the data for each household (HH), determine the average daily trips per HH (Figure 2.3).

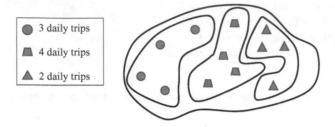

FIGURE 2.3   Zone with household data for Problem 2.3.

Solution:

The average daily trips per household (HH) is determined, as shown below:

$$DT_{avg} = \frac{\sum_{i=1}^{n} T_i H_i}{\sum_{i=1}^{n} H_i}$$ (2.6)

Where:
   DT_{avg} = the average daily trips per household (HH) in the zone
   T_i = number of daily trips per HH for subzone$_i$
   H_i = number of households (HH) in subzone$_i$

$\Rightarrow$

$$DT_{avg} = \frac{(3 \times 4) + (4 \times 4) + (2 \times 4)}{4 + 4 + 4} = 3 \text{ daily trips}$$

2.4   A travel demand survey generates the data shown in Table 2.1. To obtain this data, 25 households (HH) are interviewed. The survey data includes the number of trips produced per HH and the annual HH income (in US dollars) as well as the number of cars owned by HH. Based on the data given, develop the relationships and summary tables for this survey data using the cross-classification model (technique).

TABLE 2.1
Household-Based Travel Demand Survey Data for Problem 2.4

HH No.	Trips/HH	HH Income ($)	Cars/HH
1	4	22,000	0
2	2	11,000	0
3	3	15,000	0
4	5	30,000	1
5	8	45,000	1
6	10	62,000	2
7	12	70,000	2
8	14	80,000	3
9	8	48,000	1

(Continued)

TABLE 2.1 (CONTINUED)
Household-Based Travel Demand Survey Data for Problem 2.4

HH No.	Trips/HH	HH Income ($)	Cars/HH
10	10	60,000	2
11	15	85,000	3
12	16	86,000	4
13	14	78,000	3
14	6	32,000	1
15	6	30,000	1
16	5	28,000	1
17	4	20,000	0
18	2	12,000	0
19	2	14,000	0
20	4	22,000	0
21	4	21,000	0
22	6	28,000	1
23	7	40,000	2
24	9	50,000	2
25	10	58,000	3

Solution:

Different classes of household income are established. The number of households in each income category and car ownership is determined, as shown in Tables 2.2 and 2.3.

TABLE 2.2
Number of HHs per Income Category and Car Ownership for Problem 2.4

HH Income (1000 $)	HH Median Income (1000 $)	Number of HHs Car Ownership				Total
		0	1	2	≥ 3	
0–<20	10	4	0	0	0	4
20–<30	25	4	2	0	0	6
30–<40	35	0	3	0	0	3
40–<50	45	0	2	1	0	3
50–<60	55	0	0	1	1	2
60–<90	75	0	0	4	3	7
Total		8	7	6	4	**25**

The percentage of households per income category is determined for each car owner-ship, as summarized in Table 2.3.

TABLE 2.3
Percentage of HHs per Income Category for Each Car Ownership for Problem 2.4

HH Income (1000 $)	HH Median Income (1000 $)	% of HHs per Income Category Car Ownership				% of HHs
		0	1	2	≥3	
0–<20	10	50.0	0.0	0.0	0.0	16.0
20–<30	25	50.0	28.6	0.0	0.0	24.0
30–<40	35	0.0	42.9	0.0	0.0	12.0
40–<50	45	0.0	28.6	16.7	0.0	12.0
50–<60	55	0.0	0.0	16.7	25.0	8.0
60–<90	75	0.0	0.0	66.7	75.0	28.0
Total		100.0	100	100	100	**100.0**

In addition, the percentage of households per car ownership is determined for each income category, as summarized in Table 2.4.

TABLE 2.4
Percentage of HHs per Car Ownership in Each Income Category for Problem 2.4

HH Income (1000 $)	HH Median Income (1000 $)	% of HHs per Car Ownership Car Ownership				Total
		0	1	2	≥ 3	
0–<20	10	100.0	0.0	0.0	0.0	100.0
20–<30	25	66.7	33.3	0.0	0.0	100.0
30–<40	35	0.0	100.0	0.0	0.0	100.0
40–<50	45	0.0	66.7	33.3	0.0	100.0
50–<60	55	0.0	0.0	50.0	50.0	100.0
60–<90	75	0.0	0.0	57.1	42.9	100.0

The percentage of households per income category for each car ownership is plotted against the household median income, as shown in Figure 2.4.

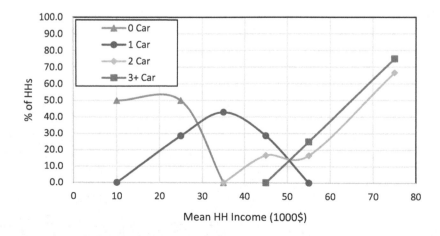

FIGURE 2.4   Percentage of households per income category and car ownership for Problem 2.4.

This figure provides a lot of information, including:

(1) For a 0-car ownership, the percentage of HHs starts high and then decreases sharply with an increase in income. This is rational as households with higher income tend to have more cars.

(2) For a 1-car ownership, the percentage of HHs increases up to a peak value and then starts to decrease at the medium-income range to the high-income range. Again, this is logical since medium- and high-income households tend to have more than one car.

(3) For a 2-car ownership, the percentage of HHs increases all the way, starting from medium-income range to high-income range. This is realistic as medium-income households typically own two cars or more. The increase in the percentage of the HHs along with an increase in income is very logical.

(4) For a 3+-car ownership, a similar trend to that of a 2-car ownership is obtained. However, the starting point of the increase occurs at a higher income range because households of higher income tend to own three or more cars.

To analyze the number of trips produced per household, per each income category and car ownership, the average number of trips per household is determined based on the household income category and car ownership, as shown in Table 2.5.

TABLE 2.5
Average Number of Trips per HH based on Income Category and Car Ownership for Problem 2.4

HH Income (1000 $)	HH Median Income (1000 $)	Average Number of Trips/HH Car Ownership				Average Trips/HH
		0	1	2	≥ 3	
0–<20	10	2.3	2.3	0.0	0.0	2.3
20–<30	25	4.0	4.5	0.0	0.0	9.5
30–<40	35	0.0	5.7	0.0	0.0	5.7
40–<50	45	0.0	6.3	7.0	0.0	15.0
50–<60	55	0.0	9.5	9.0	10.0	19.0
60–<90	75	0.0	13.0	10.7	14.8	25.4

Four curves for the 4-car ownership classes are plotted between the household median income and the average number of trips per household, as shown in Figure 2.5.

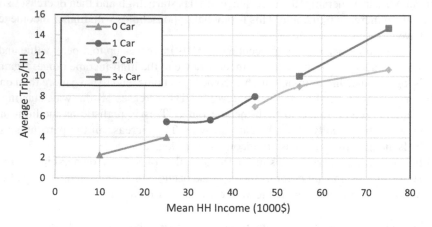

FIGURE 2.5   Average number of trips/HH versus income for Problem 2.4.

According to this figure, the relationship between the average number of trips generated per household is directly proportional to the household income. The four curves of the 4-car ownership categories (0, 1, 2, and 3⁺ cars) provide a similar trend. However, the starting and end points for the curves are different because the car ownership is also based on the household income range. Using these relationships, the number of trips generated per household can be determined using the annual household income.

2.5   Using the two figures obtained in Problem 2.4 above to determine the total number of trips generated in a zone having 150 houses and average annual income of $40,000 if the characteristics predicted in Problem 2.4 applies for this problem.

Solution:

Table 2.6 summarizes the percentage of households in each HH income category.

TABLE 2.6
Percentage of HHs in Each Income Category for Problem 2.5

HH Income (1000 $)	% of HHs
0–<20	16.0
20–<30	24.0
30–<40	12.0
40–<50	12.0
50–<60	8.0
60–<90	28.0
Total	100.0

Based on Table 2.6, the percentage of households with a HH income of $40,000 is equal to 12%. Considering car ownership and this income, 66.7% of the households in the income category of ($40,000–<$50,000) have one car and 33.3% of the households have two cars based on the results of Table 2.7 obtained from the data in Problem 2.4.

TABLE 2.7

Percentage of HHs per Car Ownership in Each Income Category for Problem 2.5

| HH Income (1000 $) | % of HHs per Car Ownership Car Ownership | | | |
	0	1	2	≥ 3
0–<20	100.0	0.0	0.0	0.0
20–<30	66.7	33.3	0.0	0.0
30–<40	0.0	100.0	0.0	0.0
40–<50	0.0	66.7	33.3	0.0
50–<60	0.0	0.0	50.0	50.0
60–<90	0.0	0.0	57.1	42.9

The average number of trips generated for each HH income category is shown in Table 2.8. In addition, the average number of trips in each income category is also determined per car ownership, as shown in Table 2.9.

TABLE 2.8

Average Number of Generated Trips in Each HH Income Category for Problem 2.5

HH Income (1000 $)	Average Trips/HH
0–<20	2.3
20–<30	4.5
30–<40	5.7
40–<50	7.7
50–<60	9.5
60–<90	13.0

TABLE 2.9

Average Number of Generated Trips per Car Ownership in Each Income Category for Problem 2.5

| HH Income (1000 $) | Average Number of Trips/HH Car Ownership | | | |
	0	1	2	≥ 3
0–<20	2.3	0.0	0.0	0.0
20–<30	4.0	5.5	0.0	0.0
30–<40	0.0	5.7	0.0	0.0
40–<50	0.0	8.0	7.0	0.0
50–<60	0.0	0.0	9.0	10.0
60–<90	0.0	0.0	10.7	14.8

Using the above data, the following computations can be performed:

$$\text{No. of HHs} = 0.12 \times 150 = 18 \text{ households}$$

66.7% of the 18 households have one car, hence:

$$\text{No. of HHs with 1 car} = 0.667 \times 18 = 12 \text{ households}$$

And 33.3% of the 18 households have two cars, therefore:

$$\text{No. of HHs with 2 cars} = 0.333 \times 18 = 6 \text{ households}$$

Based on the given income, 7.7 trips per HH is generated (eight trips for 1-car ownership and seven trips for 2-car ownership). Therefore, the total number of trips generated for this zone is equal to:

$$\text{No. of Trips Generated} = 18 \times 7.7 = 138.6 \text{ trips}$$

Among the total number of trips generated for this zone, eight trips per HH were generated for 1-car ownership and seven trips per HH were generated for 2-car ownership. Consequently:

$$\text{No. of Trips Generated} = 12 \times 8 = 96 \text{ trips for households with 1 car}$$

$$\text{No. of Trips Generated} = 6 \times 7 = 42 \text{ trips for households with 2 cars}$$

In general, the following formula can be used to determine the number of trips based on income level and car ownership:

$$P_T = \sum_{i=1}^{n} \sum_{j=1}^{m} (HH)(H_i)(HI_{ij})(T_{ij}) \qquad (2.7)$$

Where:
P_T = total number of trips generated in the zone
HH = number of the households in the zone
H_i = percentage of households in income level i
HI_{ij} = percentage of households with car ownership j in income level i
T_{ij} = number of daily trips per HH at income level i and car ownership j

In the above case, $n = 6$ (the income levels), and $m = 4$ (0, 1, 2, 3+ cars). To determine the total number of trips generated in the zone, the formula is used as described below:
$\Rightarrow$

$$P_T = \sum_{i=1}^{6} \sum_{j=1}^{4} (HH)(H_i)(HI_j)(T_{ij})$$

$$P_T = \sum_{i=1}^{6} \left[(HH)(H_i)(HI_{i1})(T_{i1}) + (HH)(H_i)(HI_{i2})(T_{i2}) \right.$$
$$\left. + (HH)(H_i)(HI_{i3})(T_{i3}) + (HH)(H_i)(HI_{i4})(T_{i4}) \right]$$

However, in this problem, only the total number of trips in the zone for income level $= \$40,000$ is required. Therefore, one income level will be used, as shown below:

$$P_{T40000} = \sum_{i=1}^{1}\left[(HH)(H_i)(HI_{11})(T_{i1})+(HH)(H_i)(HI_{12})(T_{i2})\right.$$
$$\left.+(HH)(H_i)(HI_{13})(T_{i3})+(HH)(H_i)(HI_{i4})(T_{i4})\right]$$

$$P_{T40000} = (HH)(H_1)(HI_{11})(T_{11})+(HH)(H_1)(HI_{12})(T_{12})$$
$$+(HH)(H_1)(HI_{13})(T_{13})+(HH)(H_1)(HI_{14})(T_{14})$$

$$P_{T40000} = (150)(0.12)(0.000)(0)+(150)(0.12)(0.667)(8)$$
$$+(150)(0.12)(0.333)(7)+(150)(0.12)(0.000)(0)$$
$$= 138\ trips$$

2.6 Use the relationship developed in Problem 2.4, between annual household income and average trips per household, to determine the total number of trips for an urban zone that has 200 households and an average annual income of $50,000 per household.

Solution:

The relationship developed in Problem 2.4 between annual household income and average trips per household for the 4-car ownership classes (0, 1, 2, 3⁺ cars) is shown in Figure 2.6.

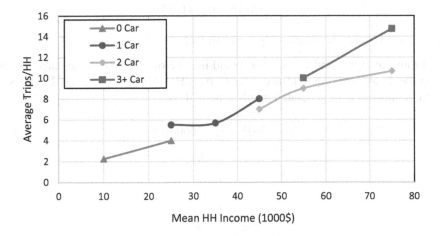

FIGURE 2.6 Average number of trips/HH versus income for Problem 2.6.

Since there is no information given on car ownership for this zone, it is more appropriate to plot another figure that illustrates the relationship between the annual household income and the average trips per household for all 4-car ownership classes. In other words, the average number of trips per household will be computed based on the average number of trips generated for an income range of the 4-car ownership classes. Table 2.10 will be used to develop this relationship, as shown in Figure 2.7.

TABLE 2.10
Average Number of Generated Trips per HH in Each Income Category for Problem 2.6

HH Income (1000 $)	Average Trips/HH
0–<20	2.3
20–<30	4.5
30–<40	5.7
40–<50	7.7
50–<60	9.5
60–<90	13.0

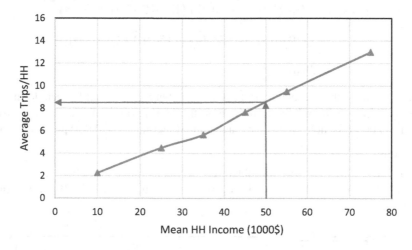

FIGURE 2.7 Mean HH income versus average trips/HH for Problem 2.6.

From this figure and using a household income of $50,000, the average number of trips per household is determined to be 8.3 trips/HH. Since the zone has 200 houses, the total number of trips generated in this zone will be:

$$P_T = (200)(8.3) = 1660 \text{ trips}$$

2.7   The relationship between the number of trips per household and the average annual household income is given in Figure 2.8 by the three logarithmic curves. The distribution of households in an urban zone is determined using a household-based travel survey, as shown in Tables 2.11 and 2.12. Determine the number of daily trips generated for low-income households if the total number of households in the zone = 80 and the average income per household = $48,000 for the low-income households.

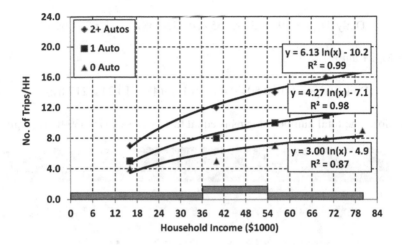

FIGURE 2.8   HH income versus number of trips/HH for Problem 2.7.

TABLE 2.11
Given Percentage of HHs per
Income Category for Problem 2.7

Income ($1000)	Households (%)
Low (< 36)	10
Medium (36–54)	50
High (> 54)	40

TABLE 2.12
Given Percentage of HHs per Auto Ownership in Each Income
Category for Problem 2.7

% Households in Each Income Category vs. Auto Ownership			
	Autos/HH		
Income	0	1	2+
Low	60	30	10
Medium	5	55	40
High	4	26	70

Solution:

For (M, 0 auto):

$$\text{Daily Trips} = 80 \times 0.50 \times 0.05 \times \left(3.00\ln(48) - 4.9\right) \cong 14 \text{ trips}$$

For (M, 1 auto):

$$\text{Daily Trips} = 80 \times 0.50 \times 0.55 \times \left(4.27\ln(48) - 7.1\right) \cong 208 \text{ trips}$$

For (M, 2+ auto):

$$\text{Daily Trips} = 80 \times 0.50 \times 0.40 \times \left(6.13\ln(48) - 10.2\right) \cong 217 \text{ trips}$$

Therefore, the total number of trips is equal to $14 + 208 + 217 = 439$ trips (Figure 2.9).

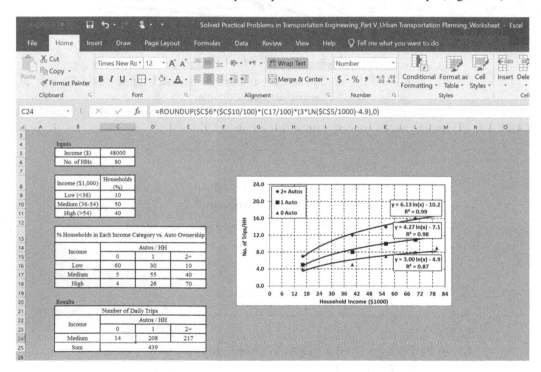

FIGURE 2.9   Image of MS Excel worksheet used for the computations of Problem 2.7.

2.8   Using Problem 2.7, estimate the number of trips produced for high-income households in this zone if the average income per HH = \$66,000 for the high-income households.

Solution:

For (H, 0 auto):

$$\text{Daily Trips} = 80 \times 0.40 \times 0.04 \times \left(3.00\ln(66) - 4.9\right) \cong 10 \text{ trips}$$

For (H, 1 auto):

$$\text{Daily Trips} = 80 \times 0.40 \times 0.26 \times \left(4.27\ln(66) - 7.1\right) \cong 90 \text{ trips}$$

For (H, 2+ auto):

$$\text{Daily Trips} = 80 \times 0.40 \times 0.70 \times (6.13\ln(66) - 10.2) \cong 347 \text{ trips}$$

Therefore, the total number of trips is equal to $10 + 90 + 347 = 447$ trips (Figure 2.10).

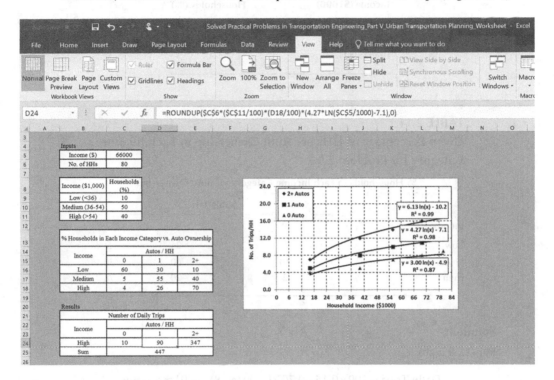

FIGURE 2.10 Image of MS Excel worksheet used for the computations of Problem 2.8.

2.9 The production rate per household is given by the relationship in Figure 2.11. Determine the number of daily trips generated for low-income households for an urban zone with the household distribution shown in Tables 2.13 and 2.14 if the total number of households in the zone is 100 and the average income per household is $18,000 for the low-income households.

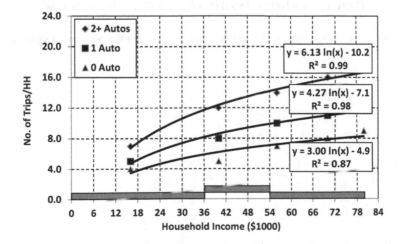

FIGURE 2.11 HH income versus number of trips/HH for Problem 2.9.

TABLE 2.13

Given Percentage of HHs per Income Category for Problem 2.9

Income ($1000)	Households (%)
Low (< 36)	15
Medium (36–54)	40
High (> 54)	45

TABLE 2.14

Given Percentage of HHs per Auto Ownership in Each Income Category for Problem 2.9

% Households in Each Income Category vs. Auto Ownership			
	Autos/HH		
Income	0	1	2+
Low	70	25	5
Medium	10	60	30
High	2	20	78

Solution:

For (L, 0 auto):

$$\text{Daily Trips} = 100 \times 0.15 \times 0.70 \times \left(3.00\ln(18) - 4.9\right) \cong 40 \text{ trips}$$

For (L, 1 auto):

$$\text{Daily Trips} = 100 \times 0.15 \times 0.25 \times \left(4.27\ln(18) - 7.1\right) \cong 20 \text{ trips}$$

For (L, 2+ auto):

$$\text{Daily Trips} = 100 \times 0.15 \times 0.05 \times \left(6.13\ln(18) - 10.2\right) \cong 6 \text{ trips}$$

Therefore, the total number of trips is equal to $40 + 20 + 6 = 66$ trips (Figure 2.12).

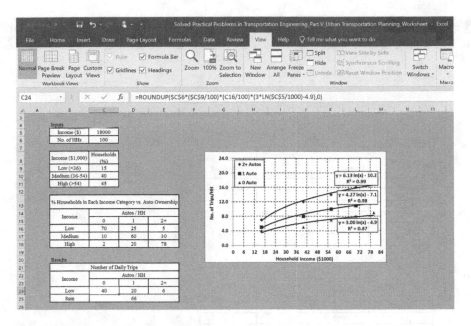

FIGURE 2.12   Image of MS Excel worksheet used for the computations of Problem 2.9.

2.10 The trip generation results shown in Table 2.15 are produced for a commercial zone based on the purpose of the trip and merchandising.

TABLE 2.15

Trip Generation Results for a Commercial Zone for Problem 2.10

Trip Purpose	Attractions per Retail Employee	Attractions per Non-Retail Employee	Attractions per Household
HBW	3.0	2.0	0
HBO	12.0	3.0	1.5
NHB	9.0	1.5	1.5

HBW = home-based work
HBO = home-based other
NHB = non-home-based

Using these results, determine the number of trips per day attracted to a commercial shopping center in the central business district (CBD) of a city that has 250 retail workers and 750 non-retail workers.

Solution:

The trip generation rates in Table 2.15 will be used in the solution of this problem as described below:

$$\text{HBW Trips} = (250)(3.0) + (750)(2.0) = 2250 \text{ HBW trips}$$

$$\text{HBO Trips} = (250)(12.0) + (750)(3.0) = 5250 \text{ HBO trips}$$

$$\text{NHB Trips} = (250)(9.0) + (750)(1.5) = 3375 \text{ NHB trips}$$

A total of 10,875 trips per day will be attracted to the shopping center (Figure 2.13).

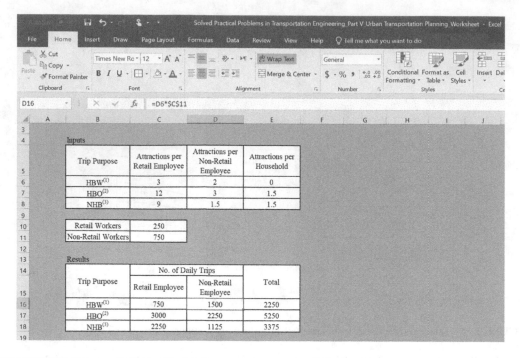

FIGURE 2.13 Image of MS Excel worksheet used for the computations of Problem 2.10.

2.11   Using the trip attraction rates (per day) in Table 2.16, estimate the total number of daily trips attracted by a retail shopping center in the downtown area with a total number of employees of 150.

TABLE 2.16
Trip Attractions Rate for a Retail Shopping Center for Problem 2.11

Trip Purpose	Attractions per Household	Attractions per Non-Retail Employee	Attractions per Downtown Retail Employee	Attractions per Other Retail Employee
HBW	–	2.0	2.0	2.0
HBO	1.5	3.0	6.0	8.0
NHB	1.5	1.0	4.0	5.0

HBW = home-based work
HBO = home-based other
NHB = non-home-based

Solution:

The trip attraction rates in Table 2.16 ("per Downtown Retail Employee" category) are used in solving this problem, as described below:

$$\text{HBW Trips} = (150)(2.0) = 300 \text{ HBW trips}$$

$$\text{HBO Trips} = (150)(6.0) = 900 \text{ HBO trips}$$

$$\text{NHB Trips} = (150)(4.0) = 600 \text{ NHB trips}$$

A total of 1800 daily trips are attracted to the downtown retail shopping center given in this problem (Figure 2.14).

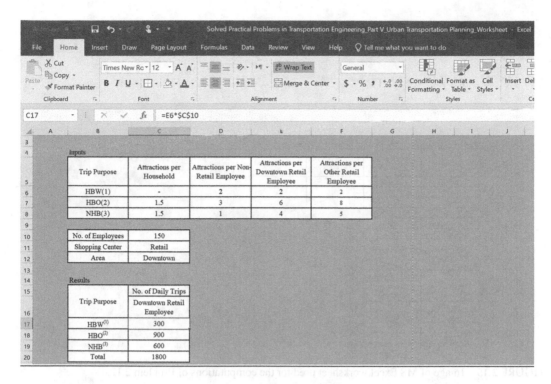

FIGURE 2.14 Image of MS Excel worksheet used for the computations of Problem 2.11.

2.12 In Problem 2.11 above, estimate the total number of daily trips attracted to a retail shopping center located in the suburbs of a city if the total number of employees of the center is 120.

Solution:

Since the retail shopping center is in the suburbs, the category "Other Retail Employee" in Table 2.16 will be used. The trip attraction rates for the category "per Other Retail Employee" are used in solving this problem as described below:

$$\text{HBW Trips} = (120)(2.0) = 240 \text{ HBW trips}$$

$$\text{HBO Trips} = (120)(8.0) = 960 \text{ HBO trips}$$

$$\text{NHB Trips} = (120)(5.0) = 600 \text{ NHB trips}$$

A total of 1800 daily trips are attracted to the suburban retail shopping center (Figure 2.15).

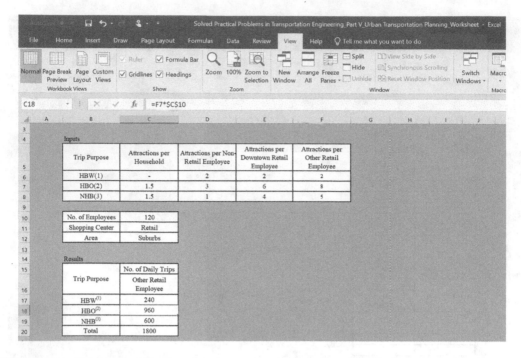

FIGURE 2.15 Image of MS Excel worksheet used for the computations of Problem 2.12.

2.13 Using the cross-classification trip production rates shown in Table 2.17, determine the number of trips generated for a rural zone with the household distribution shown in Table 2.18.

TABLE 2.17
Cross-Classification Trip Production Rates for Problem 2.13

	Number of Daily Trips per Household Persons per Household			
Auto per Household	1	2 or 3	4	≥ 5
0	0.54	1.94	4.44	6.82
1	1.32	2.89	5.39	7.77
≥ 2	1.69	3.26	5.76	8.14

*Reference: Production rates are obtained from the Oahu Metropolitan Planning Organization
https://www.oahumpo.org/resources/publications-and-reports/.

TABLE 2.18

Household Distribution based on Auto Ownership and Family Size for Problem 2.13

	Number of Households (Household) Distribution) Persons per Household			
Auto per Household	1	2 or 3	4	≥ 5
0	50	30	20	5
1	40	80	60	15
≥ 2	0	10	25	30

Solution:

The number of daily trips for each classification (category) is determined by multiplying the number of households in that category by the corresponding production rate of the category. In other words, the following formula is used:

$$P_T = \sum_{j=1}^{n} \left(N_j \times PR_j \right) \tag{2.8}$$

Where:

P_T = total number of daily trips generated in the zone
N_j = number of the households in category j
PR_j = production rate for category j
n = number of categories

In this case, there are 12 categories (classes). Therefore, 12 computations will be conducted, as shown in Table 2.19.

TABLE 2.19

Number of Daily Trips Based on Cross-Classification Rates for Problem 2.13

	Number of Daily Trips Persons per Household			
Auto per Household	1	2 or 3	4	≥ 5
0	50×0.54=27	30×1.94=58.2	20×4.44=88.8	5×6.82=34.1
1	40×1.32=52.8	80×2.89=231.2	60×5.39=323.4	15×7.77=116.6
≥ 2	0×1.69=0	10×3.26=32.6	25×5.76=144.0	30×8.14=244.2

Therefore, the total number of daily trips produced in this zone is equal to 1352.9 ≅ 1353 (Figure 2.16).

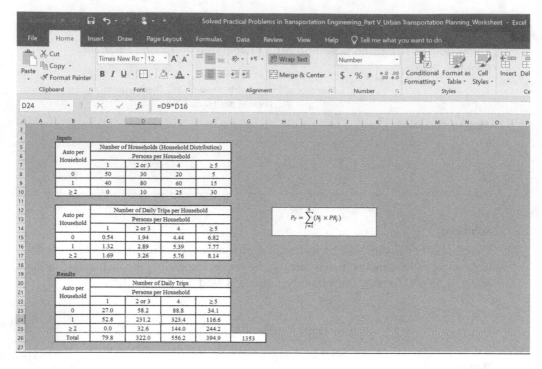

FIGURE 2.16   Image of MS Excel worksheet used for the computations of Problem 2.13.

2.14   The cross-classification trip production rates for urban areas are provided in Table 2.20. Using these rates, estimate the number of trips produced for an urban area having the household distribution shown in Table 2.21.

TABLE 2.20
Cross-Classification Trip Production Rates for Problem 2.14

	Number of Daily Trips per Household Persons per Household			
Auto per Household	1	2 or 3	4	≥ 5
0	0.57	2.07	4.57	6.95
1	1.45	3.02	5.52	7.90
≥ 2	1.82	3.39	5.89	8.27

*Reference: Production rates are obtained from the Oahu Metropolitan Planning Organization https://www.oahumpo.org/resources/publications-and-reports/.

TABLE 2.21
Household Distribution Based on Auto Ownership and Family Size for Problem 2.14

| | Number of Households (Household) Distribution) Persons per Household | | | |
| | 1 | 2 or 3 | 4 | ≥ 5 |
Auto per Household				
0	90	60	25	0
1	40	150	100	30
≥ 2	0	10	35	50

Solution:

The following formula is used to estimate the total number of trips produced in the area:

$$P_T = \sum_{j=1}^{n} \left(N_j \times PR_j \right)$$

For each category, the number of households is multiplied by the corresponding production rate, as illustrated in Table 2.22.

TABLE 2.22
Number of Daily Trips Based on Cross-Classification Rates for Problem 2.14

| | Number of Daily Trips Persons per Household | | | |
| | 1 | 2 or 3 | 4 | ≥ 5 |
Auto per Household				
0	$90 \times 0.57 = 51.3$	$60 \times 2.07 = 124.2$	$25 \times 4.57 = 114.3$	$0 \times 6.95 = 0.0$
1	$40 \times 1.45 = 58.0$	$150 \times 3.02 = 453.0$	$100 \times 5.52 = 552.0$	$30 \times 7.90 = 237.0$
≥ 2	$0 \times 1.82 = 0.0$	$10 \times 3.39 = 33.9$	$35 \times 5.89 = 206.2$	$50 \times 8.27 = 413.5$

Therefore, the total number of daily trips produced in this zone is equal to 2243.3 $\cong$ 2243 (Figure 2.17).

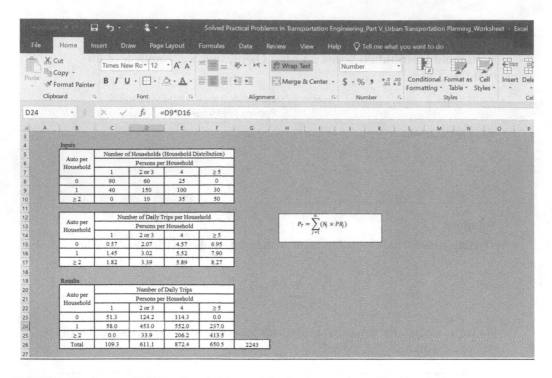

FIGURE 2.17   Image of MS Excel worksheet used for the computations of Problem 2.14.

2.15   The cross-classification trip production rates for sub-urban areas are provided in Table 2.23. Using these rates, estimate the number of trips produced for a sub-urban area with the following household distribution shown in Table 2.24.

TABLE 2.23

Cross-Classification Trip Production Rates for Problem 2.15

Auto per Household	Number of Daily Trips per Household Persons per Household			
	1	2 or 3	4	≥ 5
0	0.97	2.54	5.04	7.42
1	1.92	3.49	5.99	8.37
≥ 2	2.29	3.86	6.36	8.74

* Reference: Production rates are obtained from the Oahu Metropolitan Planning Organization
https://www.oahumpo.org/resources/publications-and-reports/.

TABLE 2.24

Household Distribution Based on Auto Ownership and Family Size for Problem 2.15

	Number of Households (Household) Distribution) Persons per Household			
Auto per Household	1	2 or 3	4	≥ 5
0	70	50	25	0
1	40	100	80	20
≥ 2	0	10	30	40

Solution:

The following formula is used to estimate the total number of trips produced in the area:

$$P_T = \sum_{j=1}^{n} \left(N_j \times PR_j \right)$$

For each category, the number of households is multiplied by the corresponding production rate, as illustrated in Table 2.25.

TABLE 2.25

Number of Daily Trips Based on Cross-Classification Rates for Problem 2.15

	Number of Daily Trips Persons per Household			
Auto per Household	1	2 or 3	4	≥ 5
0	$90 \times 0.57 = 51.3$	$60 \times 2.07 = 124.2$	$25 \times 4.57 = 114.3$	$0 \times 6.95 = 0.0$
1	$40 \times 1.45 = 58.0$	$150 \times 3.02 = 453.0$	$100 \times 5.52 = 552.0$	$30 \times 7.90 = 237.0$
≥ 2	$0 \times 1.82 = 0.0$	$10 \times 3.39 = 33.9$	$35 \times 5.89 = 206.2$	$50 \times 8.27 = 413.5$

Therefore, the total number of daily trips produced in this zone is equal to 1972.3 $\cong$ 1972 (Figure 2.18).

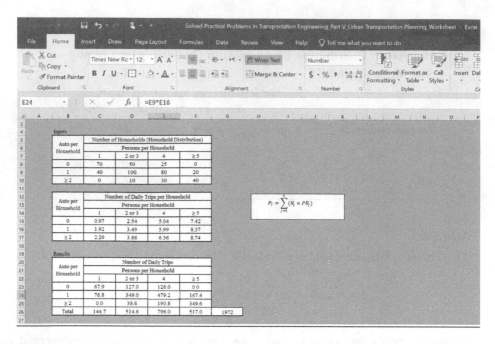

FIGURE 2.18 Image of MS Excel worksheet used for the computations of Problem 2.15.

2.16   A multiple linear regression model is developed to predict the number of daily trips per household based on three independent variables: the household size, the annual household income, and the number of family members with an age of 18 or more years. The regression model is given by the mathematical expression shown below. Use this model to determine the number of daily trips for a household with a size of six members, four of which have an age of 18 or more years, if the annual household income is $50,000.

Solution:

$$\text{DT}_{\text{HH}} = 0.15\left(\text{HH Size}\right) + 2 \times 10^{-5}\left(\text{HH Income}\right) + 0.25\left(\text{FM}_{18}\right) \tag{2.9}$$

Where:
   DT_{HH} = the total number of daily trips per household
   HH Size = the household size (number of family members)
   HH Income = the annual household income ($)
   FM_{18} = the number of family members with age of 18 or more years

$\Rightarrow$

$$\text{DT}_{\text{HH}} = 0.15\left(6\right) + 2 \times 10^{-5}\left(50000\right) + 0.25\left(4\right) = 2.9 \text{ daily trips}$$

2.17   The number of daily shopping trips per household is given by the multiple linear regression model shown in the mathematical expression below. Determine the shopping trips for a household with a size of three members and an annual income of $48,000 if the distance to the nearest shopping center is 5 miles.

$$T_{\text{Shopping}} = 0.2\left(\text{HH Size}\right) + 1.5 \times 10^{-5}\left(\text{HH Income}\right) - 0.05\left(D\right) \tag{2.10}$$

Where:

T_{Shopping} = the number of daily shopping trips per household
HH Size = the household size (number of family members)
HH Income = the annual household income ($)
D = the distance (in miles) to the nearest shopping center

Solution:

The number of daily shopping trips for this household is determined using the given model.

⇒

$$T_{\text{Shopping}} = 0.2(3) + 1.5 \times 10^{-5}(48000) - 0.05(5) = 1.07 \text{ shopping trips}$$

2.18   The number of daily social/recreational trips (T) per HH is given using the multiple linear regression model given by the mathematical expression below. Determine T for a particular HH with an annual salary of $40,000 and a size of four members.

$$T = 0.03(\text{HH Size}) + 2 \times 10^{-5}(\text{HH Income}) \qquad (2.11)$$

Where:

T = number of social/recreational trips per household in the peak hour
HH Size = the household size (number of family members)
HH Income = the annual household income ($)

Solution:

The number of daily social/recreational trips for this household is determined using the given model.

⇒

$$T = 0.03(4) + 2 \times 10^{-5}(40000) = 0.92 \cong 1 \text{ trip}$$

2.19   The number of shopping trips (T) per HH during the peak hour period is given by the multiple linear regression model shown below. The regression model is a function of HH size, the annual HH income, and the number of shops in the neighborhood. Determine T for a particular HH with a size of five members and an annual salary of $60,000 if there are five shops in the neighborhood.

$$T = 0.15 + 0.12(\text{HH Size}) + 1.2 \times 10^{-5}(\text{HH Income})$$

$$- 0.10(\text{No. of Shops in the Neighborhood}) \qquad (2.12)$$

Where:

T = number of shopping trips per household in the peak hour
HH Size = the household size (number of family members)
HH Income = the annual household income ($)

Solution:

The number of peak hour shopping trips for this household is determined using the given model.

⇒

$$T = 0.15 + 0.12(5) + 1.2 \times 10^{-5}(60000) - 0.10(5) = 0.97 \cong 1 \text{ trip}$$

Notice the significance of the signs of the coefficients for the different independent variables. For instance, the −ve sign for the "No. of Shops in the Neighborhood" variable shows that this variable is inversely contributing to trip generation; i.e., as the number of shops in the neighborhood increases, the number of shopping trips per HH decreases. This is rational, because the existence of shops in the neighborhood discourages residents from making shopping trips outside the neighborhood.

The coefficient for the independent variable (HH Income) is 1.2×10^{-5} (the least among the coefficients), which indicates that the contribution of this variable to trip generation is less significant than the other variables. On the other hand, the coefficient for the independent variable (HH Size) is 0.12 (the maximum among the coefficients), which indicates that the contribution of this variable to trip generation is more significant than the other variables.

2.20   The zonal number of trip ends for residential areas is given by the following non-linear regression model. Determine the number of trip ends generated for a new development zone with 500 units (homes).

$$T = 14.984 \left(\text{No. of HHs} \right)^{0.92} \tag{2.13}$$

Where:
   T = number of trip ends in the zone
   No. of HHs = the number of households (homes) in the zone

Solution:
The number of trip ends in the zone is estimated using the given non-linear regression model, as shown below:

$$T = 14.984 \left(500 \right)^{0.92} = 4557 \text{ trip ends}$$

2.21   The linear regression model shown below estimates the zonal number of trip ends for industrial areas as a function of the total area of the zone. Determine the number of trip ends in a new industrial zone with a total area of 500,000 m².

$$T = 0.0417 \left(A_{\text{Total}} \right) \tag{2.14}$$

Where:
   T = number of trip ends in the zone
   A_{Total} = the total area of the industrial zone

Solution:
The number of trip ends in this industrial zone is estimated using the given linear regression model, as shown below:

$$T = 0.0417 \left(500000 \right) = 20850 \text{ trip ends}$$

2.22 The data set shown in Table 2.26 is collected using a zone-based travel survey for a study area with ten traffic analysis zones (TAZs). Use this data set to develop a linear regression model to predict the trip production for the study area.

TABLE 2.26
Zone-Based Travel Survey Data for Problem 2.22

TAZ	No. of Trips	No. of Cars/Zone
1	500	180
2	450	140
3	620	240
4	750	250
5	840	280
6	910	310
7	250	90
8	280	140
9	330	170
10	950	430

Solution:

For a linear regression model of the form:

$$y = a_0 + a_1 x \tag{2.15}$$

The regression coefficients (a_1 and a_0) are determined using the following expressions:

$$a_1 = \frac{\sum x_i y_i - \frac{1}{n}\sum x_i \sum y_i}{\sum x_i^2 - \frac{1}{n}\left(\sum x_i\right)^2} \tag{2.16}$$

Where:
a_1 = the slope of the linear regression model
n = the number of data points (or number of TAZs in this case)
x = the number of cars in the zone
y = the number of trips generated in the zone

$$a_0 = \frac{\sum y_i}{n} - a_1 \frac{\sum x_i}{n} = \bar{y} - a_1 \bar{x} \tag{2.17}$$

Where:
a_0 = the intercept of the linear regression model
n = the number of data points (or the number of TAZs in this case)
x = number of autos in the zone
y = number of trips generated in the zone
$\bar{x}$ = the average of the number of cars per zone
$\bar{y}$ = the average of the number of trips per zone

$$r^2 = \frac{S_t - S_r}{S_t} \tag{2.18}$$

Where:

r^2=the coefficient of determination of the regression model

S_t=sum of squares of errors around the mean

S_r=sum of squares of errors (residuals) around the regression line (model)

S_t and S_r are given in the two expressions shown in the following two formulas, respectively:

$$S_t = \sum_{i=1}^{n}\left(y_i - \bar{y}\right)^2 \tag{2.19}$$

$$S_r = \sum_{i=1}^{n}\left(y_i - y_{i\text{-predicted}}\right)^2 \tag{2.20}$$

Where:

y_i=number of trips generated in zone i

$\bar{y}$ =the average of the number of trips per zone

$y_{i\text{-predicted}}$=the predicted number of trips in zone i based on the linear regression model

n=the number of data points (or the number of TAZs)

Table 2.27 shows the computations required to determine the regression coefficients (a_1 and a_0) and the coefficient of determination (r^2) of the linear regression model.

TABLE 2.27

Regression Analysis for a Trip Production Linear Regression Model for Problem 2.22

TAZ	No. of Trips (y)	No. of Cars/Zone (x)	x^2	xy	$y_{predicted}$	$(y-y_{average})^2$	$(y-y_{predicted})^2$
1	500	180	32,400	90,000	483.0	7744	287.6
2	450	140	19,600	63,000	385.4	19,044	4172.7
3	620	240	57,600	148,800	629.5	1024	90.2
4	750	250	62,500	187,500	653.9	26,244	9234.3
5	840	280	78,400	235,200	727.1	63,504	12,739.1
6	910	310	96,100	282,100	800.4	103,684	12,020.9
7	250	90	8100	22,500	263.4	114,244	178.4
8	280	140	19,600	39,200	385.4	94,864	11,109.9
9	330	170	28,900	56,100	458.6	66,564	16,546.0
10	950	430	184,900	408,500	1093.3	131,044	20,526.5
Sum	5880	2230	588,100	1,532,900		627,960	86,905.5
Average	588.0	223.0					

$$a_1 = \frac{\sum x_i y_i - \dfrac{1}{n}\sum x_i \sum y_i}{\sum x_i^2 - \dfrac{1}{n}\left(\sum x_i\right)^2}$$

$\Rightarrow$

$$a_1 = \frac{1532900 - \dfrac{1}{10}(2230)(5880)}{588100 - \dfrac{1}{10}(2230)^2} = 2.441$$

$$a_0 = \frac{\sum y_i}{n} - a_1 \frac{\sum x_i}{n} = \bar{y} - a_1 \bar{x}$$

$\Rightarrow$

$$a_0 = \frac{5880}{10} - (2.441)\frac{2230}{10} = 43.67$$

$$S_t = \sum_{i=1}^{n}(y_i - \bar{y})^2$$

$\Rightarrow$

$$S_t = 627960$$

$$S_r = \sum_{i=1}^{n}(y_i - y_{i\text{-predicted}})^2$$

$\Rightarrow$

$$S_r = 86905.5$$

$$r^2 = \frac{S_t - S_r}{S_t}$$

$\Rightarrow$

$$r^2 = \frac{627960 - 86905.5}{627960} = 0.86$$

The linear regression model for the trip production (T) is given by the following expression and is shown in Figure 2.19. The MS Excel worksheet used to perform the computations of the regression analysis in this problem is shown in Figure 2.20.

$$T = 43.67 + 2.441\,(\text{No. of Cars})$$

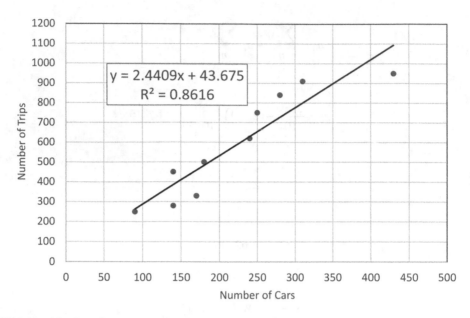

FIGURE 2.19   Number of cars versus number of produced trips in the study area for Problem 2.22.

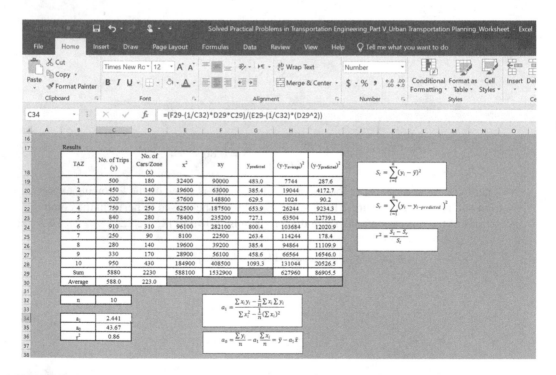

FIGURE 2.20   Image of MS Excel worksheet used for the regression analysis of Problem 2.22.

TRIP DISTRIBUTION

2.23 Table 2.28 shows the number of home-based shopping productions for a study area with three traffic zones. The three zones generate 150, 150, and 300 productions for zones 1, 2, and 3, respectively. Each of the three zones has a shopping center that attracts 200 attractions. The factors F_{11}, F_{21}, and F_{31} from zones 1, 2, and 3, respectively, to zone 1 are equal to 0.3 due to the difficulty and high cost of parking for zone 1. The factors F_{12}, F_{22}, and F_{32} from zones 1, 2, and 3, respectively, to zone 2 are equal to 0.5 due to a moderate parking cost. On the other hand, the factors F_{13}, F_{23}, and F_{33} from zones 1, 2, and 3, respectively, to zone 3 are equal to 1.0 due to parking availability in this zone and the accessibility to the zone. Determine the trip distribution (zone-to-zone trips) using the gravity model if the socioeconomic adjustment factor (K) is equal to 1.0.

TABLE 2.28

Home-Based Productions and Attractions for a Study Area with Three Traffic Zones for Problem 2.23

Zone	Productions	Attractions
1	150	200
2	150	200
3	300	200
Total	600	600

Solution:

The gravity model given by the mathematical expression of the following formula is used to estimate the zone-to-zone trips (trip distribution):

$$T_{ij} = P_i \left[\frac{A_j F_{ij} K_{ij}}{\sum_j A_j F_{ij} K_{ij}} \right] \quad (2.21)$$

Where:

T_{ij} = the number of trips produced in zone i and attracted to zone j
P_i = the total number of trips produced in zone i
A_j = the number of trips attracted to zone j
F_{ij} = a value which is an inverse function of travel time
K_{ij} = socioeconomic adjustment factor for interchange ij

Sample Calculation:
The number of zone 1 to zone 2 trips is determined as below:

$$T_{ij} = P_i \left[\frac{A_j F_{ij} K_{ij}}{\sum_j A_j F_{ij} K_{ij}} \right]$$

$\Rightarrow$

$$T_{1-2} = P_1 \left[\frac{A_2 F_{12} K_{12}}{\sum_j A_j F_{1j} K_{1j}} \right] = P_1 \left[\frac{A_2 F_{12} K_{12}}{A_1 F_{11} K_{11} + A_2 F_{12} K_{12} + A_3 F_{13} K_{13}} \right]$$

$\Rightarrow$

$$T_{1-2} = 150 \left[\frac{(200)(0.5)(1.0)}{(200)(0.3)(1.0)+(200)(0.5)(1.0)+(200)(1.0)(1.0)} \right]$$

$$= 41.7 \cong 42 \text{ trips}$$

The results of the zone-to-zone trips are all summarized in Table 2.29. The MS Excel worksheet used to perform the computations of this problem is shown in Figure 2.21.

TABLE 2.29
Computed Trip Distribution (Zone-to-Zone Trips) by the Gravity Model for the Study Area for Problem 2.23

Zone	1	2	3	Computed P	Given P
1	25.0	41.7	83.3	150.0	150
2	25.0	41.7	83.3	150.0	150
3	50.0	83.3	166.7	300.0	300
Computed A	100.0	166.7	333.3	600.0	600
Given A	200	200	200	600	

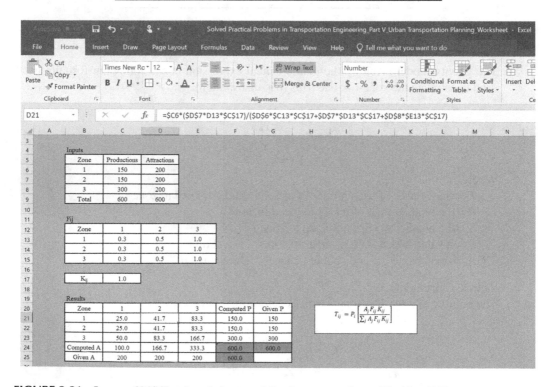

FIGURE 2.21 Image of MS Excel worksheet used for the computations of Problem 2.23.

2.24   A study area with three zones has 1000 productions and 1000 attractions, as shown in Table 2.30. The travel time between zones and the friction factor are given in Tables 2.31 and 2.32, respectively. Determine the number of zone-to-zone trips (trip distribution) using the gravity model after two iterations. Assume that the socioeconomic adjustment factor $(K_{ij}) = 1.0$.

TABLE 2.30

Productions and Attractions for a Study Area with Three Traffic Zones for Problem 2.24

Zone	Productions	Attractions
1	250	200
2	350	300
3	400	500
Total	1000	1000

TABLE 2.31

Travel Times (in Minutes) between Zones for the Study Area for Problem 2.24

Zone	Travel Time (minutes)		
	1	2	3
1	6	4	2
2	2	8	3
3	1	3	5

TABLE 2.32

Friction Factor Values for the Study Area for Problem 2.24

Travel Time (minutes)	1	2	3	4	5	6	7	8
F Value	82	52	50	41	39	26	20	13

Solution:

The gravity model given below is used to estimate the zone-to-zone trips (trip distribution):

$$T_{ij} = P_i \left[\frac{A_j F_{ij} K_{ij}}{\sum_j A_j F_{ij} K_{ij}} \right]$$

Sample Calculation:
After the first iteration, the number of zone 2 to zone 3 trips is determined as below:

$$T_{ij} = P_i \left[\frac{A_j F_{ij} K_{ij}}{\sum_j A_j F_{ij} K_{ij}} \right]$$

⇒

$$T_{2-3} = P_2 \left[\frac{A_3 F_{23} K_{23}}{\sum_j A_j F_{2j} K_{2j}} \right] = P_2 \left[\frac{A_3 F_{23} K_{23}}{A_1 F_{21} K_{21} + A_2 F_{22} K_{22} + A_3 F_{23} K_{23}} \right]$$

Based on the zone-to-zone travel times given in the problem, the factors F_{ij} are determined from the relationship between the travel time and F. The values of the F_{ij} are summarized in Table 2.33.

TABLE 2.33
Friction Factor Values between Zones for Problem 2.24

Zone	Friction Factor, F_{ij}		
	1	2	3
1	26	41	52
2	52	13	50
3	82	50	39

⇒

$$T_{2-3} = 350 \left[\frac{(500)(50)(1.0)}{(200)(52)(1.0) + (300)(13)(1.0) + (500)(50)(1.0)} \right]$$

$$= 222.6 \cong 223 \text{ trips}$$

The results of the zone-to-zone trips after the first iteration are all summarized in Table 2.34. The MS Excel worksheet used to perform the computations of the first iteration results in this problem is shown in Figure 2.22.

TABLE 2.34
Computed Trip Distribution (Zone-to-Zone Trips) by the Gravity Model for the Study Area for Problem 2.24

Zone	1	2	3	Computed P	Given P
1	29.9	70.7	149.4	250.0	250
2	92.6	34.7	222.6	350.0	350
3	128.9	117.9	153.2	400.0	400
Computed A	251.4	223.3	525.3	1000.0	1000
Given A	200	300	500	1000	

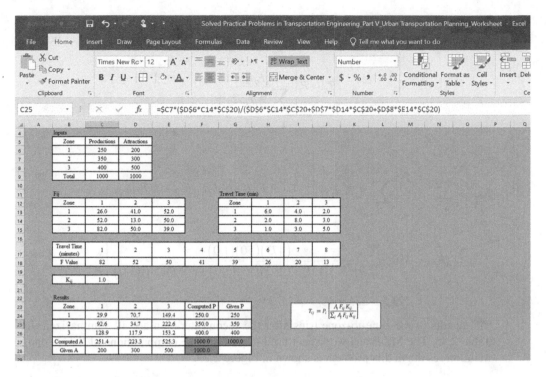

The formula bar shows: C25 | fx | =$C7*($D$6*C14*$C$20)/($D$6*$C14*$C$20+$D$7*$D14*$C$20+$D$8*$E14*$C$20)

Inputs

Zone	Productions	Attractions
1	250	200
2	350	300
3	400	500
Total	1000	1000

Fij

Zone	1	2	3
1	26.0	41.0	52.0
2	52.0	13.0	50.0
3	82.0	50.0	39.0

Travel Time (min)

Zone	1	2	3
1	6.0	4.0	2.0
2	2.0	8.0	3.0
3	1.0	3.0	5.0

Travel Time (minutes)	1	2	3	4	5	6	7	8
F Value	82	52	50	41	39	26	20	13

K_{ij}	1.0

Results

Zone	1	2	3	Computed P	Given P
1	29.9	70.7	149.4	250.0	250
2	92.6	34.7	222.6	350.0	350
3	128.9	117.9	153.2	400.0	400
Computed A	251.4	223.3	525.3	1000.0	1000.0
Given A	200	300	500	1000.0	

$$T_{ij} = P_i \left[\frac{A_j F_{ij} K_{ij}}{\sum_j A_j F_{ij} K_{ij}} \right]$$

FIGURE 2.22 Image of MS Excel worksheet used for estimating zone-to-zone trips by the gravity model after the first iteration in Problem 2.24.

The adjusted values for the attraction factors of the three zones that will be used in the second iteration are determined using the following formula:

$$A_{jk} = \frac{A_j}{C_{j(k-1)}} A_{j(k-1)} \tag{2.22}$$

Where:

A_{jk} = adjusted attraction factor for attraction zone j, iteration k
C_{jk} = actual attractions for zone j, iteration k
A_j = desired attractions for attraction zone j
j = attraction zone number (1, 2, ... , n)
n = number of zones
k = iteration number (1, 2, ... , m)
m = number of iterations

$\Rightarrow$

$$A_1 = \frac{200}{252} \times 200 = 159$$

$$A_2 = \frac{300}{224} \times 300 = 402$$

$$A_3 = \frac{500}{524} \times 500 = 477$$

Sample Calculation:
After the second iteration, the number of zone 2 to zone 3 trips is determined as below:

$$T_{ij} = P_i \left[\frac{A_j F_{ij} K_{ij}}{\sum_j A_j F_{ij} K_{ij}} \right]$$

$\Rightarrow$

$$T_{2-3} = P_2 \left[\frac{A_3 F_{23} K_{23}}{\sum_j A_j F_{2j} K_{2j}} \right] = P_2 \left[\frac{A_3 F_{23} K_{23}}{A_1 F_{21} K_{21} + A_2 F_{22} K_{22} + A_3 F_{23} K_{23}} \right]$$

$\Rightarrow$

$$T_{2-3} = 350 \left[\frac{(477)(50)(1.0)}{(159)(52)(1.0) + (402)(13)(1.0) + (477)(50)(1.0)} \right]$$

$$= 223.5 \cong 224 \text{ trips}$$

The results of the zone-to-zone trips after the second iteration are all summarized in Table 2.35. The MS Excel worksheet used to perform the computations of the second iteration results in this problem is shown in Figure 2.23.

TABLE 2.35
Second Iteration-Computed Trip Distribution (Zone-to-Zone Trips) by the Gravity Model for the Study Area for Problem 2.24

Zone	1	2	3	Computed P	Given P
1	22.8	90.7	136.5	250.0	250
2	77.5	49.0	223.5	350.0	350
3	100.8	155.4	143.8	400.0	400
Computed A	201.0	295.1	503.9	1000.0	1000
Given A	200	300	500	1000	

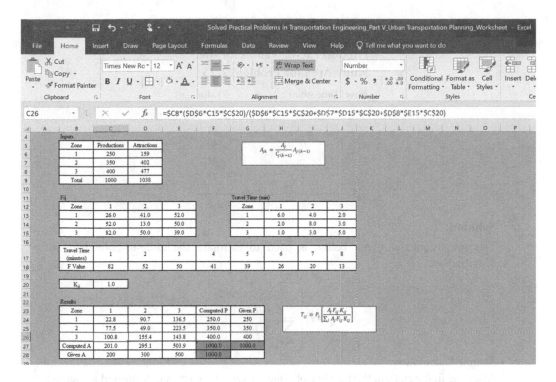

FIGURE 2.23 Image of MS Excel worksheet used for estimating zone-to-zone trips by the gravity model after the second iteration in Problem 2.24.

2.25 A two-zone system with a total of 500 productions is shown in Table 2.36. Zones 1 and 2 generate 300 and 200 productions, respectively, and attract 250 attractions per each zone. The average travel times between the zones and the relationship between the travel time and the friction factor (F) are given in Tables 2.37 and 2.38. Assume that the socioeconomic adjustment factor (K_{ij}) = 1.0. Determine the number of zone-to-zone trips using the gravity model through only one iteration.

TABLE 2.36
Productions and Attractions for a Study Area with Two Traffic Zones for Problem 2.25

Zone	Productions	Attractions
1	300	250
2	200	250
Total	500	500

TABLE 2.37
Travel Times (in Minutes) between Zones for the Study Area for Problem 2.25

Zone	1	2
1	3	5
2	5	2

TABLE 2.38
Friction Factor Values for the
Study Area for Problem 2.25

Travel Time (min)	F Value
1	82
2	52
3	50
4	41
5	39

Solution:

The gravity model given below is used to estimate the zone-to-zone trips (trip distribution):

$$T_{ij} = P_i \left[\frac{A_j F_{ij} K_{ij}}{\sum_j A_j F_{ij} K_{ij}} \right]$$

Sample Calculation:
After the first iteration, the number of zone 1 to zone 2 trips is determined as below:

$$T_{ij} = P_i \left[\frac{A_j F_{ij} K_{ij}}{\sum_j A_j F_{ij} K_{ij}} \right]$$

$\Rightarrow$

$$T_{1-2} = P_1 \left[\frac{A_2 F_{12} K_{12}}{\sum_j A_j F_{1j} K_{1j}} \right] = P_1 \left[\frac{A_2 F_{12} K_{12}}{A_1 F_{11} K_{11} + A_2 F_{12} K_{12}} \right]$$

Based on the zone-to-zone travel times given in the problem, the factors F_{ij} are determined from the relationship between the travel time and F. The values of the F_{ij} are summarized in Table 2.39.

TABLE 2.39
Friction Factor Values between
Zones for Problem 2.25

Zone	Friction Factor, F_{ij}	
	1	2
1	50	39
2	39	52

⇒

$$T_{1-2} = 300 \left[\frac{(250)(39)(1.0)}{(250)(50)(1.0) + (250)(39)(1.0)} \right] = 131.5 \cong 132 \text{ trips}$$

The procedure and the results of the zone-to-zone trips after the first iteration are all summarized in Tables 2.40 and 2.41, respectively. The MS Excel worksheet used to perform the computations of this problem is shown in Figure 2.24.

TABLE 2.40
Computations of the Zone-to-Zone Trips by the Gravity Model for Problem 2.25

Zone-to-Zone Trips	Procedure	Value
T_{1-1}	$T_{1-1} = 300 \left[\dfrac{(250)(50)}{(250)(50) + (250)(39)} \right] =$	169
T_{1-2}	$T_{1-2} = 300 \left[\dfrac{(250)(39)}{(250)(50) + (250)(39)} \right] =$	131
T_{2-1}	$T_{2-1} = 200 \left[\dfrac{(250)(39)}{(250)(39) + (250)(52)} \right] =$	86
T_{2-2}	$T_{2-2} = 200 \left[\dfrac{(250)(52)}{(250)(39) + (250)(52)} \right] =$	114
	Total	500

TABLE 2.41
Computed Trip Distribution (Zone-to-Zone Trips) by the Gravity Model for the Study Area for Problem 2.25

Zone	1	2	Computed P	Given P
1	169	131	300	300
2	86	114	200	200
Computed A	255	245	500	500
Given A	250	250	500	

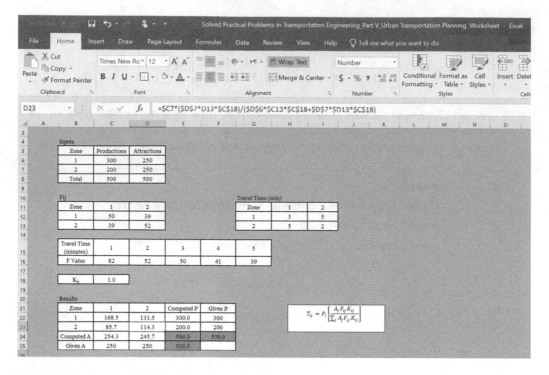

FIGURE 2.24 Image of MS Excel worksheet used for estimating zone-to-zone trips by the gravity model after the first iteration in Problem 2.25.

2.26   For Problem 2.25 above, determine the number of zone-to-zone trips (trip distribution) using the gravity model after the second iteration (one more iteration).

Solution:

The adjusted values for the attraction factors of the two zones that will be used in the second iteration are determined using the following formula:

$$A_{jk} = \frac{A_j}{C_{j(k-1)}} A_{j(k-1)}$$

⇒

$$A_1 = \frac{250}{255} \times 250 = 245$$

$$A_2 = \frac{250}{245} \times 250 = 255$$

Sample Calculation:
After the second iteration, the number of zone 1 to zone 2 trips is determined as below:

$$T_{ij} = P_i \left[\frac{A_j F_{ij} K_{ij}}{\sum_j A_j F_{ij} K_{ij}} \right]$$

$$\Rightarrow$$

$$T_{1-2} = P_1 \left[\frac{A_2 F_{12} K_{12}}{\sum_j A_j F_{1j} K_{1j}} \right] = P_1 \left[\frac{A_2 F_{12} K_{12}}{A_1 F_{11} K_{11} + A_2 F_{12} K_{12}} \right]$$

Using the F_{ij} values shown in Table 2.39 (used earlier in Problem 2.25),

$$\Rightarrow$$

$$T_{1-2} = 300 \left[\frac{(255)(39)(1.0)}{(245)(50)(1.0) + (255)(39)(1.0)} \right] = 134.4 \cong 134 \text{ trips}$$

The results of the zone-to-zone trips after the second iteration are all summarized in Table 2.42. The MS Excel worksheet used to perform the computations of this problem is shown in Figure 2.25.

TABLE 2.42

Second Iteration-Computed Trip Distribution (Zone-to-Zone Trips) by the Gravity Model for the Study Area for Problem 2.26

Zone	1	2	Computed P	Given P
1	165.6	134.4	300.0	300
2	83.8	116.2	200.0	200
Computed A	249.3	250.7	500.0	500
Given A	200	300	500	

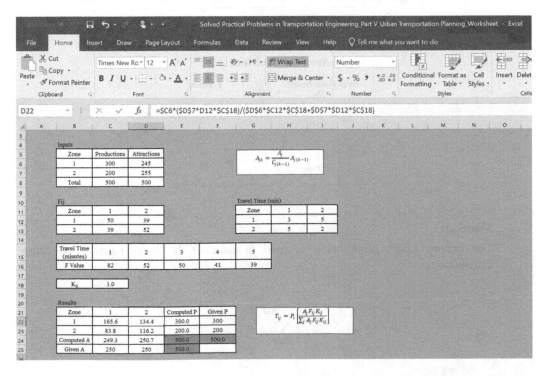

FIGURE 2.25 Image of MS Excel worksheet used for estimating zone-to-zone trips by the gravity model for Problem 2.26.

2.27   A two-zone system with a total of 600 productions is shown in Table 2.43; 400 and
        200 productions are generated for zones 1 and 2, respectively. Each zone contains a
        business center that attracts 300 attractions. The average travel times between the
        zones and the relationship between the travel time and the friction factor (F) are given
        in Tables 2.44 and 2.45. Assume that the socioeconomic adjustment factor (K_{ij}) = 1.0.
        Determine the number of zone-to-zone trips using the gravity model through only one
        iteration.

TABLE 2.43
Productions and Attractions for a Study Area with Three Traffic Zones for Problem 2.27

Zone	Productions	Attractions
1	400	300
2	200	300
Total	600	600

TABLE 2.44
Travel Times between Zones for the Study Area for Problem 2.27

Zone	1	2
1	2	4
2	4	3

TABLE 2.45
Friction Factor Values for the Study Area for Problem 2.27

Travel Time (min)	F Value
1	82
2	52
3	50
4	41
5	39

Solution:

The gravity model given below is used to estimate the zone-to-zone trips (trip
distribution):

$$T_{ij} = P_i \left[\frac{A_j F_{ij} K_{ij}}{\sum_j A_j F_{ij} K_{ij}} \right]$$

Sample Calculation:

After the first iteration, the number of zone 2 to zone 1 trips is determined as below:

$$T_{ij} = P_i \left[\frac{A_j F_{ij} K_{ij}}{\sum_j A_j F_{ij} K_{ij}} \right]$$

$\Rightarrow$

$$T_{2\text{-}1} = P_2 \left[\frac{A_1 F_{21} K_{21}}{\sum_j A_j F_{2j} K_{2j}} \right] = P_2 \left[\frac{A_1 F_{21} K_{21}}{A_1 F_{21} K_{21} + A_2 F_{22} K_{22}} \right]$$

The zone-to-zone travel times given in the problem are used to determine the factors F_{ij} from the relationship between the travel time and F. The values of the F_{ij} are summarized in the table below:

	Friction Factor, F_{ij}	
Zone	1	2
1	52	41
2	41	50

$\Rightarrow$

$$T_{2\text{-}1} = 200 \left[\frac{(300)(41)(1.0)}{(300)(41)(1.0) + (300)(50)(1.0)} \right] = 90.1 \cong 90 \text{ trips}$$

The procedure and the results of the zone-to-zone trips after the first iteration are all summarized in Tables 2.46 and 2.47, respectively. Figure 2.26 shows a screen image of the MS Excel worksheet used to perform the computations of this problem.

TABLE 2.46

Computations of the Zone-to-Zone Trips by the Gravity Model for Problem 2.27

Zone-to-Zone Trips	Procedure	Value
$T_{1\text{-}1}$	$T_{1\text{-}1} = 400 \left[\dfrac{(300)(52)(1)}{(300)(52)(1) + (300)(41)(1)} \right] =$	223.7
$T_{1\text{-}2}$	$T_{1\text{-}2} = 400 \left[\dfrac{(300)(41)(1)}{(300)(52)(1) + (300)(41)(1)} \right] =$	176.3
$T_{2\text{-}1}$	$T_{2\text{-}1} = 200 \left[\dfrac{(300)(41)(1)}{(300)(41)(1) + (300)(50)(1)} \right] =$	90.1
$T_{2\text{-}2}$	$T_{2\text{-}2} = 200 \left[\dfrac{(300)(50)(1)}{(300)(41)(1) + (300)(50)(1)} \right] =$	109.9
Total		600.0

TABLE 2.47
Computed Trip Distribution (Zone-to-Zone Trips) by the Gravity Model for the Study Area for Problem 2.27

Zone	1	2	Computed P	Given P
1	223.7	176.3	400.0	400
2	90.1	109.9	200.0	200
Computed A	313.8	286.2	600.0	600
Given A	300	300	600	

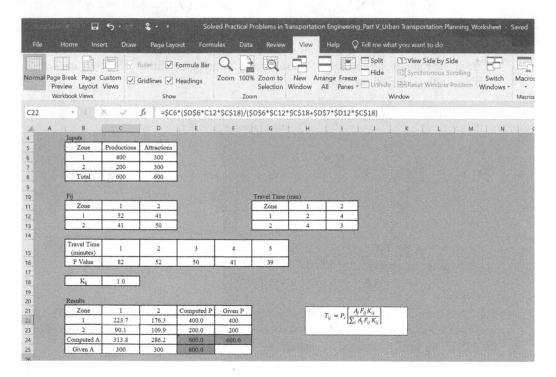

FIGURE 2.26 Image of MS Excel worksheet used for estimating zone-to-zone trips by the gravity model after the first iteration in Problem 2.27.

2.28   The trip productions and attractions for a two-zone study area are given in Table 2.48. If the travel times between the zones are equal and $K_{ij} = 1.0$; then using the gravity model, determine the zone-to-zone trips.

TABLE 2.48
Productions and Attractions for a Study Area
with Three Traffic Zones for Problem 2.28

Zone	Productions	Attractions
1	120	220
2	280	180
Total	400	400

Solution:

The gravity model given in the following formula is used to estimate the zone-to-zone trips:

$$T_{ij} = P_i \left[\frac{A_j F_{ij} K_{ij}}{\sum_j A_j F_{ij} K_{ij}} \right]$$

Sample Calculation:
The number of zone 1 to zone 2 trips is determined as below:

$$T_{ij} = P_i \left[\frac{A_j F_{ij} K_{ij}}{\sum_j A_j F_{ij} K_{ij}} \right]$$

$\Rightarrow$

$$T_{1\text{-}2} = P_1 \left[\frac{A_2 F_{12} K_{12}}{\sum_j A_j F_{1j} K_{1j}} \right] = P_1 \left[\frac{A_2 F_{12} K_{12}}{A_1 F_{11} K_{11} + A_2 F_{12} K_{12}} \right]$$

Since the zone-to-zone travel time is equal, the factor F_{ij} is the same between the zone $= F$.

$\Rightarrow$

$$T_{1\text{-}2} = 120 \left[\frac{(180)(F)(1.0)}{(220)(F)(1.0) + (180)(F)(1.0)} \right] = 54 \text{ trips}$$

The procedure and the results of the zone-to-zone trips after the first iteration are all summarized in Tables 2.49 and 2.50, respectively. Figure 2.27 shows a screen image of the MS Excel worksheet used to perform the computations of this problem.

TABLE 2.49

Computations of the Zone-to-Zone Trips by the Gravity Model for Problem 2.28

Zone-to-Zone Trips	Procedure	Value
T_{1-1}	$T_{1\text{-}1} = 120 \left[\dfrac{(220)(F)(1)}{(220)(F)(1) + (180)(F)(1)} \right] =$	66
T_{1-2}	$T_{1\text{-}2} = 120 \left[\dfrac{(180)(F)(1)}{(220)(F)(1) + (180)(F)(1)} \right] =$	54
T_{2-1}	$T_{2\text{-}1} = 280 \left[\dfrac{(220)(F)(1)}{(220)(F)(1) + (180)(F)(1)} \right] =$	154
T_{2-2}	$T_{2\text{-}2} = 280 \left[\dfrac{(180)(F)(1)}{(220)(F)(1) + (180)(F)(1)} \right] =$	126
Total		400

TABLE 2.50

Computed Trip Distribution (Zone-to-Zone Trips) by the Gravity Model for the Study Area for Problem 2.28

Zone	1	2	Computed P	Given P
1	66	54	120	120
2	154	126	280	280
Computed A	220	180	400	400
Given A	220	180	400	

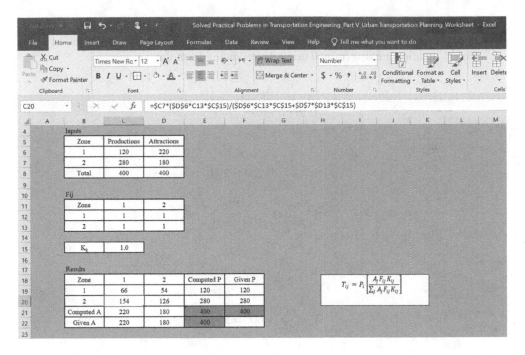

FIGURE 2.27 Image of MS Excel worksheet used for estimating zone-to-zone trips by the gravity model after the first iteration in Problem 2.28.

2.29   In Table 2.51, the productions and attractions for a four-zone system used in the first iteration of a trip distribution method (gravity model) along with the resulting productions and attractions are shown. Determine the adjusted productions and attractions for each of the four zones in the second iteration. The F_{ij} between zones is shown in Table 2.52. Assume $K_{ij} = 1.0$.

TABLE 2.51

Productions and Attractions after the First Iteration for a Study Area with Four Traffic Zones for Problem 2.29

Zone	1	2	3	4
Productions	150	200	350	500
Attractions	200	100	300	600
$P_{\text{First Iteration}}$	150	200	350	500
$A_{\text{First Iteration}}$	197	99	298	606

TABLE 2.52

Friction Factor Values between Zones for Problem 2.29

	Friction Factor, F_{ij}			
Zone	1	2	3	4
1	52	50	41	39
2	50	52	41	39
3	41	41	52	39
4	39	39	39	52

Solution:

The adjusted values for the attraction factors of the four zones that will be used in the second iteration are determined using the following formula:

$$A_{jk} = \frac{A_j}{C_{j(k-1)}} A_{j(k-1)}$$

$\Rightarrow$

$$A_1 = \frac{200}{197} \times 200 = 203$$

$$A_2 = \frac{100}{99} \times 100 = 101$$

$$A_3 = \frac{300}{298} \times 300 = 302$$

$$A_4 = \frac{600}{606} \times 600 = 594$$

The gravity model given in the following formula is used to estimate the zone-to-zone trips:

$$T_{ij} = P_i \left[\frac{A_j F_{ij} K_{ij}}{\sum_j A_j F_{ij} K_{ij}} \right]$$

Sample Calculation:

The number of zone 3 to zone 4 trips is determined as below:

$$T_{ij} = P_i \left[\frac{A_j F_{ij} K_{ij}}{\sum_j A_j F_{ij} K_{ij}} \right]$$

$\Rightarrow$

$$T_{3\text{-}4} = P_3 \left[\frac{A_4 F_{34} K_{34}}{\sum_j A_j F_{3j} K_{3j}} \right]$$

$$= P_3 \left[\frac{A_4 F_{34} K_{34}}{A_1 F_{31} K_{31} + A_2 F_{32} K_{32} + A_3 F_{33} K_{33} + A_4 F_{34} K_{34}} \right]$$

$\Rightarrow$

$$T_{3\text{-}4} = 350 \left[\frac{(594)(39)(1.0)}{(203)(41)(1.0) + (101)(41)(1.0) + (302)(52)(1.0) + (594)(39)(1.0)} \right]$$

$$= 157.9 \text{ trips} \cong 158 \text{ trips}$$

The results of the zone-to-zone trips after the second iteration are summarized in Table 2.53. Figure 2.28 shows a screen image of the MS Excel worksheet used to perform the computations of this problem.

TABLE 2.53

Second Iteration-Computed Trip Distribution (Zone-to-Zone Trips) by the Gravity Model for the Study Area for Problem 2.29

Zone	1	2	3	4	Computed P	Given P
1	30.9	14.8	36.3	67.9	150	150
2	39.8	20.7	48.6	90.9	200	200
3	56.7	28.3	107.1	157.9	350	350
4	72.5	36.2	108.0	283.3	500	500
Computed A	200	100	300	600	1200	1200
Given A	203	101	302	594	1200	

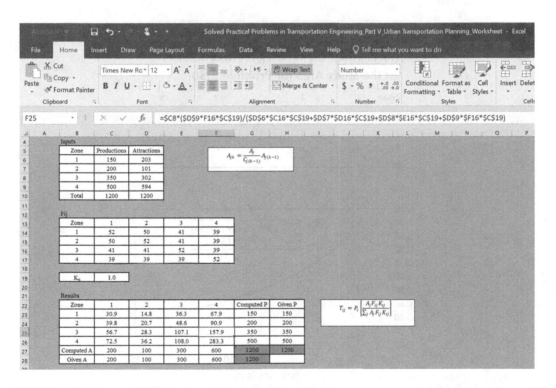

FIGURE 2.28 image of MS Excel worksheet used for estimating zone-to-zone trips by the gravity model for Problem 2.29.

2.30 The trip productions and attractions, as well as the travel times between the zones for a three-zone study area, are given in Tables 2.54 and 2.55, respectively. The relationship between the travel time and the friction factor between the zones is given by the curve in Figure 2.29. If the socioeconomic adjustment factor $(K_{ij}) = 1.0$, use the gravity model to determine the zone-to-zone trips after two iterations.

TABLE 2.54

Productions and Attractions for a Study Area with Three Traffic Zones for Problem 2.30

Zone	Productions	Attractions
1	120	150
2	180	100
3	100	150
Total	400	400

TABLE 2.55

Travel Times (in Minutes) between Zones for the Study Area for Problem 2.30

Zone	1	2	3
1	2	4	6
2	4	2	6
3	6	6	2

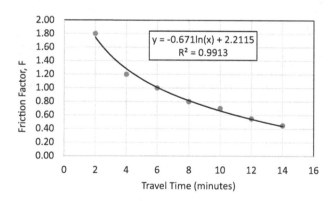

FIGURE 2.29 Relationship between travel time and friction factor (F) for Problem 2.30.

Solution:

The gravity model given in the following expression is used to compute the zone-to-zone trips:

$$T_{ij} = P_i \left[\frac{A_j F_{ij} K_{ij}}{\sum_j A_j F_{ij} K_{ij}} \right]$$

Sample Calculation:

The number of zone 1 to zone 3 trips is determined as below:

$$T_{ij} = P_i \left[\frac{A_j F_{ij} K_{ij}}{\sum_j A_j F_{ij} K_{ij}} \right]$$

$\Rightarrow$

$$T_{1\text{-}3} = P_1 \left[\frac{A_3 F_{13} K_{13}}{\sum_j A_j F_{1j} K_{1j}} \right] = P_1 \left[\frac{A_3 F_{13} K_{13}}{A_1 F_{11} K_{11} + A_2 F_{12} K_{12} + A_3 F_{13} K_{13}} \right]$$

Based on the travel times between the zones given in this problem and using the relationship between the travel time and the friction factor (F), the friction factors between the zones (F_{ij}) are computed as shown below:

Sample Calculation (for the friction factor):

$$F_{ij} = -0.671 \ln\left(\text{Travel Time} \right) + 2.2115 \qquad (2.23)$$

Between zone 1 and zone 3, the travel time = 6 minutes, therefore:

$$F_{ij} = -0.671 \ln\left(\text{Travel Time} \right) + 2.2115$$

$\Rightarrow$

$$F_{1\text{-}3} = -0.671 \ln\left(6 \right) + 2.2115 = 1.01$$

The values of the friction factors between the three zones are summarized in Table 2.56.

TABLE 2.56

Friction Factor Values for the Study Area for Problem 2.30

	Friction Factor, F_{ij}		
Zone	1	2	3
1	1.75	1.28	1.01
2	1.28	1.75	1.01
3	1.01	1.01	1.75

$\Rightarrow$

$$T_{1\text{-}3} = 120 \left[\frac{(150)(1.01)(1.0)}{(150)(1.75)(1.0) + (100)(1.28)(1.0) + (150)(1.01)(1.0)} \right]$$

$$= 33.5 \text{ trips} \cong 34 \text{ trips}$$

The procedure and the results of the zone-to-zone trips after the first iteration are summarized in Tables 2.57 and 2.58, respectively. The MS Excel worksheet used to perform the computations of the first iteration results in this problem is shown in Figure 2.30.

TABLE 2.57

Computations of the Zone-to-Zone Trips by the Gravity Model for Problem 2.30

Zone-to-Zone Trips	Procedure	Value
T_{1-1}	$T_{1-1} = 120 \left[\dfrac{(150)(1.7)(1)}{(150)(1.7)(1)+(100)(1.3)(1)++(150)(1.0)(1)} \right]$	58.1
T_{1-2}	$T_{1-2} = 120 \left[\dfrac{(100)(1.3)(1)}{(150)(1.7)(1)+(100)(1.3)(1)++(150)(1.0)(1)} \right]$	28.3
T_{1-3}	$T_{1-3} = 120 \left[\dfrac{(150)(1.0)(1)}{(150)(1.7)(1)+(100)(1.3)(1)++(150)(1.0)(1)} \right]$	33.5
T_{2-1}	$T_{2-1} = 180 \left[\dfrac{(150)(1.3)(1)}{(150)(1.3)(1)+(100)(1.7)(1)++(150)(1.0)(1)} \right]$	66.7
T_{2-2}	$T_{2-2} = 180 \left[\dfrac{(100)(1.7)(1)}{(150)(1.3)(1)+(100)(1.7)(1)++(150)(1.0)(1)} \right]$	60.8
T_{2-3}	$T_{2-3} = 180 \left[\dfrac{(150)(1.0)(1)}{(150)(1.3)(1)+(100)(1.7)(1)++(150)(1.0)(1)} \right]$	52.6
T_{3-1}	$T_{3-1} = 100 \left[\dfrac{(150)(1.0)(1)}{(150)(1.0)(1)+(100)(1.0)(1)++(150)(1.7)(1)} \right]$	29.4
T_{3-2}	$T_{3-2} = 100 \left[\dfrac{(100)(1.0)(1)}{(150)(1.0)(1)+(100)(1.0)(1)++(150)(1.7)(1)} \right]$	19.6
T_{3-3}	$T_{3-3} = 100 \left[\dfrac{(150)(1.7)(1)}{(150)(1.0)(1)+(100)(1.0)(1)++(150)(1.7)(1)} \right]$	51.0
Total		400

TABLE 2.58

First Iteration-Computed Trip Distribution (Zone-to-Zone Trips) by the Gravity Model for the Study Area for Problem 2.30

Zone	1	2	3	Computed P	Given P
1	58.1	28.4	33.5	120	120
2	66.8	60.7	52.6	180	180
3	29.4	19.6	50.9	100	100
Computed A	154	109	137	400	400
Given A	150	100	150	400	

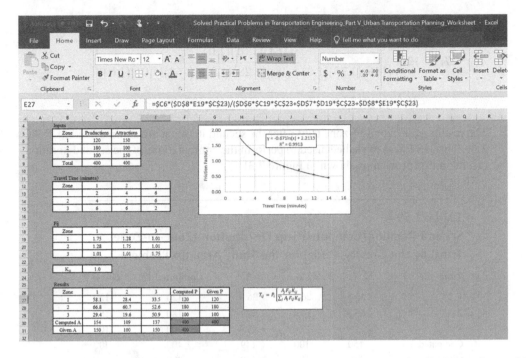

FIGURE 2.30 Image of MS Excel worksheet used for estimating zone-to-zone trips by the gravity model after the first iteration in Problem 2.30.

The adjusted values for the attraction factors of the three zones that will be used in the second iteration are determined using the following formula:

$$A_{jk} = \frac{A_j}{C_{j(k-1)}} A_{j(k-1)}$$

⇒

$$A_1 = \frac{150}{154} \times 150 = 146$$

$$A_2 = \frac{100}{109} \times 100 = 92$$

$$A_3 = \frac{150}{137} \times 150 = 164$$

Sample Calculation:
After the second iteration, the number of zone 1 to zone 3 trips is determined, as shown below:

$$T_{ij} = P_i \left[\frac{A_j F_{ij} K_{ij}}{\sum_j A_j F_{ij} K_{ij}} \right]$$

⇒

$$T_{1\text{-}3} = P_1 \left[\frac{A_3 F_{13} K_{13}}{\sum_j A_j F_{1j} K_{1j}} \right] = P_1 \left[\frac{A_3 F_{13} K_{13}}{A_1 F_{11} K_{11} + A_2 F_{12} K_{12} + A_3 F_{13} K_{13}} \right]$$

$\Rightarrow$

$$T_{1-3} = 120 \left[\frac{(164)(1.01)(1.0)}{(146)(1.75)(1.0) + (92)(1.28)(1.0) + (164)(1.01)(1.0)} \right]$$

$$= 36.9 \text{ trips} \cong 37 \text{ trips}$$

The results of the zone-to-zone trips after the second iteration are summarized in Table 2.59. The MS Excel worksheet used to perform the computations of the second iteration results in this problem is shown in Figure 2.31.

TABLE 2.59
Second Iteration-Computed Trip Distribution (Zone-to-Zone Trips) by the Gravity Model for the Study Area for Problem 2.30

Zone	1	2	3	Computed *P*	Given *P*
1	56.8	26.3	36.9	120	120
2	65.5	56.4	58.1	180	180
3	27.9	17.6	54.4	100	100
Computed *A*	150	100	149	400	400
Given *A*	150	100	150	400	

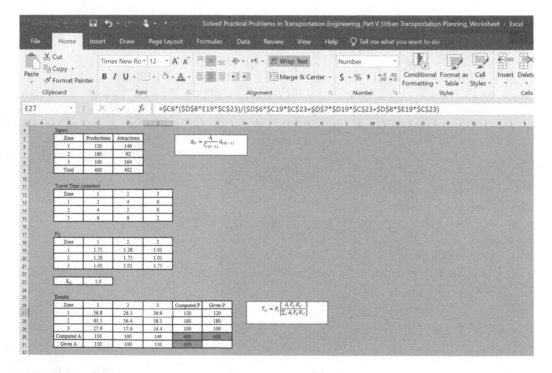

FIGURE 2.31 Image of MS Excel worksheet used for estimating zone-to-zone trips by the gravity model after the second iteration in Problem 2.30.

2.31 A study area consists of three zones (1, 2, and 3). The current number of trips between the zones is shown in Table 2.60. In five years, it is predicted that the number of trips in each zone will be increased to the values shown in the last column of the table. Determine the number of trips between zones (trip distribution) after five years using the Fratar model after only one iteration.

TABLE 2.60
Present Trip Matrix and Trips in Five Years for a Three-Zone Study Area for Problem 2.31

Zone	1	2	3	Present Trip Generation	Trip Generation in Five Years	Growth Factor
1	–	430	170	600	750	1.25
2	430	–	330	760	1064	1.40
3	170	330	–	500	550	1.10
Total	600	760	500			

Solution:

The Fratar model given in the following expression is used to compute the zone-to-zone trips:

$$T_{ij} = \left(t_i G_i\right)\left[\frac{t_{ij} G_j}{\sum_j t_{ij} G_j}\right] \qquad (2.24)$$

Where:

T_{ij} = the number of trips produced in zone i and attracted to zone j (the number of trips estimated from zone i to zone j)

t_i = the number of trips produced in zone i (trip generation in zone i)

$T_i = t_i G_i$ = the number of trips produced in zone i in the future (future trip generation in zone i)

t_{ij} = the number of trips between zone i and zone j

G_j = the growth factor for zone j

Sample Calculation:
The number of trips between zone 1 and zone 2 is determined as below:

$$T_{ij} = \left(t_i G_i\right)\left[\frac{t_{ij} G_j}{\sum_j t_{ij} G_j}\right]$$

$\Rightarrow$

$$T_{1\text{-}2} = \left(t_1 G_1\right)\left[\frac{t_{12} G_2}{\sum_j t_{1j} G_j}\right] = \left(t_1 G_1\right)\left[\frac{t_{12} G_2}{t_{12} G_2 + t_{13} G_3}\right]$$

$\Rightarrow$

$$T_{1\text{-}2} = \left(600 \times 1.25\right)\left[\frac{(430)(1.40)}{(430)(1.40) + (170)(1.10)}\right]$$

$$= 572.2 \text{ trips} \cong 572 \text{ trips}$$

And:

$$T_{2\text{-}1} = \left(t_2 G_2\right)\left[\dfrac{t_{21}G_1}{\displaystyle\sum_j t_{2j}G_j}\right] = \left(t_2 G_2\right)\left[\dfrac{t_{21}G_1}{t_{21}G_1 + t_{23}G_3}\right]$$

$\Rightarrow$

$$T_{2\text{-}1} = \left(760 \times 1.40\right)\left[\dfrac{(430)(1.25)}{(430)(1.25)+(330)(1.10)}\right]$$

$$= 635.1 \text{ trips} \cong 635 \text{ trips}$$

$\Rightarrow$

The average value of $T_{1\text{-}2}$ is equal to:

$$T_{1\text{-}2} = \dfrac{572.2 + 635.1}{2} = 603.7 \text{ trips} \cong 604 \text{ trips}$$

The procedure to estimate the number of trips between the zones and the final results are summarized in Tables 2.61 through 2.63, respectively. The MS Excel worksheet used to perform the computations of this problem is shown in Figure 2.32.

TABLE 2.61

Computations of the Zone-to-Zone Trips by the Fratar Model for Problem 2.31

Zone-to-Zone Trips	Procedure	Value
$T_{1\text{-}2}$	$T_{1\text{-}2} = \left(600 \times 1.25\right)\left[\dfrac{(430)(1.40)}{(430)(1.40)+(170)(1.10)}\right] =$	572.2
$T_{2\text{-}1}$	$T_{2\text{-}1} = \left(760 \times 1.40\right)\left[\dfrac{(430)(1.25)}{(430)(1.25)+(330)(1.10)}\right] =$	635.1
Average $T_{1\text{-}2}$		603.7
$T_{1\text{-}3}$	$T_{1\text{-}3} = \left(600 \times 1.25\right)\left[\dfrac{(170)(1.10)}{(430)(1.40)+(170)(1.10)}\right] =$	177.8
$T_{3\text{-}1}$	$T_{3\text{-}1} = \left(500 \times 1.10\right)\left[\dfrac{(170)(1.25)}{(170)(1.25)+(330)(1.40)}\right] =$	173.3
Average $T_{1\text{-}3}$		175.5
$T_{2\text{-}3}$	$T_{2\text{-}3} = \left(760 \times 1.40\right)\left[\dfrac{(330)(1.10)}{(430)(1.25)+(330)(1.10)}\right] =$	428.9
$T_{3\text{-}2}$	$T_{3\text{-}2} = \left(500 \times 1.10\right)\left[\dfrac{(330)(1.40)}{(170)(1.25)+(330)(1.40)}\right] =$	376.7
Average $T_{2\text{-}3}$		402.8

TABLE 2.62
Computed Trip Distribution (Zone-to-Zone Trips) by the Fratar Model for the Study Area for Problem 2.31

				Future Trip Generation	
Zone	**1**	**2**	**3**	**Computed Total Trips**	**Actual Total Trips**
1	–	572	178	750	750
2	635	–	429	1064	1064
3	173	377	–	550	550
Total	808	949	607		

TABLE 2.63
Average Computed Trip Distribution (Zone-to-Zone) Trips by the Fratar Model for the Study Area for Problem 2.31

				Future Trip Generation	
Zone	**1**	**2**	**3**	**Computed Total Trips**	**Actual Total Trips**
1	–	603.7	175.5	779	750
2	603.7	–	402.8	1006	1064
3	175.5	402.8	–	578	550
Total	779	1006	578		

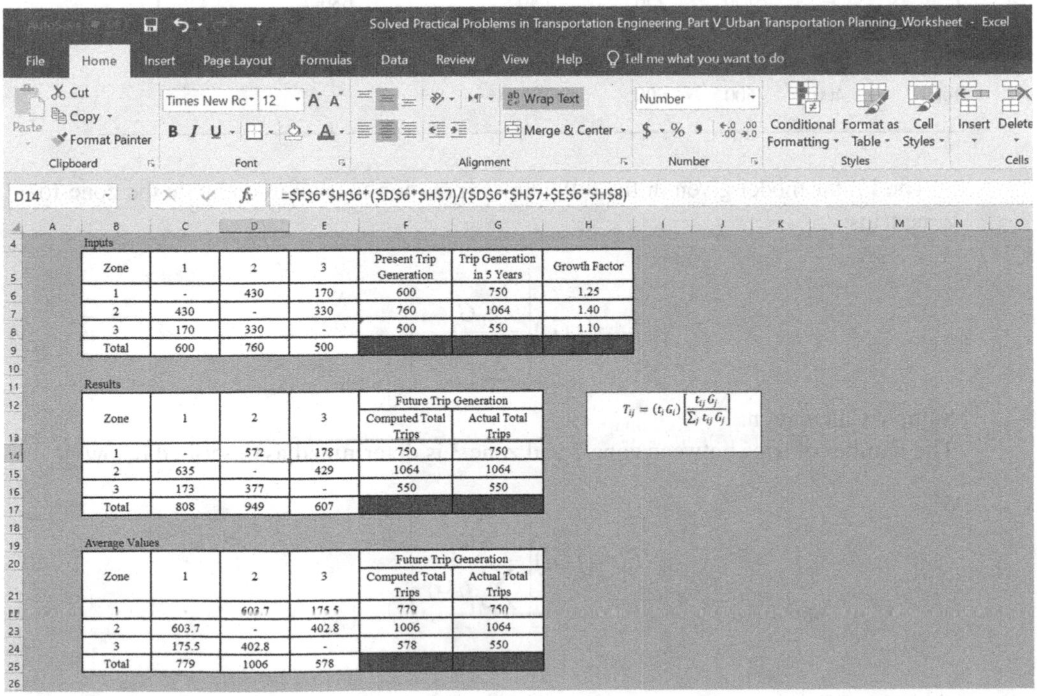

Solved Practical Problems in Transportation Engineering_Part V_Urban Transportation Planning_Worksheet - Excel

D14 fx =$F$6*$H$6*($D$6*$H$7)/($D$6*$H$7+$E$6*$H$8)

Inputs

Zone	1	2	3	Present Trip Generation	Trip Generation in 5 Years	Growth Factor
1	-	430	170	600	750	1.25
2	430	-	330	760	1064	1.40
3	170	330	-	500	550	1.10
Total	600	760	500			

Results

				Future Trip Generation	
Zone	1	2	3	Computed Total Trips	Actual Total Trips
1	-	572	178	750	750
2	635	-	429	1064	1064
3	173	377	-	550	550
Total	808	949	607		

$$T_{ij} = (t_i G_i)\left[\frac{t_{ij} G_j}{\sum_j t_{ij} G_j}\right]$$

Average Values

				Future Trip Generation	
Zone	1	2	3	Computed Total Trips	Actual Total Trips
1	-	603.7	175.5	779	750
2	603.7	-	402.8	1006	1064
3	175.5	402.8	-	578	550
Total	779	1006	578		

FIGURE 2.32 Image of MS Excel worksheet used for estimating zone-to-zone trips distribution by the Fratar model for Problem 2.31.

2.32   The trips between zones and the trips after five years for a three-zone study area are shown in Table 2.64. Using the Fratar model and only one iteration, estimate the zone-to-zone trips (trip distribution) after five years.

TABLE 2.64
Present Trip Matrix and Trips in Five Years for a Three-Zone Study Area for Problem 2.32

Zone	1	2	3	Trips after Five Years
1	–	150	250	680
2	150	–	350	700
3	250	350	–	720
Total	400	500	600	

Solution:

The present trip generation (productions) and the growth factor are computed for each zone based on the given data. The results are shown in Table 2.65.

TABLE 2.65
Calculated Present Trip Productions and Growth Factors for Problem 2.32

Zone	1	2	3	Present Trip Generation	Trip Generation after Five Years	Growth Factor
1	–	150	250	400	680	1.7
2	150	–	350	500	700	1.4
3	250	350	–	600	720	1.2
Total	400	500	600			

The Fratar model given in the following expression is used to compute the zone-to-zone trips:

$$T_{ij} = \left(t_i G_i \right) \left[\frac{t_{ij} G_j}{\sum_j t_{ij} G_j} \right]$$

Sample Calculation:
The number of trips between zone 2 and zone 3 is determined as described below:

$$T_{ij} = \left(t_i G_i \right) \left[\frac{t_{ij} G_j}{\sum_j t_{ij} G_j} \right]$$

But:

$$(t_iG_i) = T_i = \text{future trip generation in zone } i$$

$\Rightarrow$

$$T_{2\text{-}3} = (t_2G_2)\left[\frac{t_{23}G_3}{\sum_j t_{2j}G_j}\right] = (T_2)\left[\frac{t_{23}G_3}{t_{21}G_1 + t_{23}G_3}\right]$$

$\Rightarrow$

$$T_{2\text{-}3} = (700)\left[\frac{(350)(1.2)}{(150)(1.7)+(350)(1.2)}\right] = 435.6 \text{ trips} \cong 436 \text{ trips}$$

And:

$$T_{3\text{-}2} = (t_3G_3)\left[\frac{t_{32}G_2}{\sum_j t_{3j}G_j}\right] = (T_3)\left[\frac{t_{32}G_2}{t_{31}G_1 + t_{32}G_2}\right]$$

$\Rightarrow$

$$T_{3\text{-}2} = (720)\left[\frac{(350)(1.4)}{(250)(1.7)+(350)(1.4)}\right] = 385.6 \text{ trips} \cong 386 \text{ trips}$$

$\Rightarrow$
The average value of $T_{2\text{-}3}$ is equal to:

$$T_{2\text{-}3} = \frac{435.6 + 385.6}{2} = 410.6 \text{ trips} \cong 411 \text{ trips}$$

The procedure to estimate the number of trips between the zones and the final results are summarized in Tables 2.66 through 2.68, respectively. The MS Excel worksheet used to perform the computations of this problem is shown in Figure 2.33.

TABLE 2.66

Computations of the Zone-to-Zone Trips by the Fratar Model for Problem 2.32

Zone-to-Zone Trips	Procedure	Value
T_{1-2}	$T_{1\text{-}2} = (680)\left[\dfrac{(150)(1.4)}{(150)(1.4)+(250)(1.2)}\right] =$	280.0
T_{2-1}	$T_{2\text{-}1} = (700)\left[\dfrac{(150)(1.7)}{(150)(1.7)+(350)(1.2)}\right] =$	264.4
Average T_{1-2}		272.2
T_{1-3}	$T_{1\text{-}3} = (680)\left[\dfrac{(250)(1.2)}{(150)(1.4)+(250)(1.2)}\right] =$	400.0
T_{3-1}	$T_{3\text{-}1} = (720)\left[\dfrac{(250)(1.7)}{(250)(1.7)+(350)(1.4)}\right] =$	334.4
Average T_{1-3}		367.2
T_{2-3}	$T_{2\text{-}3} = (700)\left[\dfrac{(350)(1.2)}{(150)(1.7)+(350)(1.2)}\right] =$	435.6
T_{3-2}	$T_{3\text{-}2} = (720)\left[\dfrac{(350)(1.4)}{(250)(1.7)+(350)(1.4)}\right] =$	385.6
Average T_{2-3}		410.6

TABLE 2.67

Iteration-Computed Trip Distribution (Zone-to-Zone Trips) by the Fratar Model for the Study Area for Problem 2.32

Zone	1	2	3	Future Trip Generation	
				Computed Total Trips	Actual Total Trips
1	–	280.0	400.0	680	680
2	264.4	–	435.6	700	700
3	334.4	385.6	–	720	720
Total	599	666	836		

TABLE 2.68

Average Computed Trip Distribution (Zone-to-Zone) Trips by the Fratar Model for the Study Area for Problem 2.32

Zone	1	2	3	Future Trip Generation	
				Computed Total Trips	Actual Total Trips
1	–	272.2	367.2	639	680
2	272.2	–	410.6	683	700
3	367.2	410.6	–	778	720
Total	639	683	778		

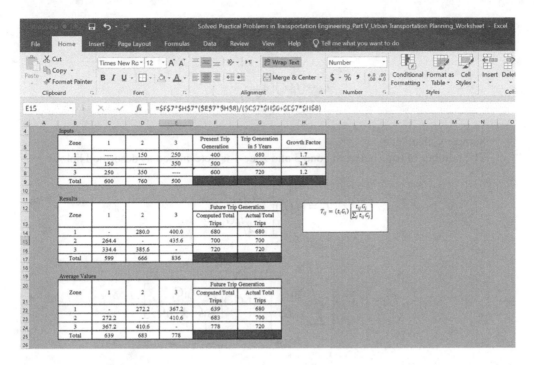

FIGURE 2.33 Image of MS Excel worksheet used for estimating zone-to-zone trips distribution by the Fratar model for Problem 2.32.

2.33   If the number of present trips between zones and the growth factors for a three-zone study area are as shown in Table 2.69, estimate the future zone-to-zone trips (trip distribution) using the Fratar model and after two iterations.

TABLE 2.69
Present Trip Matrix and Growth Factors for a Three-Zone Study Area for Problem 2.33

Zone	1	2	3	Growth Factor
1	–	100	300	1.2
2	100	–	200	1.5
3	300	200	–	2.0
Total	400	300	500	

Solution:

The present trip generation (productions) and future trips are computed for each zone based on the given data. The results are shown in Table 2.70.

TABLE 2.70
Calculated Present and Future Trip Generation Results for Problem 2.33

Zone	1	2	3	Present Trip Generation	Future Trip Generation	Growth Factor
1	–	100	300	400	480	1.2
2	100	–	200	300	450	1.5
3	300	200	–	500	1000	2.0
Total	400	300	500			

The Fratar model given in the following expression is used to compute the zone-to-zone trips:

$$T_{ij} = \left(t_i G_i\right)\left[\frac{t_{ij}G_j}{\sum_j t_{ij}G_j}\right]$$

Sample Calculation:
The number of trips between zone 3 and zone 1 is determined as described below:

$$T_{ij} = \left(t_i G_i\right)\left[\frac{t_{ij}G_j}{\sum_j t_{ij}G_j}\right]$$

$\Rightarrow$

$$T_{3\text{-}1} = \left(t_3 G_3\right)\left[\frac{t_{31}G_1}{\sum_j t_{3j}G_j}\right] = \left(t_3 G_3\right)\left[\frac{t_{31}G_1}{t_{31}G_1 + t_{32}G_2}\right]$$

$\Rightarrow$

$$T_{3\text{-}1} = \left(500 \times 2.0\right)\left[\frac{(300)(1.2)}{(300)(1.2)+(200)(1.5)}\right] = 545.5 \text{ trips} \cong 546 \text{ trips}$$

And:

$$T_{1\text{-}3} = \left(t_1 G_1\right)\left[\frac{t_{13}G_3}{\sum_j t_{1j}G_j}\right] = \left(t_1 G_1\right)\left[\frac{t_{13}G_3}{t_{12}G_2 + t_{13}G_3}\right]$$

$\Rightarrow$

$$T_{1\text{-}3} = \left(400 \times 1.2\right)\left[\frac{(300)(2.0)}{(100)(1.5)+(300)(2.0)}\right] = 384.0 \text{ trips}$$

$\Rightarrow$

The average value of T_{3-1} is equal to:

$$T_{3\text{-}1} = \frac{546+384}{2} = 465 \text{ trips}$$

The procedure to estimate the number of trips between the zones and the final results are summarized in Tables 2.71 through 2.73, respectively. Figure 2.34 shows a screen image of the MS Excel worksheet used to perform the computations of the first iteration results in this problem.

TABLE 2.71

Computations of the Zone-to-Zone Trips by the Fratar Model for Problem 2.33

Zone-to-Zone Trips	Procedure	Value
T_{1-2}	$T_{1-2} = (400 \times 1.2)\left[\dfrac{(100)(1.5)}{(100)(1.5)+(300)(2.0)}\right] =$	96.0
T_{2-1}	$T_{2-1} = (300 \times 1.5)\left[\dfrac{(100)(1.2)}{(100)(1.2)+(200)(2.0)}\right] =$	103.8
Average T_{1-2}		99.9
T_{1-3}	$T_{1-3} = (400 \times 1.2)\left[\dfrac{(300)(2.0)}{(100)(1.5)+(300)(2.0)}\right] =$	384.0
T_{3-1}	$T_{3-1} = (500 \times 2.0)\left[\dfrac{(300)(1.2)}{(300)(1.2)+(200)(1.5)}\right] =$	545.5
Average T_{1-3}		464.7
T_{2-3}	$T_{2-3} = (300 \times 1.5)\left[\dfrac{(200)(2.0)}{(100)(1.2)+(200)(2.0)}\right] =$	346.2
T_{3-2}	$T_{3-2} = (500 \times 2.0)\left[\dfrac{(200)(1.5)}{(300)(1.2)+(200)(1.5)}\right] =$	454.5
Average T_{2-3}		400.3

TABLE 2.72

First Iteration-Computed Trip Distribution (Zone-to-Zone Trips) by the Fratar Model for the Study Area for Problem 2.33

Zone	1	2	3	Future Trip Generation	
				Computed Total Trips	Actual Total Trips
1	–	96.0	384.0	480	480
2	103.8	–	346.2	450	450
3	545.5	454.5	–	1000	1000
Total	649	551	730		

TABLE 2.73

First Iteration-Average Computed Trip Distribution (Zone-to-Zone) Trips by the Fratar Model for the Study Area for Problem 2.33

Zone	1	2	3	Future Trip Generation	
				Computed Total Trips	Actual Total Trips
1	–	99.9	464.7	565	480
2	99.9	–	400.3	500	450
3	464.7	400.3	–	865	1000
Total	565	500	865		

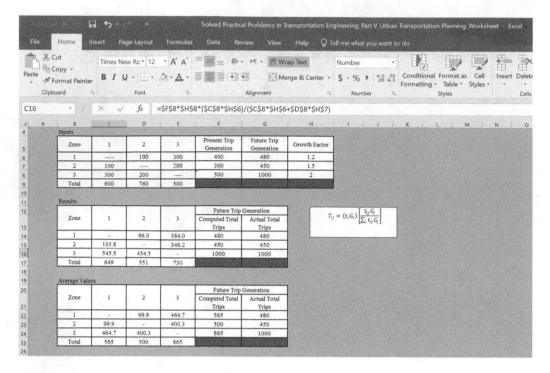

FIGURE 2.34 Image of MS Excel worksheet used for estimating zone-to-zone trips by the Fratar model for Problem 2.33.

In the second iteration, new growth factors are estimated based on the estimated future trip generation and the actual future trip generation. Therefore, the results in Table 2.74 are obtained.

TABLE 2.74
Second Iteration Growth Factors for Problem 2.33

| Zone | 1 | 2 | 3 | Future Trip Generation | | |
				Computed Total Trips	Actual Total Trips	Growth Factor
1	–	99.9	464.7	565	480	0.85
2	99.9	–	400.3	500	450	0.90
3	464.7	400.3	–	865	1000	1.16
Total	565	500	865			

Following the same procedure performed in the first iteration and using the Fratar model given in the following formula, the zone-to-zone trips are estimated.

$$T_{ij} = \left(t_i G_i\right)\left[\frac{t_{ij}G_j}{\displaystyle\sum_j t_{ij}G_j}\right]$$

But:

$$\left(t_i G_i\right) = T_i = \text{future trip generation in zone } i$$

Sample Calculation:

The number of trips between zone 3 and zone 1 is determined as below:

$$T_{ij} = (t_i G_i) \left[\frac{t_{ij} G_j}{\sum_j t_{ij} G_j} \right]$$

$\Rightarrow$

$$T_{3\text{-}1} = (t_3 G_3) \left[\frac{t_{31} G_1}{\sum_j t_{3j} G_j} \right] = (T_3) \left[\frac{t_{31} G_1}{t_{31} G_1 + t_{32} G_2} \right]$$

$\Rightarrow$

$$T_{3\text{-}1} = (1000) \left[\frac{(464.7)(0.85)}{(464.7)(0.85) + (400.3)(0.90)} \right] \cong 523.0 \text{ trips}$$

And:

$$T_{1\text{-}3} = (t_1 G_1) \left[\frac{t_{13} G_3}{\sum_j t_{1j} G_j} \right] = (T_1) \left[\frac{t_{13} G_3}{t_{12} G_2 + t_{13} G_3} \right]$$

$\Rightarrow$

$$T_{1\text{-}3} = (480) \left[\frac{(464.7)(1.16)}{(99.9)(0.90) + (464.7)(1.16)} \right] \cong 411 \text{ trips}$$

$\Rightarrow$

The average value of T_{3-1} is equal to:

$$T_{3\text{-}1} = \frac{523 + 411}{2} = 467 \text{ trips}$$

The final trip distribution results of the second iteration are summarized in Tables 2.75 and 2.76. Figure 2.35 shows a screen image of the MS Excel worksheet used to perform the computations of the second iteration results in this problem.

TABLE 2.75

Second Iteration-Computed Trip Distribution (Zone-to-Zone Trips) by the Fratar Model for the Study Area for Problem 2.33

| Zone | 1 | 2 | 3 | Future Trip Generation | |
				Computed Total Trips	Actual Total Trips
1	–	68.8	411.2	480	480
2	69.8	–	380.2	450	450
3	523.1	476.9	–	1000	1000
Total	593	546	791		

TABLE 2.76

Second Iteration-Average Computed Trip Distribution (Zone-to-Zone) Trips by the Fratar Model for the Study Area for Problem 2.33

				Future Trip Generation	
Zone	1	2	3	Computed Total Trips	Actual Total Trips
1	–	69.3	467.2	536	480
2	69.3	–	428.5	498	450
3	467.2	428.5	–	896	1000
Total	536	498	896		

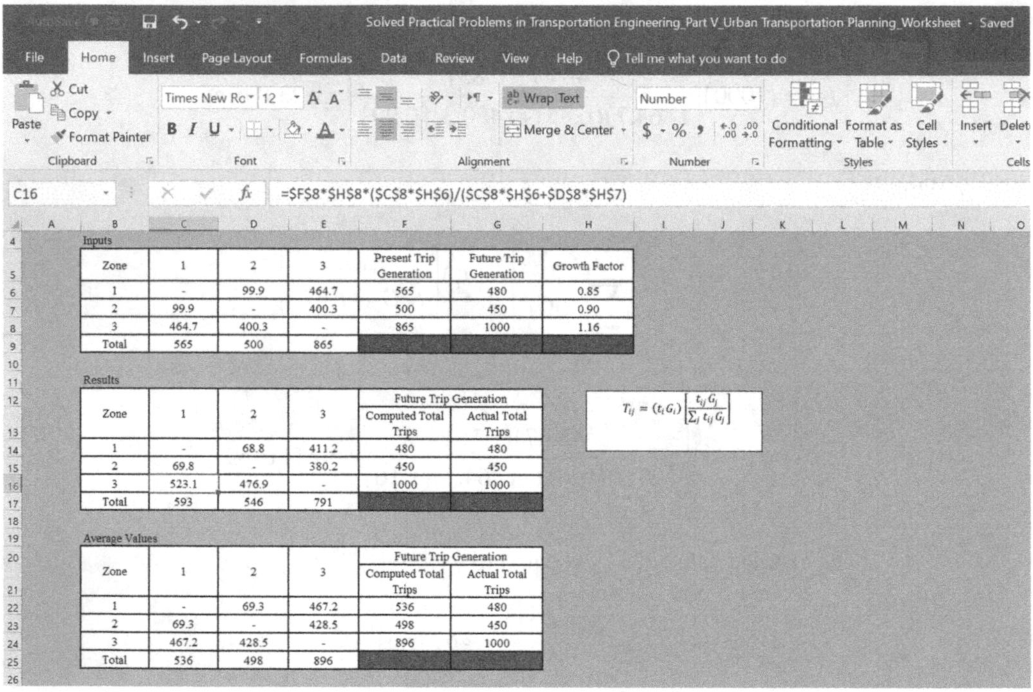

FIGURE 2.35 Image of MS Excel worksheet used for estimating zone-to-zone trips by the Fratar model after the second iteration for Problem 2.33.

The iterative process can be performed several times until the future estimated trips are close enough to the actual future trips originally given in the problem. In each iteration, new growth factors are estimated based on the future estimated trips in the previous iteration and the actual future trips (given originally in the problem).

TRANSPORT MODE CHOICE

2.34 In a residential zone, the average number of cars per household is 0.40 and the residential density is 10,000 person/mile² (3861 person/km²). Use the relationship between the transit mode split and the urban travel factor (UTF) shown in Figure 2.36 to determine the percentage of residents who are expected to use transit.

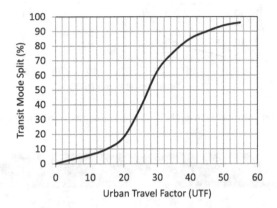

FIGURE 2.36 Urban travel factor (UTF) versus transit mode split for Problem 2.34. (From Garber/Hoel. *Traffic and Highway Engineering, 4E.* ©2009 Cengage Learning, a part of Cengage, Inc. Reproduced by permission. www.cengage.com/permissions.)

Solution:

The urban travel factor is given by the following formula:

$$\text{UTF} = \frac{1}{1000}\left(\frac{\text{Residential Density}}{\text{Car Ownership}}\right) \tag{2.25}$$

Where:

UTF (urban travel factor) = the product of the residential density and the households per car (the inverse of car ownership measured by cars/HH)

Residential Density = the density of the residential area (person/mile²)

Car Ownership = the number of cars per household in the study area (cars/HH)

⇒

$$\text{UTF} = \frac{1}{1000}\left(\frac{10000}{0.40}\right) = 25.0$$

Using the relationship between the transit mode split and the urban travel factor in the figure above; at UTF = 25.0, the transit mode split = 37%. Therefore, the percentage of residents who are expected to use the transit mode is 37% (Figure 2.37). The MS Excel worksheet used to conduct the computations of this problem is shown in Figure 2.38.

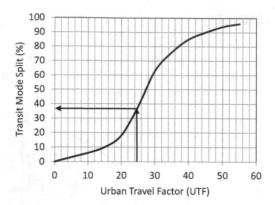

FIGURE 2.37 Using the urban travel factor (UTF) to determine the transit mode split for Problem 2.34.

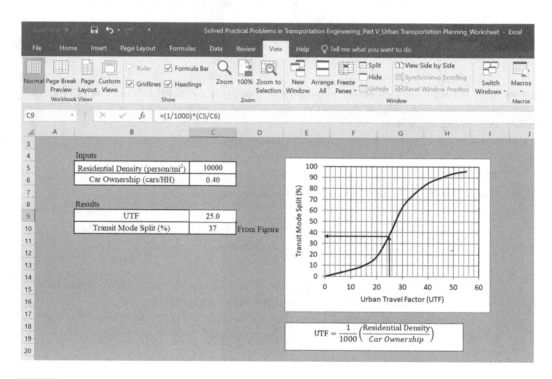

FIGURE 2.38 Image of MS Excel worksheet used for determining the transit mode split using the UTF for Problem 2.34.

2.35 A survey conducted for a medium-income residential area shows that the average car (auto) ownership for a residential area is 1.0 car/HH and that 85% of the residents use for their own cars for trips. Use the relationship between the transit mode split and the urban travel factor (UTF) in Problem 2.34 to estimate the residential density of the area.

Solution:

Since 85% of the residents use their own cars; therefore, the transit mode split is determined as:

$$\text{transit mode split} = 100 - 85 = 15\%.$$

Using the relationship between the transit mode split and the urban travel factor, at a transit mode split = 15%, the UTF $\cong$ 18 (Figure 2.39). Hence, the residential density of the area is estimated using the UTF formula, as shown below:

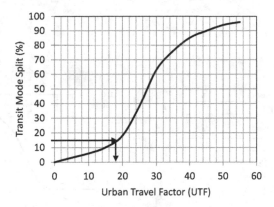

FIGURE 2.39 Using the urban travel factor (UTF) and transit mode split relationship for Problem 2.35.

$$\text{UTF} = \frac{1}{1000} \left(\frac{\text{Residential Density}}{\text{Car Ownership}} \right)$$

$\Rightarrow$

$$18 = \frac{1}{1000} \left(\frac{\text{Residential Density}}{1.0} \right)$$

$\Rightarrow$

$$\text{Residential Density} = 18000 \text{ person/mi}^2 \left(6950 \text{ person/km}^2 \right)$$

The MS Excel worksheet shown in Figure 2.40 is used to perform the computations of this problem.

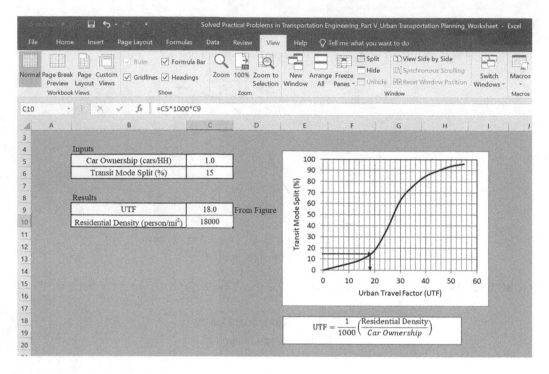

FIGURE 2.40   Image of MS Excel worksheet used for the computations of Problem 2.35.

2.36   A high-income residential area with a residential density of 8000 person/mile² (3089 person/km²) and an average car ownership of two cars/HH. Estimate the total number of trips made by private cars if the total number of production trips in the area is 8000 trips/day using the UTF method.

Solution:

The UTF formula given below will be used:

$$UTF = \frac{1}{1000}\left(\frac{\text{Residential Density}}{\text{Car Ownership}}\right)$$

⇒

$$UTF = \frac{1}{1000}\left(\frac{8000}{2.0}\right) = 4.0$$

Using the relationship between the transit mode split and the urban travel factor, at UTF =4, the transit mode split=3% (Figure 2.41). Therefore:

The percentage of residents expected to use their own cars = 100 − 3 = 97%.

The number of trip made by private cars = $\dfrac{97}{100}(8000) = 7760$ trips

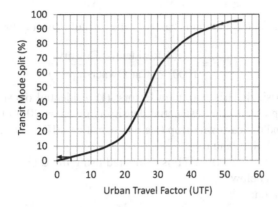

FIGURE 2.41 Using the urban travel factor (UTF) to determine the transit mode split for Problem 2.36.

The MS Excel worksheet shown in Figure 2.42 is used to perform the computations of this problem.

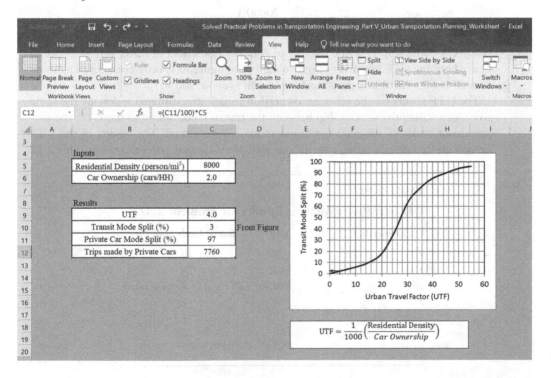

FIGURE 2.42 Image of MS Excel worksheet used for determining the transit mode split using the UTF for Problem 2.36.

2.37   Perform a sensitivity analysis for the impact of the car ownership factor (variable) on the transit mode split (the percentage of residents that will use the transit mode) in the urban travel factor (UTF) method for residential zones with a residential density = 20,000 person/mile^2 (7722 person/km^2). What conclusions can be made based on this analysis?

Solution:

The urban travel factor method is based on the formula given below:

$$UTF = \frac{1}{1000}\left(\frac{\text{Residential Density}}{\text{Car Ownership}}\right)$$

This formula basically has two main factors: (1) residential density, and (2) car ownership. In this problem, the residential density is given as 20,000 person/mile^2 (7722 person/km^2), and the car ownership (the number of cars/HH) will change according to the values: 1, 2, 3, 4, 5, and 6.

Based on these values, the UTF is computed using the UTF formula. A sample calculation is shown below:

Sample Calculation:

$$UTF = \frac{1}{1000}\left(\frac{\text{Residential Density}}{\text{Car Ownership}}\right)$$

$\Rightarrow$

$$UTF = \frac{1}{1000}\left(\frac{20000}{1}\right) = 20.0$$

When car ownership is 0 (the number of cars per household is zero), this is a special case where the UTF has a very high value. In the figure, it is only considered as more than 60 (>60). Therefore, the transit mode split is about 100%.

The UTF results for the other car ownership values are estimated following the same procedure. Using the relationship between the transit mode split and the UTF, the percentage of residents who use the transit mode is determined. The results of the UTF along with the corresponding transit mode split (%) values are summarized in Table 2.77.

TABLE 2.77
UTF Sensitivity Analysis Results for Problem 2.37

Car Ownership	UTF	Transit Mode Split (%)
0	> 60	100
1	20.0	18
2	10.0	7
3	6.7	5
4	5.0	4
5	4.0	3
6	3.3	2

Based on the sensitivity analysis results, it can be concluded that the impact of car ownership has a significant effect on transit mode split when the value ≤1 car/HH. However, when the value ≥2 cars/HH, the effect is minimal and the difference between the transit mode split values is very low. The change in the transit mode split with the change in car ownership is plotted in Figure 2.43 and this supports the above conclusions. The MS Excel worksheet shown in Figure 2.44 is used to perform the computations of this problem.

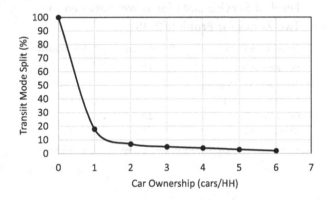

FIGURE 2.43 Car ownership versus transit mode split from the sensitivity analysis for Problem 2.37.

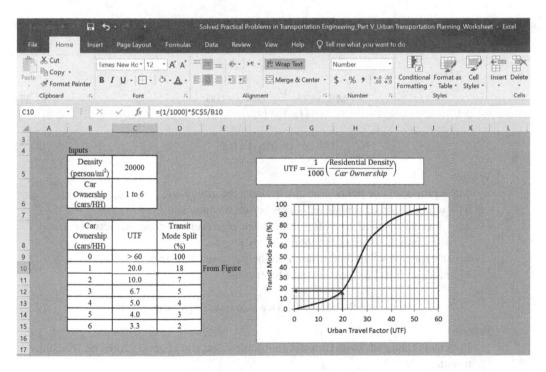

FIGURE 2.44 Image of MS Excel worksheet used for the sensitivity analysis of car ownership on transit mode split for Problem 2.37.

2.38   Data on the level-of-service (LOS) parameters for travel between a residential area and a commercial zone is developed, as shown in Table 2.78. Determine the percentage of trips made by transit and auto using trip interchange models if the average income in the residential area is $30,000. Use a value of 2.0 for exponent b. Assume that the year has 120,000 working minutes.

TABLE 2.78
Level-of-Service Data for Travel between the Two Zones for Problem 2.38

Travel LOS Data	Private Auto	Public Transit
Distance (mile)	15	12
Trip Fuel Cost ($/mile)	0.22	0.14
Waste Time (min)	7	11
Other Cost ($)	2.3	0
Speed (mph)	45	30

Solution:

The trip interchange model expressed in the following formula is used:

$$MS_a = \frac{(I_{ija})^{-b}}{(I_{ija})^{-b} + (I_{ijt})^{-b}} \times 100 \qquad (2.26)$$

Or:

$$MS_a = \frac{(I_{ijt})^{b}}{(I_{ija})^{b} + (I_{ijt})^{b}} \times 100 \qquad (2.27)$$

$$MS_t = (1 - MS_a) \times 100 \qquad (2.28)$$

Where:

   MS_t = proportion of public transit trips between zone i and j
   MS_a = proportion of private auto trips between zone i and j
   I_{ijm} = a value referred to as the impedance of travel of mode m, between i and j (which is a measure of the total trip cost)
   b = an exponent value that is based on the purpose of the trip
   m = mode of travel (a for auto and t for transit)

$$I_{ijm} = (\text{In-vehicle time}) + (2.5 \times \text{waste time}) + (3 \times \text{trip cost}) \qquad (2.29)$$

Where:

   In-vehicle time and waste time are in minutes, and trip cost is in units of $/income/ minute.

$\Rightarrow$

$$I_{1\text{-}2a} = \left(\frac{15}{45} \times 60\right) + (2.5 \times 7) + \left[\frac{3(0.22 \times 15 + 2.3)}{30000/120000}\right]$$

$$= 104.7 \text{ equivalent minutes}$$

$$I_{1\text{-}2t} = \left(\frac{12}{30} \times 60\right) + \left(2.5 \times 11\right) + \left[\frac{3\left(0.14 \times 12\right)}{30000 / 120000}\right]$$

$$= 71.7 \text{ equivalent minutes}$$

$\Rightarrow$

$$MS_a = \frac{\left(I_{1\text{-}2a}\right)^{-b}}{\left(I_{1\text{-}2a}\right)^{-b} + \left(I_{1\text{-}2t}\right)^{-b}} \times 100$$

$\Rightarrow$

$$MS_a = \frac{\left(104.7\right)^{-2}}{\left(104.7\right)^{-2} + \left(71.7\right)^{-2}} \times 100 = 31.9\%$$

$$MS_t = \left(1 - MS_a\right) \times 100$$

$\Rightarrow$

$$MS_t = 100 - 31.9 = 68.1\%$$

The MS Excel worksheet shown in Figure 2.45 is used to perform the computations of this problem.

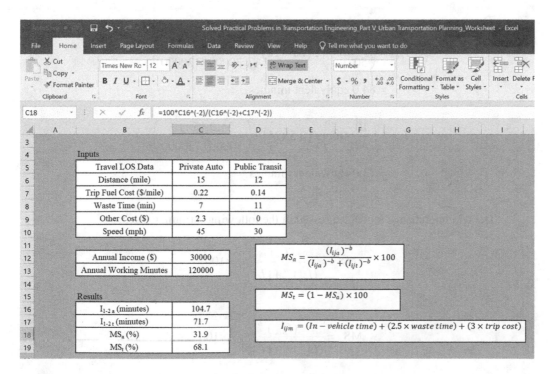

FIGURE 2.45 Image of MS Excel worksheet used for estimating transit mode split by trip interchange models for Problem 2.38.

2.39   The travel data between residential zone R and business zone B are shown in Table 2.79. If an exponent value of 2.0 is used for work travel, the median annual income is $30,000, and the travel impedance function for transportation mode m is given as:

TABLE 2.79
Travel Data between Two Zones for Problem 2.39

Parameter	Auto	Transit
Distance (mile)	15	10
Trip cost per mile ($)	0.20	0.15
Excess time (min)	12	16
Speed (mph)	50	40

$$I_{ijm} = \left(\text{In-vehicle time}\right) + \left(2.5 \times \text{waste time}\right) + \left(3 \times \text{trip cost}\right);$$

then using the trip interchange model, estimate the percentage of work trips by auto and the percentage of work trips by transit.

Solution:

The travel impedance function for transportation mode m given in the problem will be used to determine the equivalent minutes for each mode, as shown below:

$$I_{ijm} = \left(\text{In-vehicle time}\right) + \left(2.5 \times \text{waste time}\right) + \left(3 \times \text{trip cost}\right)$$

$\Rightarrow$

$$I_{RBa} = \left(\frac{15}{50} \times 60\right) + \left(2.5 \times 12\right) + \left(\frac{3 \times 0.20 \times 15}{30000/120000}\right)$$

$$= 84.0 \text{ equivalent minutes}$$

$$I_{RBt} = \left(\frac{10}{40} \times 60\right) + \left(2.5 \times 16\right) + \left(\frac{3 \times 0.15 \times 10}{30000/120000}\right)$$

$$= 73.0 \text{ equivalent minutes}$$

The trip interchange model expressed in the following formula is used:

$$MS_a = \frac{\left(I_{ija}\right)^{-b}}{\left(I_{ija}\right)^{-b} + \left(I_{ijt}\right)^{-b}} \times 100$$

$\Rightarrow$

$$MS_a = \frac{\left(I_{RBa}\right)^{-b}}{\left(I_{RBa}\right)^{-b} + \left(I_{RBt}\right)^{-b}} \times 100$$

$\Rightarrow$

$$MS_a = \frac{\left(84.0\right)^{-2}}{\left(84.0\right)^{-2} + \left(73.0\right)^{-2}} \times 100 = 43.0\%$$

$$MS_t = (1 - MS_a) \times 100$$

$$\Rightarrow$$

$$MS_t = 100 - 43.0 = 57.0\%$$

The MS Excel worksheet shown in Figure 2.46 is used to perform the computations of this problem.

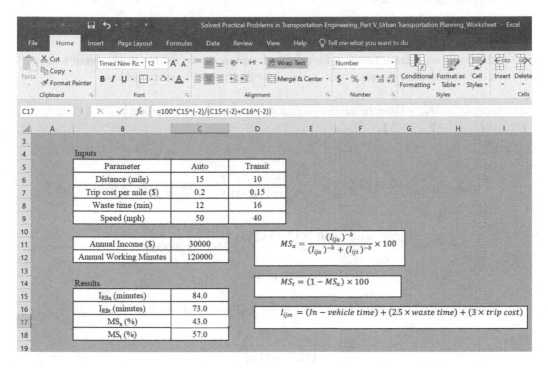

FIGURE 2.46 Image of MS Excel worksheet used for estimating transit mode split by trip interchange models for Problem 2.39.

2.40 A road transport authority in a certain region conducts a study to improve the percentage of work trips made by metro (public transit). The travel data between residential area R and business area B is shown in Table 2.80.

TABLE 2.80
Travel Data between Two Zones for Problem 2.40

Variable	Auto	Transit (Metro)
Distance (mile)	20	15
Trip cost per mile ($)	0.25	0.15
Waste time (min)	8	12
Speed (mph)	50	30

If an exponent value of 2.0 is used for work travel, the median annual income is $25,000, and the travel impedance function for transportation mode m is given as:

$$I_{ijm} = (\text{In-vehicle time}) + (2.5 \times \text{waste time}) + (3 \times \text{trip cost}).$$

The study considers two options to encourage residents to use the metro: (1) alternative #1: increase the speed of the metro from 30 to 40 mph, (2) alternative #2: use another shorter route for the metro with a length of 10 miles. Determine the effect of the two alternatives on the percentage of work trips done by the metro (public transit) using the trip interchange models.

Solution:

Existing Conditions:

the travel impedance function for transportation mode m given in the problem will be used to determine the equivalent minutes for each mode, as shown below:

$$I_{ijm} = (\text{In-vehicle time}) + (2.5 \times \text{waste time}) + (3 \times \text{trip cost})$$

$\Rightarrow$

$$I_{\text{RB auto}} = \left(\frac{20}{50} \times 60\right) + (2.5 \times 8) + \left(\frac{3 \times 0.25 \times 20}{25000/120000}\right) = 116.0 \text{ equivalent minutes}$$

$$I_{\text{RB metro}} = \left(\frac{15}{30} \times 60\right) + (2.5 \times 12) + \left(\frac{3 \times 0.15 \times 15}{25000/120000}\right)$$

$$= 92.4 \text{ equivalent minutes}$$

The trip interchange model expressed in the following formula is used:

$$MS_a = \frac{(I_{ija})^{-b}}{(I_{ija})^{-b} + (I_{ijt})^{-b}} \times 100$$

$\Rightarrow$

$$MS_a = \frac{(I_{\text{RB auto}})^{-b}}{(I_{\text{RB auto}})^{-b} + (I_{\text{RB metro}})^{-b}} \times 100$$

$\Rightarrow$

$$MS_a = \frac{(116.0)^{-2}}{(116.0)^{-2} + (92.4)^{-2}} \times 100 = 38.8\%$$

$$MS_t = (1 - MS_a) \times 100$$

$\Rightarrow$

$$MS_t = 100 - 38.8 = 61.2\%$$

Alternative #1 (speed of metro $= 40$ mph):

$$I_{ijm} = (\text{In-vehicle time}) + (2.5 \times \text{waste time}) + (3 \times \text{trip cost})$$

$\Rightarrow$

$$I_{\text{RB auto}} = 116.0 \text{ equivalent minutes} \left(\text{no change}\right)$$

$$I_{\text{RB metro}} = \left(\frac{15}{40} \times 60\right) + \left(2.5 \times 12\right) + \left(\frac{3 \times 0.15 \times 15}{25000/120000}\right)$$

$$= 84.9 \text{ equivalent minutes}$$

The trip interchange model expressed in the following formula is used:

$$MS_a = \frac{\left(I_{ija}\right)^{-b}}{\left(I_{ija}\right)^{-b} + \left(I_{ijt}\right)^{-b}} \times 100$$

$\Rightarrow$

$$MS_a = \frac{\left(I_{\text{RB auto}}\right)^{-b}}{\left(I_{\text{RB auto}}\right)^{-b} + \left(I_{\text{RB metro}}\right)^{-b}} \times 100$$

$\Rightarrow$

$$MS_a = \frac{(116.0)^{-2}}{(116.0)^{-2} + (84.9)^{-2}} \times 100 = 34.9\%$$

$$MS_t = \left(1 - MS_a\right) \times 100$$

$\Rightarrow$

$$MS_t = 100 - 34.9 = 65.1\%$$

Therefore, an increase of about 4% in the percentage of work trips done by metro is the result of increasing the metro speed from 30 to 40 mph.
Alternative #2 (distance used by metro = 10 miles):

$$I_{ijm} = \left(\text{In-vehicle time}\right) + \left(2.5 \times \text{waste time}\right) + \left(3 \times \text{trip cost}\right)$$

$\Rightarrow$

$$I_{\text{RB auto}} = 116.0 \text{ equivalent minutes} \left(\text{no change}\right)$$

$$I_{\text{RB metro}} = \left(\frac{10}{30} \times 60\right) + \left(2.5 \times 12\right) + \left(\frac{3 \times 0.15 \times 10}{25000/120000}\right)$$

$$= 71.6 \text{ equivalent minutes}$$

The trip interchange model expressed in the following formula is used:

$$MS_a = \frac{\left(I_{ija}\right)^{-b}}{\left(I_{ija}\right)^{-b} + \left(I_{ijt}\right)^{-b}} \times 100$$

$\Rightarrow$

$$MS_a = \frac{\left(I_{RB\,auto}\right)^{-b}}{\left(I_{RB\,auto}\right)^{-b} + \left(I_{RB\,metro}\right)^{-b}} \times 100$$

$\Rightarrow$

$$MS_a = \frac{(116.0)^{-2}}{(116.0)^{-2} + (71.6)^{-2}} \times 100 = 27.6\%$$

$$MS_t = \left(1 - MS_a\right) \times 100$$

$\Rightarrow$

$$MS_t = 100 - 27.6 = 72.4\%$$

In conclusion, an increase of more than 11% in the percentage of work trips using the metro is the result of reducing the metro route length from 15 to 10 miles. Figure 2.47 shows a screen image of the MS Excel worksheet used to perform the computations of this problem.

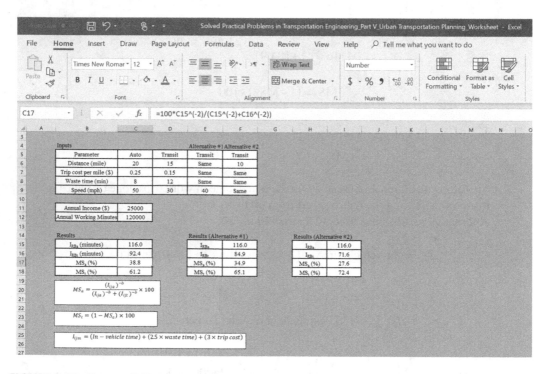

FIGURE 2.47 Image of MS Excel worksheet used for estimating transit mode split by trip interchange models for different alternatives in Problem 2.40.

2.41 The travel impedance function values for trips between two zones are 44 and 30 minutes for auto mode and transit mode, respectively. If an exponent value of 1.5 is used for travel between the two zones, using the trip interchange model, estimate the percentage of trips using an auto mode and using a transit mode.

Solution:

The trip interchange model shown below is used:

$$MS_a = \frac{\left(I_{ija}\right)^{-b}}{\left(I_{ija}\right)^{-b} + \left(I_{ijt}\right)^{-b}} \times 100$$

$$\Rightarrow$$

$$MS_{auto} = \frac{\left(I_{1\text{-}2\,auto}\right)^{-b}}{\left(I_{1\text{-}2\,auto}\right)^{-b} + \left(I_{1\text{-}2\,transit}\right)^{-b}} \times 100$$

$$\Rightarrow$$

$$MS_{auto} = \frac{(44)^{-1.5}}{\left(44\right)^{-1.5} + \left(30\right)^{-1.5}} \times 100 = 36.0\%.$$

$$MS_t = \left(1 - MS_a\right) \times 100$$

$$\Rightarrow$$

$$MS_{transit} = 100 - 36.0 = 64.0\%$$

The MS Excel worksheet shown in Figure 2.48 is used to perform the computations of this problem.

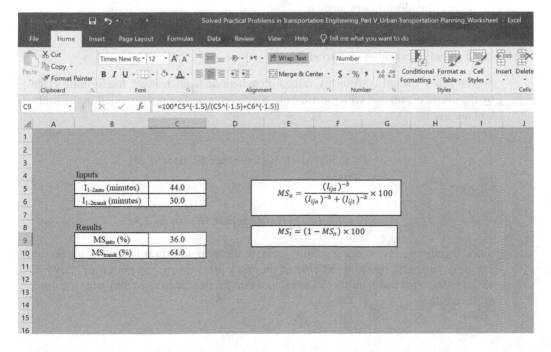

FIGURE 2.48 Image of MS Excel worksheet used for estimating transit mode split by trip interchange models for Problem 2.41.

2.42 The utility functions for a public transit (bus) mode and a private auto mode for travel between two zones are given below. The values of the travel parameters for the two modes are also provided in Table 2.81. Determine the percentage of trips by auto and transit using the logit model.

TABLE 2.81
Values of Travel Parameters for Auto
and Transit Modes for Problem 2.42

Travel Parameter	Auto	Transit
T_t (minutes)	15	20
T_w (minutes)	5	3
C (cents)	250	120

Auto: $U_A = -0.50 - 0.30T_t - 0.10T_w - 0.008C$
Transit: $U_T = -0.10 - 0.30T_t - 0.10T_w - 0.008C$

T_t = travel time (minutes)
T_w = waiting time (minutes)
C = cost (cents)

Solution:

The utility/disutility of each mode is determined using the formula below:

$$U_m = \sum_{i=1}^{n} a_i X_i \tag{2.30}$$

Where:
    U_m = utility/disutility of mode m
    n = number of attributes
    X_i = attribute value (travel time, waiting time, cost, ... etc.)
    a_i = coefficient value for attribute i

To determine the utility/disutility value for each mode (auto, transit), the formula given by the above expression with the attribute coefficients and values provided in the problem are used:
$\Rightarrow$

$$U_{\text{auto}} = -0.50 - 0.30(15) - 0.10(5) - 0.008(250) = -7.50$$

And:

$$U_{\text{transit}} = -0.10 - 0.30(20) - 0.10(8) - 0.008(120) = -7.86$$

Since the sign is −ve, it is a disutility for both the auto and transit modes. In general, disutility functions measure the cost of the mode (travel time, waiting time, cost, ... etc.). On the other hand, utility functions measure the degree of satisfaction for the users of the mode.

According to the logit model, the choice of a specific mode (auto, or transit) is based on the probability distribution. Therefore, for the transport modes (auto, transit), the logit model is given in the following formula as the probability of selecting the auto mode:

$$P(\text{auto}) = \frac{e^{U_{\text{auto}}}}{e^{U_{\text{auto}}} + e^{U_{\text{transit}}}} \tag{2.31}$$

Similarly, the probability for selecting the transit mode can be expressed by the logit model as below:

$$P(\text{transit}) = \frac{e^{U_{\text{transit}}}}{e^{U_{\text{auto}}} + e^{U_{\text{transit}}}} = 1 - P(\text{auto}) \tag{2.32}$$

Where:
   P (auto) = the probability of selecting the auto mode
   P (transit) = the probability of selecting the transit mode
   U_{auto} = the utility/disutility value for the auto mode
   U_{transit} = the utility/disutility value for the transit mode

If there are more than two transport modes available, the logit model is expressed in the following formula:

$$P(m) = \frac{e^{U_m}}{\sum\limits_m e^{U_m}} \tag{2.33}$$

Where:
   $P(m)$ = the probability of selecting the mode, m
   U_m = the utility/disutility value for the mode, m

$\Rightarrow$

$$P(\text{auto}) = \frac{e^{-7.50}}{e^{-7.50} + e^{-7.86}} = 0.59$$

And:

$$P(\text{transit}) = \frac{e^{-7.86}}{e^{-7.50} + e^{-7.86}} = 0.41$$

Or:

$$P(\text{transit}) = 1 - P(\text{auto}) = 1 - 0.59 = 0.41$$

Therefore, 59% will use the auto mode and 41% will use the transit mode (Figure 2.49).

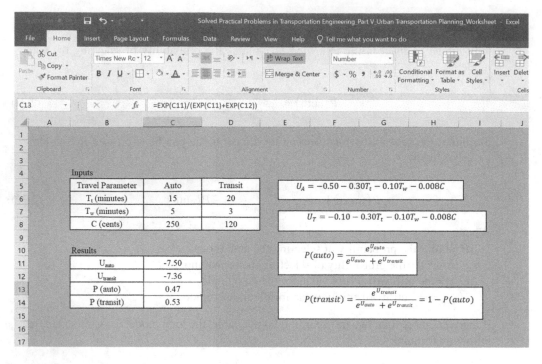

FIGURE 2.49 Image of MS Excel worksheet used for estimating transit mode split by the logit model for Problem 2.42.

2.43 Three transport modes are available for a residential zone in the downtown area of a city: the auto, the bus, and the metro. If the utility functions for the three modes along with the values of the travel attributes are given below, determine the predicted percentage of trips by each mode using the logit model (Table 2.82).

Auto: $U_{\text{auto}} = -0.42 - 0.28T_t - 0.10T_w - 0.010C$
Bus: $U_{\text{bus}} = -0.08 - 0.32T_t - 0.12T_w - 0.006C$
Metro: $U_{\text{metro}} = -0.08 - 0.36T_t - 0.14T_w - 0.006C$

T_t = travel time (minutes)
T_w = waiting time (minutes)
C = cost (cents)

TABLE 2.82
Values of Travel Parameters for Three Transport Modes for Problem 2.43

Travel Attribute	Auto	Bus	Metro
T_t (minutes)	25	40	32
T_w (minutes)	5	10	12
C (cents)	400	200	350

Solution:

The utility/disutility value for each mode (auto, bus, metro) is estimated using the formulas and the attribute coefficients and values provided in the problem:

$$U_{auto} = -0.42 - 0.28T_t - 0.10T_w - 0.010C$$

$\Rightarrow$

$$U_{auto} = -0.42 - 0.28(25) - 0.10(5) - 0.010(400) = -11.92$$

$$U_{bus} = -0.08 - 0.32T_t - 0.12T_w - 0.006C$$

$\Rightarrow$

$$U_{bus} = -0.08 - 0.32(40) - 0.12(10) - 0.006(200) = -15.28$$

$$U_{metro} = -0.08 - 0.36T_t - 0.14T_w - 0.006C$$

$\Rightarrow$

$$U_{metro} = -0.08 - 0.36(32) - 0.14(12) - 0.006(350) = -15.38$$

The logit model described by the probability function shown below is used:

$$P(m) = \frac{e^{U_m}}{\sum_m e^{U_m}}$$

$\Rightarrow$

$$P(auto) = \frac{e^{U_{auto}}}{e^{U_{auto}} + e^{U_{bus}} + e^{U_{metro}}}$$

$\Rightarrow$

$$P(auto) = \frac{e^{-11.92}}{e^{-11.92} + e^{-15.28} + e^{-15.38}} = 0.938$$

$$P(bus) = \frac{e^{U_{bus}}}{e^{U_{auto}} + e^{U_{bus}} + e^{U_{metro}}}$$

$\Rightarrow$

$$P(bus) = \frac{e^{-15.28}}{e^{-11.92} + e^{-15.28} + e^{-15.38}} = 0.033$$

$$P(metro) = \frac{e^{U_{metro}}}{e^{U_{auto}} + e^{U_{bus}} + e^{U_{metro}}}$$

$\Rightarrow$

$$P(metro) = \frac{e^{-15.38}}{e^{-11.92} + e^{-15.28} + e^{-15.38}} = 0.029$$

In summary, 93.8% will use the auto, 3.3% will use the bus, and 2.9% will use the metro (Figure 2.50).

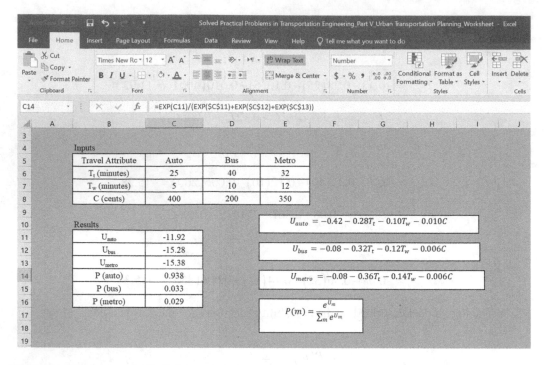

FIGURE 2.50   Image of MS Excel worksheet used for estimating transit and auto mode split by the logit model for Problem 2.43.

2.44   A new metro line is proposed by a road transport authority in an urban area. It is required to estimate the percentage of travelers who will prefer using the new metro line overusing the auto mode. The attribute (disutility) values for the two modes (auto, metro) are given in Table 2.83. The average annual household income in the area is $45,000.

TABLE 2.83

Values of Travel Attributes for Two Transport Modes for Problem 2.44

Travel Attribute	Auto	Metro
Travel Time (minutes)	30	40
Waste Time (minutes)	5	10
Cost (cents)	480	400

Solution:

Since this is a new proposed metro system, the data about the system and the attribute coefficients are not available. In this case, the general utility function with the following expression is used:

$$U_m = a_0 + a_1\left(T_{\text{in-vehicle}}\right) + a_2\left(T_{\text{out-of-vehicle}}\right) + a_3\left(C_{\text{out-of-pocket}}\right) + \ldots \qquad (2.34)$$

Where:
   U_m = utility/disutility of mode m
   $a_0, a_1, a_2, a_3, \ldots$ = values of coefficients for attributes, respectively
   $T_{\text{in-vehicle}}$ = in-vehicle travel time (minute)
   $T_{\text{out-of-vehicle}}$ = out-of-vehicle travel time (minute)
   $C_{\text{out-of-pocket}}$ = out-of-pocket cost (cents)

In this case and since there are only three attributes for the mode, the formula for U_m becomes:

$$U_m = a_0 + a_1\left(T_{\text{in-vehicle}}\right) + a_2\left(T_{\text{out-of-vehicle}}\right) + a_3\left(C_{\text{out-of-pocket}}\right) \tag{2.35}$$

The values of a_0, a_1, a_2, a_3 are calibrated using the approach in the National Cooperative Highway Research Program (NCHRP) Report No. 365.

$a_0 = 0$ in this case since it a constant that reflects mode service improvement
$a_1 = -0.025$
$a_2 = -0.050$

a_3 is given by the following formula:

$$a_3 = \frac{a_1}{\dfrac{\left(R_{\text{TTER}}\right)\left(I_{\text{HH}}\right)}{1248}} \tag{2.36}$$

Where:
   R_{TTER} = ratio of one-hour travel time over hourly employment rate (a typical value of 0.3 is used when the data is not available)
   I_{HH} = average annual household income in the area ($)
   $(1/1248)$ = a conversion factor that converts $/year to cents/minute (in this case, the year is considered to have 124,800 working minutes)

$\Rightarrow$

$$a_3 = \frac{-0.025}{\dfrac{(0.3)(45000)}{1248}} = -0.00231$$

$\Rightarrow$

$$U_{\text{auto}} = 0 - 0.025(30) - 0.050(5) - 0.00231(480) = -2.109$$

$$U_{\text{metro}} = 0 - 0.025(40) - 0.050(10) - 0.00231(400) = -2.424$$

The percentage of travelers who will use an auto mode is computed using the logit model as described below:

$$P(\text{auto}) = \frac{e^{U_{\text{auto}}}}{e^{U_{\text{auto}}} + e^{U_{\text{transit}}}}$$

$\Rightarrow$

$$P(\text{auto}) = \frac{e^{-2.109}}{e^{-2.109} + e^{-2.424}} = 0.58$$

And the percentage of travelers who will use the metro is also computed using the logit model as follows:

$$P(\text{transit}) = \frac{e^{U_{\text{transit}}}}{e^{U_{\text{auto}}} + e^{U_{\text{transit}}}} = 1 - P(\text{auto})$$

$$\Rightarrow$$

$$P(\text{metro}) = \frac{e^{-2.424}}{e^{-2.109} + e^{-2.424}} = 0.42$$

Therefore, 42% of travelers will use the new proposed metro line with the given travel attributes (Figure 2.51).

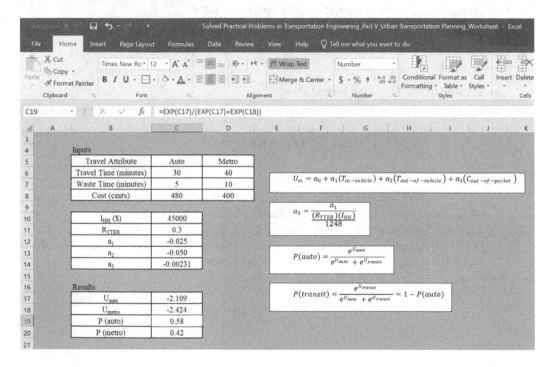

FIGURE 2.51 Image of MS Excel worksheet used for estimating transit and auto mode split by the logit model for Problem 2.44.

2.45   In the previous problem (Problem 2.44), if the metro line offers a Wi-Fi service for the metro users, determine the expected percentage of travelers that will use the metro line given that the new Wi-Fi service will add a +ve coefficient (a_0) with a value of 0.411 to the utility function of the metro mode.

Solution:

The utility/disutility function values for the two modes (auto and metro) are calculated, as shown below:

$$U_m = a_0 + a_1\left(T_{\text{in-vehicle}}\right) + a_2\left(T_{\text{out-of-vehicle}}\right) + a_3\left(C_{\text{out-of-pocket}}\right)$$

$$U_{\text{auto}} = 0 - 0.025(30) - 0.050(5) - 0.00231(480) = -2.109$$

(no change; it is the same value as in Problem 2.44).

$$U_{\text{metro}} = 0.411 - 0.025(40) - 0.050(10) - 0.00231(400) = -2.013$$

$$P(\text{auto}) = \frac{e^{U_{\text{auto}}}}{e^{U_{\text{auto}}} + e^{U_{\text{transit}}}}$$

$$\Rightarrow$$

$$P(\text{auto}) = \frac{e^{-2.109}}{e^{-2.109} + e^{-2.013}} = 0.48$$

$$P(\text{transit}) = \frac{e^{U_{\text{transit}}}}{e^{U_{\text{auto}}} + e^{U_{\text{transit}}}} = 1 - P(\text{auto})$$

$$\Rightarrow$$

$$P(\text{metro}) = \frac{e^{-2.013}}{e^{-2.109} + e^{-2.013}} = 0.52$$

Hence, as a result of the new Wi-Fi service offered on the metro line for travelers, the percentage of travelers who will use the metro line increased by 10% from 42 to 52% (based on the logit model) (Figure 2.52).

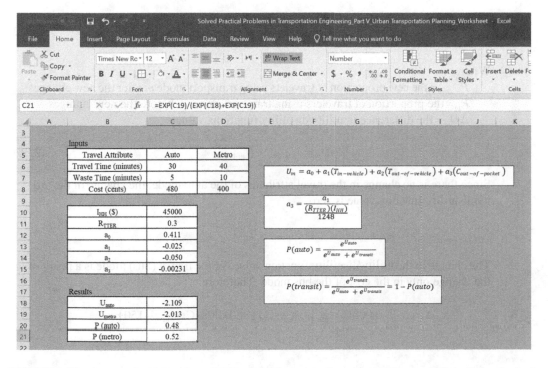

FIGURE 2.52 Image of MS Excel worksheet used for estimating transit and auto mode split by the logit model for Problem 2.45.

2.46   A municipality in a big city desires to increase the public transit (bus) mode split over the private auto mode split. The transportation engineers in the municipality are planning to increase the percentage of travelers using the public transit mode from 35% to 50% by reducing the trip cost for the current bus line, which will encourage travelers to use it more frequently. The utility functions for auto and bus modes are given below:

$$U_{auto} = -0.50 - 0.30(T_{travel}) - 0.10(T_{waiting}) - 0.008(Cost)$$

$$U_{bus} = -0.10 - 0.30(T_{travel}) - 0.10(T_{waiting}) - 0.008(Cost)$$

Where:
   T_{travel} = travel time (minutes)
   $T_{waiting}$ = waiting time (minutes)
   Cost is in cents

What should the new cost of the bus be to achieve the planning objectives of the municipality transportation engineers if the current cost for the public transit (bus) mode is 150 cents? Do you think this will be a feasible solution? And why? What other recommendations can the transportation engineers implement to increase the public transit (bus) mode split?

Solution:

The incremental logit model expressed in the following formula is used to solve this problem:

$$P_{m\text{-after}} = \frac{P_m e^{\Delta U_m}}{\sum\limits_{m} P_m e^{\Delta U_m}} \tag{2.37}$$

Where:
   $P_{m\text{-new}}$ = the new proportion of travelers using transport mode m after the new changes
   P_m = the proportion of travelers using transport mode m before the new changes
   ΔU_m = the difference in utility/disutility functions values

$$\Delta U_m = U_{m\text{-after}} - U_{m\text{-before}} \tag{2.38}$$

Since there is no change in the travel attributes of the auto mode, the change in the utility/disutility function value is zero.

$$\Delta U_{auto} = 0$$

The change in the utility/disutility function value for the transit (bus) mode is a result of the change only in the cost of the bus mode. Therefore:

$$\Delta U_{bus} = -0.008(Cost_{new} - Cost_{old}) = -0.008(Cost_{new} - 150)$$

$\Rightarrow$

$$\Delta U_{bus} = -0.008 Cost_{new} + 1.2$$

$$P_{m\text{-after}} = \frac{P_m e^{\Delta U_m}}{\sum_m P_m e^{\Delta U_m}}$$

$\Rightarrow$

$$P_{\text{bus-after}} = \frac{P_{\text{bus}} e^{\Delta U_{\text{bus}}}}{P_{\text{bus}} e^{\Delta U_{\text{bus}}} + P_{\text{auto}} e^{\Delta U_{\text{auto}}}}$$

$\Rightarrow$

$$0.50 = \frac{0.35 e^{\Delta U_{\text{bus}}}}{0.35 e^{\Delta U_{\text{bus}}} + 0.65 e^0}$$

$\Rightarrow$

$$\frac{1}{0.50} = 1 + \frac{0.65 e^0}{0.35 e^{\Delta U_{\text{bus}}}}$$

Or:

$$\frac{1}{\dfrac{1}{0.50} - 1} = \frac{0.35 e^{\Delta U_{\text{bus}}}}{0.65 e^0}$$

$\Rightarrow$

$$1 = \frac{0.35 e^{\Delta U_{\text{bus}}}}{0.65}$$

$\Rightarrow$

$$\Delta U_{\text{bus}} = 0.619$$

But:

$$\Delta U_{\text{bus}} = -0.008 \text{Cost}_{\text{new}} + 1.2$$

$\Rightarrow$

$$\text{Cost}_{\text{new}} = 72.6 \text{ cents}$$

This solution may not be feasible because to achieve the planning objectives, the cost should be reduced significantly (to approximately half the original cost) and that is not reasonable and may create some financial troubles. Another solution might be reducing the travel time by either introducing a new bus system with higher speed or constructing a new short-cut route. Overall, the three alternatives have to be studied carefully and a cost-effectiveness study should be conducted to select the most feasible solution (Figure 2.53).

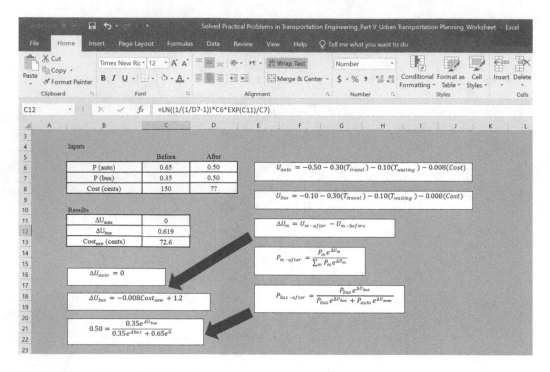

FIGURE 2.53 Image of MS Excel worksheet used for the analysis by the incremental logit model for Problem 2.46.

2.47 In the above problem (Problem 2.46), develop a relationship between the change in travel time and the percentage split for public transit (bus) mode using the incremental logit model (at a travel time of 60 minutes, percentage transit split = 20%). Tabulate your computations and draw the relationship using a scatter plot with smooth lines.

Solution:

Using the incremental logit model expressed in the formula below:

$$P_{m\text{-after}} = \frac{P_m e^{\Delta U_m}}{\sum_m P_m e^{\Delta U_m}}$$

At $T_{\text{travel}} = 60$ minutes, P (bus) = 0.20.

Based on the change in the travel time of the bus mode, the change in the percentage split of the bus mode (and hence the auto mode) is estimated using the incremental logit model. An increment of 5 minutes is used in the change of travel time. In other words, the travel time decreases by 5 minutes from 60 minutes. Eleven travel time values (from 60 to 10 minutes) are therefore used.

Sample Calculation:
Since there is no change in the travel attributes of the auto mode,

$$\Delta U_{\text{auto}} = 0$$

The change in the utility/disutility function value for the transit (bus) mode is a result of the change only in the travel time of the bus mode. Therefore:

$$\Delta U_{\text{bus}} = -0.30\left(T_{\text{travel-new}} - T_{\text{travel-old}}\right) = -0.30\left(55 - 60\right) = 1.5$$

$$P_{m\text{-after}} = \frac{P_m e^{\Delta U_m}}{\sum_m P_m e^{\Delta U_m}}$$

$\Rightarrow$

$$P_{\text{bus-after}} = \frac{P_{\text{bus}} e^{\Delta U_{\text{bus}}}}{P_{\text{bus}} e^{\Delta U_{\text{bus}}} + P_{\text{auto}} e^{\Delta U_{\text{auto}}}}$$

$\Rightarrow$

$$P_{\text{bus-after}} = \frac{0.20 e^{1.5}}{0.20 e^{1.5} + 0.80 e^0} = 0.53$$

$\Rightarrow$

$$P_{\text{auto-after}} = 1 - P_{\text{bus-after}} = 1 - 0.53 = 0.47$$

The results at the other travel time values can be computed using the incremental logit model following the same exact procedure. The results are tabulated below (Table 2.84):

TABLE 2.84

The Percentage Split of the Bus and Auto Modes after the Change in the Bus Travel Time for Problem 2.47

T_{travel} (minutes)	P (bus)	P (auto)
60	0.20	0.80
55	0.53	0.47
50	0.83	0.17
45	0.96	0.04
40	0.99	0.01
35	1.00	0.00
30	1.00	0.00
25	1.00	0.00
20	1.00	0.00
15	1.00	0.00
10	1.00	0.00

Figure 2.54 illustrates the relationship between the travel time for the transit (bus) mode and the transit (bus) split. Figure 2.55 shows the MS Excel worksheet used for developing the relationship between the bus travel time and mode split by the incremental logit model.

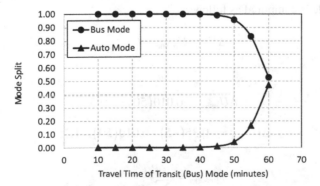

FIGURE 2.54 Relationship between bus travel time and mode split for Problem 2.47.

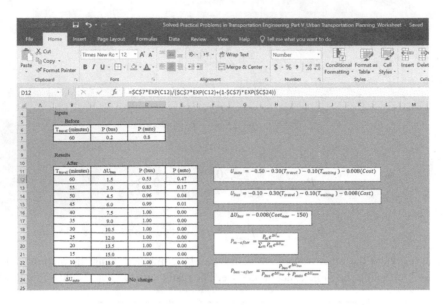

FIGURE 2.55 Image of MS Excel worksheet used for developing a relationship between bus travel time and mode split by the incremental logit model for Problem 2.47.

2.48   The utility function for both auto and public transit modes in a specific urban area is given below:

$$U = -0.35T_1 - 0.15T_2 - 0.010C$$

Where:
   T_1 = travel time (minutes)
   T_2 = waiting time (minutes)
   C = cost (cents)

The transportation authority in that area is planning to increase the percentage of travelers using the public transit by reducing the travel time of the transit mode. At the same time, the trip cost for both modes is raised due to an increase in oil prices. If the reduction in travel time is 5 minutes and the increase in cost is 50 cents, use the incremental logit model to determine the new split for public transit knowing that the original transit split is 40%.

Solution:

The change in the utility/disutility function value for both the auto mode and the public transit mode is computed as follows:

$$\Delta U_m = U_{m\text{-after}} - U_{m\text{-before}}$$

$\Rightarrow$

$$\Delta U_{\text{auto}} = -0.010(50) = -0.5$$

$$\Delta U_{\text{transit}} = -0.35(-5) - 0.010(50) = 1.25$$

$$P_{m\text{-after}} = \frac{P_m e^{\Delta U_m}}{\sum_m P_m e^{\Delta U_m}}$$

$$\Rightarrow$$

$$P_{\text{transit-after}} = \frac{P_{\text{transit}} e^{\Delta U_{\text{transit}}}}{P_{\text{transit}} e^{\Delta U_{\text{transit}}} + P_{\text{auto}} e^{\Delta U_{\text{auto}}}}$$

$$\Rightarrow$$

$$P_{\text{transit-after}} = \frac{0.40 e^{1.25}}{0.40 e^{1.25} + 0.60 e^{-0.5}} = 0.79$$

$$\Rightarrow$$

$$P_{\text{auto-after}} = 1 - P_{\text{transit-after}} = 1 - 0.79 = 0.21$$

Therefore, 79% of travelers are expected to use public transit after these changes. Figure 2.56 shows the MS Excel worksheet used for quantifying the effect of the change in travel time and cost on mode split by the incremental logit model.

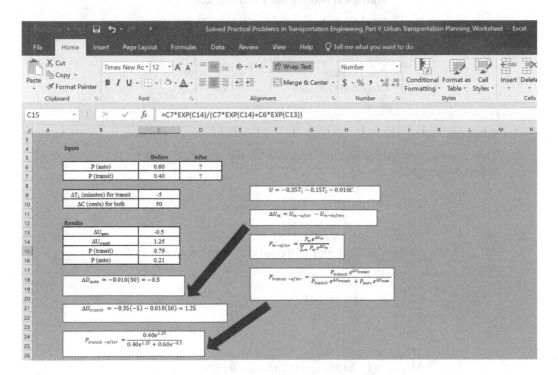

FIGURE 2.56 Image of MS Excel worksheet used for quantifying the effect of the change in travel time and cost on mode split by the incremental logit model for Problem 2.48.

2.49   A transportation engineer has conducted a mode choice study in an urban zone. Survey data on the travelers in that area has been collected, as shown in Table 2.85. The engineer intends to use the logit model in the study based on the data collected. The study shows that the percentage of travelers who use the transit mode is 20%. Based on this mode split, the utility function used in the study has been calibrated:

$$U_m = -0.15T_{travel} - 0.20T_{waste} - 0.008C + k$$

A Wi-Fi service has been added to the transit mode resulting in a mode split of 30% and 70% for the transit mode and the auto mode, respectively. Estimate the value of the calibration constant for the transit mode ($k_{transit}$) based on this change.

TABLE 2.85
Values of Travel Attributes for Two
Transport Modes for Problem 2.49

Travel Attribute	Auto	Transit
Travel Time, T_{travel} (minutes)	25	40
Waste Time, T_{waste} (minutes)	8	12
Cost, C (cents)	400	200
Calibration Constant, k	−0.40	−0.35

Solution:

The change in the utility/disutility function for the auto mode and for the transit mode is first determined:

$$\Delta U_m = U_{m\text{-after}} - U_{m\text{-before}}$$

⇒

$\Delta U_{Auto} = 0$ since there is no change in the service of the auto mode.
And:

$$\Delta U_{transit} = U_{transit\text{-after}} - U_{transit\text{-before}}$$

⇒

$$\Delta U_{transit} = \left(-0.15T_{travel} - 0.20T_{waste} - 0.008C + k\right)_{after}$$

$$-\left(-0.15T_{travel} - 0.20T_{waste} - 0.008C + k\right)_{before}$$

Since the improvement in the service affects only the constant k and none of the other travel attributes (travel time, waste time, cost) have been changed, the change in the utility/disutility function for the transit mode is determined as follows:

$$\Delta U_{transit} = k_{after} - k_{before} = k_{after} - \left(-0.35\right) = k_{after} + 0.35$$

The incremental logit mode expressed in the following formula is used to solve this problem.

$$P_{m\text{-after}} = \frac{P_m e^{\Delta U_m}}{\sum_m P_m e^{\Delta U_m}}$$

$\Rightarrow$

$$P_{transit\text{-after}} = \frac{P_{transit} e^{\Delta U_{transit}}}{P_{transit} e^{\Delta U_{transit}} + P_{auto} e^{\Delta U_{auto}}}$$

$\Rightarrow$

$$0.30 = \frac{0.20 e^{(k_{after} + 0.35)}}{0.20 e^{(k_{after} + 0.35)} + 0.80 e^0}$$

Solving for k_{after} yields the following value:

$$k_{after} = 0.189$$

Figure 2.57 shows the MS Excel worksheet used for estimating the transit mode calibration constant after adding a Wi-Fi service to the mode by the incremental logit model.

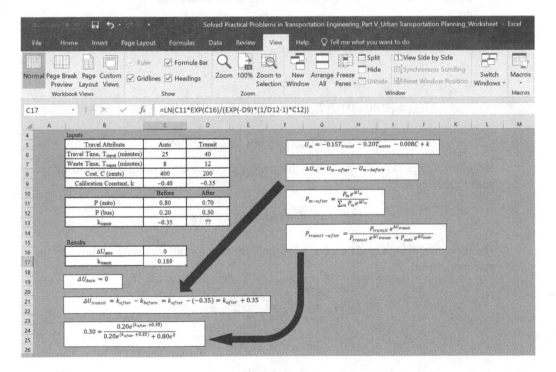

FIGURE 2.57 Image of MS Excel worksheet used for estimating the transit mode calibration constant after adding a Wi-Fi service to the mode by the incremental logit model for Problem 2.49.

2.50 The utility functions for auto and bus modes for a study area are given below:

$$U_{\text{auto}} = -0.03(T_{\text{travel}}) - 0.06(T_{\text{waste}}) - 0.003(C)$$

$$U_{\text{bus}} = 0.40 - 0.03(T_{\text{travel}}) - 0.06(T_{\text{waste}}) - 0.003(C)$$

T_{travel} = time of travel (minutes)
T_{waste} = waste time (minutes)
C = trip cost (cents)

Using these functions, the logit model predicts that 40% of the travelers use the auto mode and 60% of them use the bus mode. If the cost (cents) for the bus line was increased from 80 to 140 cents, what would the new modal split be for the bus mode after this change?

Solution:

The change in the utility/disutility function for the auto mode and for the transit mode is first determined:

$$\Delta U_m = U_{m\text{-after}} - U_{m\text{-before}}$$

Since there is no change in the value of the travel attributes of the auto mode ⇒ $\Delta U_{\text{Auto}} = 0$, there is no change in the service of the auto mode.
And:

$$\Delta U_{\text{bus}} = U_{\text{bus-after}} - U_{\text{bus-before}}$$

⇒

$$\Delta U_{\text{bus}} = \left(0.40 - 0.03(T_{\text{travel}}) - 0.06(T_{\text{waste}}) - 0.003(C)\right)_{\text{after}}$$

$$- \left(0.40 - 0.03(T_{\text{travel}}) - 0.06(T_{\text{waste}}) - 0.003(C)\right)_{\text{before}}$$

Since the change is on the cost only, the change in the utility/disutility function for the bus mode becomes:

$$\Delta U_{\text{bus}} = -0.003(140 - 80) = -0.18$$

The incremental logit mode expressed in the following formula is used to determine the new mode split:

$$P_{m\text{-after}} = \frac{P_m e^{\Delta U_m}}{\sum_m P_m e^{\Delta U_m}}$$

⇒

$$P_{\text{bus-after}} = \frac{P_{\text{bus}} e^{\Delta U_{\text{bus}}}}{P_{\text{bus}} e^{\Delta U_{\text{bus}}} + P_{\text{auto}} e^{\Delta U_{\text{auto}}}}$$

⇒

$$P_{\text{bus-after}} = \frac{0.60 e^{-0.18}}{0.60 e^{-0.18} + 0.40 e^{0}} = 0.56$$

Therefore, an increase in the trip cost for the bus mode will result in a decrease in the percentage of travelers expected to use the bus mode from 60 to 56%.

The MS Excel worksheet used to conduct the computations of this problem is shown in Figure 2.58.

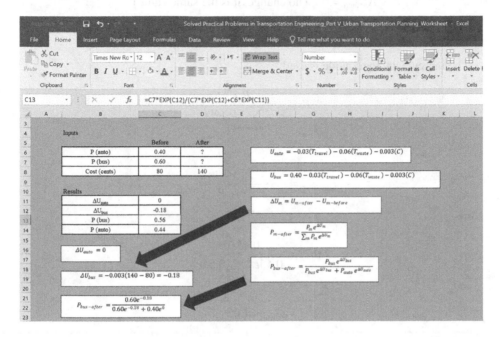

FIGURE 2.58 Image of MS Excel worksheet used for quantifying the effect of the change in the cost for the bus line on mode split by the incremental logit model for Problem 2.50.

2.51 The utility functions for auto and bus modes for a study area are given below. Using these functions, the logit model predicts that 55% of the travelers use the auto mode and 45% of them use the bus mode. A new service improvement to the bus line has been implemented. If the modal split for the bus line after the improvement becomes 70%, determine the constant in the utility function for the bus mode (A_{bus}) that reflects this change.

$$U_{auto} = -0.03\left(T_{travel}\right) - 0.06\left(T_{waste}\right) - 0.003\left(C\right)$$

$$U_{bus} = -0.03\left(T_{travel}\right) - 0.06\left(T_{waste}\right) - 0.003\left(C\right)$$

T_{travel} = time of travel (minutes)
T_{waste} = waste time (minutes)
C = trip cost (cents)

Solution:

Before improvement:

$$U_{auto} = -0.03\left(T_{travel}\right) - 0.06\left(T_{waste}\right) - 0.003\left(C\right)$$

$$U_{bus} = -0.03\left(T_{travel}\right) - 0.06\left(T_{waste}\right) - 0.003\left(C\right)$$

After improvement:

$$U_{bus} = A_{bus} - 0.03\left(T_{travel}\right) - 0.06\left(T_{waste}\right) - 0.003\left(C\right)$$

$$\Delta U_m = U_{m\text{-after}} - U_{m\text{-before}}$$

$\Rightarrow$

$$\Delta U_{\text{auto}} = 0 \quad \left(\text{no change; it is the same value}\right)$$

$$\Delta U_{\text{bus}} = A_{\text{bus}}$$

The incremental logit mode is used:

$$P_{m\text{-after}} = \frac{P_m e^{\Delta U_m}}{\displaystyle\sum_m P_m e^{\Delta U_m}}$$

$\Rightarrow$

$$P_{\text{bus-after}} = \frac{P_{\text{bus}} e^{\Delta U_{\text{bus}}}}{P_{\text{bus}} e^{\Delta U_{\text{bus}}} + P_{\text{auto}} e^{\Delta U_{\text{auto}}}}$$

$\Rightarrow$

$$0.70 = \frac{0.45 e^{A_{\text{bus}}}}{0.45 e^{A_{\text{bus}}} + 0.55 e^0}$$

$\Rightarrow$

$$A_{\text{bus}} = 1.05$$

The MS Excel worksheet used to perform the computations of this problem is shown in Figure 2.59.

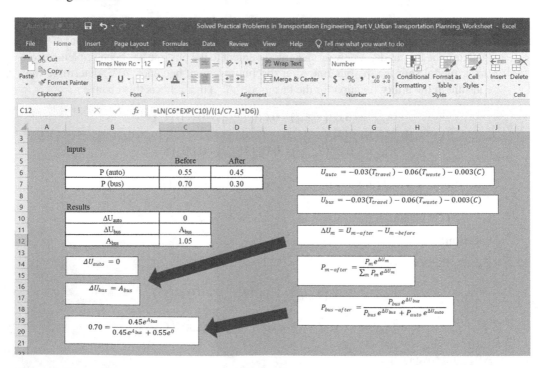

FIGURE 2.59 Image of MS Excel worksheet used for determining utility function constant after implementing a new service improvement to the bus line by the incremental logit model for Problem 2.51.

2.52 The transportation authority in an urban area desires to increase the number of travelers using the public transit (bus) mode over those using the private auto mode. The authority is planning to increase the percentage of travelers using the public transit mode from 20 to 50% by reducing the travel time for the current bus line through diverting the bus to a new short-cut route that will encourage travelers to use it more frequently. The utility functions for auto and transit modes are given below:

Private auto: $U_{auto} = -0.50 - 0.30T_1 - 0.10T_2 - 0.008C$

Public transit (bus): $U_{bus} = -0.10 - 0.30T_1 - 0.10T_2 - 0.008C$

$T_1 = $ travel time (minutes)

$T_2 = $ waiting time (minutes)

$C = $ cost (cents)

What should the new travel time of the new bus route be to achieve the planning objectives of the transportation authority if the current travel time for the public transit (bus) mode is 35 minutes?

Solution:

$$\Delta U_m = U_{m\text{-after}} - U_{m\text{-before}}$$

$\Rightarrow$

Since there is no change in the private auto mode,

$$\Delta U_{auto} = 0$$

Since there is only a change in the travel time of the bus mode,

$$\Delta U_{bus} = -0.30\left(T_{1\text{-new}} - T_{1\text{-old}}\right)$$

$\Rightarrow$

$$\Delta U_{bus} = -0.30\left(T_{1\text{-new}} - 35\right) = -0.30T_{1\text{-new}} + 10.5$$

Using the incremental logit mode shown in the expression below:

$$P_{m\text{-after}} = \frac{P_m e^{\Delta U_m}}{\sum_m P_m e^{\Delta U_m}}$$

$\Rightarrow$

$$P_{bus\text{-after}} = \frac{P_{bus} e^{\Delta U_{bus}}}{P_{bus} e^{\Delta U_{bus}} + P_{auto} e^{\Delta U_{auto}}}$$

$\Rightarrow$

$$0.50 = \frac{0.20 e^{\Delta U_{bus}}}{0.80 e^{0} + 0.20 e^{\Delta U_{bus}}}$$

$\Rightarrow$

$$\frac{1}{0.50} = 1 + \frac{0.80 e^{0}}{0.20 e^{\Delta U_{bus}}}$$

Rearranging yields:

$$\frac{1}{\dfrac{1}{0.50} - 1} = \frac{0.20e^{\Delta U_{bus}}}{0.80e^0}$$

Or:

$$1.0 = \frac{0.20e^{\Delta U_{bus}}}{0.80}$$

$\Rightarrow$

$$\Delta U_{bus} = 1.3863$$

But:

$$\Delta U_{bus} = -0.30T_{1\text{-new}} + 10.5$$

$\Rightarrow$

$$T_{1\text{-new}} = 30.4 \text{ minutes}$$

Therefore, the travel time of the new bus route should be 30.4 minutes to achieve the planning objectives of the transportation authority in this area.

Figure 2.60 shows the MS Excel worksheet used for estimating the new travel time for the bus line by the incremental logit model.

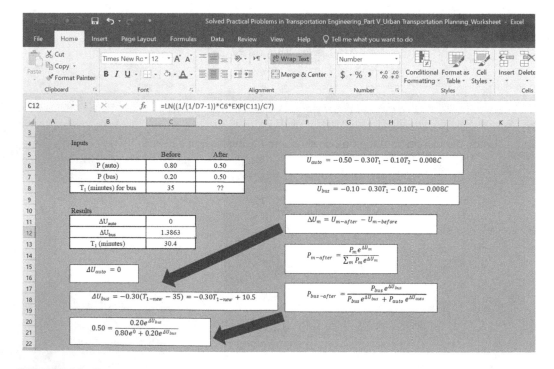

FIGURE 2.60 Image of MS Excel worksheet used for estimating the new travel time for the bus line by the incremental logit model for Problem 2.51.

TRAFFIC ASSIGNMENT

2.53 Determine the shortest travel path from node 1 (home node) to all other nodes (zones) in the following 12-node network with travel times on each link shown for each node (zone) pair. The travel times are in minutes (Figure 2.61).

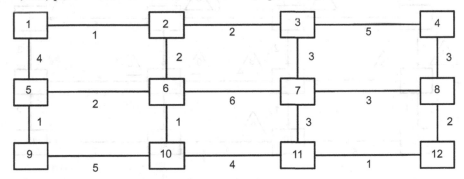

FIGURE 2.61 A 12-node network for Problem 2.54.

Solution:

Step 1: the travel times from node 1 to the nearest nodes (nodes 2 and 5) are determined (Figure 2.62).

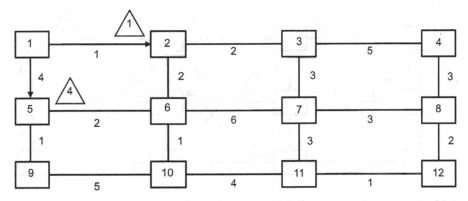

FIGURE 2.62 Travel times from node 1 to nearest nodes (nodes 2 and 5) for Problem 2.53.

Step 2: the travel times from the node with the shortest travel time (node 2) to the nearest nodes (nodes 3 and 6) are determined (Figure 2.63).

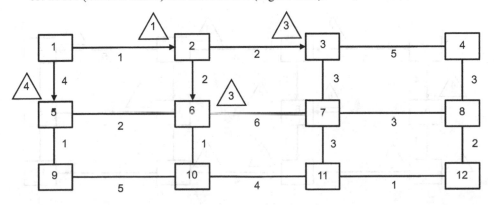

FIGURE 2.63 Travel times from the node with the shortest travel time (node 2) to the nearest nodes (nodes 3 and 6) for Problem 2.53.

Step 3: the travel times from the node with the shortest travel time (nodes 3 and 6 since they are equal) to the nearest nodes (nodes 4 and 7 for node 3 and nodes 5, 7, and 10 for node 6) are determined (Figure 2.64).

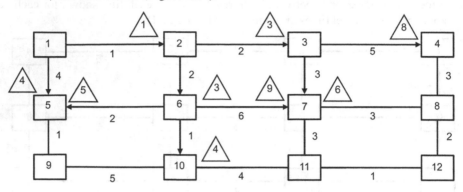

FIGURE 2.64 Travel times from the node with the shortest travel time (nodes 3 and 6) to the Nearest nodes (nodes 4 and 7 for node 3 and nodes 5, 7, and 10 for node 6) for Problem 2.53.

Step 4: eliminate link 6 to 7 and link 6 to 5 because the accumulative travel time to node 7 through link 3–7 is shorter than that through link 6–7 and the accumulative travel time to node 5 through link 1–5 is shorter than that through link 6–5 (Figure 2.65).

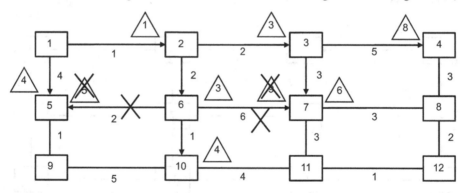

FIGURE 2.65 Elimination of link 6 to 7 and link 6 to 5 for Problem 2.53.

Step 5: the travel times from the node with the shortest travel time (nodes 5 and 10 since they are equal) to the nearest nodes (nodes 6 and 9 for node 5 and nodes 9 and 11 for node 10) (Figure 2.66).

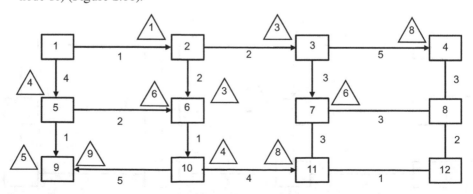

FIGURE 2.66 Travel times from the node with the shortest travel time (nodes 5 and 10) to the nearest nodes (nodes 6 and 9 for node 5 and nodes 9 and 11 for node 10) for Problem 2.53.

Step 6: eliminate link 5 to 6 and link 10 to 9 because the accumulative travel time to node 6 through link 2–6 is shorter than that through link 5–6 and the accumulative travel time to node 9 through link 5–9 is shorter than that through link 10–9 (Figure 2.67).

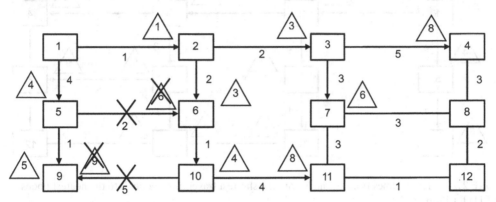

FIGURE 2.67 Elimination of link 5 to 6 and link 10 to 9 to for Problem 2.53.

Step 7: the travel times from the node with the shortest travel time (node 9) to the nearest nodes (node 10) (Figure 2.68).

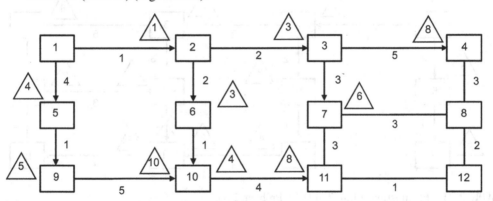

FIGURE 2.68 Travel times from the node with the shortest travel time (node 9) to the nearest nodes (node 10) for Problem 2.53.

Step 8: eliminate link 9 to 10 because the accumulative travel time to node 10 through link 6–10 is shorter than that through link 9–10 (Figure 2.69).

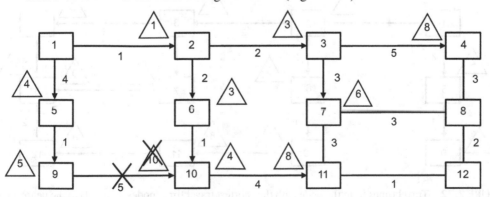

FIGURE 2.69 Elimination of link 9 to 10 for Problem 2.53.

Step 9: the travel times from the node with the shortest travel time (node 7) to the nearest nodes (nodes 8 and 11) (Figure 2.70).

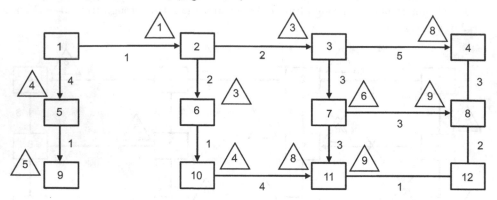

FIGURE 2.70 Travel times from the node with the shortest travel time (node 7) to the nearest nodes (nodes 8 and 11) for Problem 2.53.

Step 10: eliminate link 7 to 11 because the accumulative travel time to node 11 through link 10–11 is shorter than that through link 7–11 (Figure 2.71).

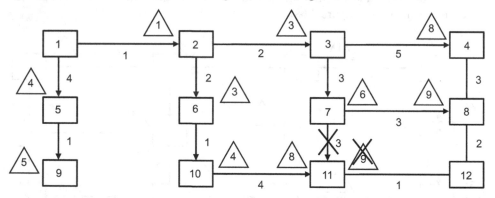

FIGURE 2.71 Elimination of link 7 to 11 for Problem 2.53.

Step 11: the travel times from the node with the shortest travel time (nodes 4 and 11 since they are equal) to the nearest nodes (node 8 for node 4 and node 12 for node 11) (Figure 2.72).

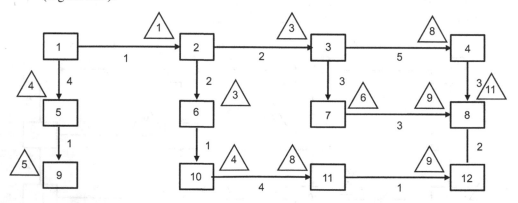

FIGURE 2.72 Travel times from the node with the shortest travel time (nodes 4 and 11) to the nearest nodes (node 8 for node 4 and node 12 for node 11) for Problem 2.53.

Step 12: eliminate link 4 to 8 because the accumulative travel time to node 8 through link 7–8 is shorter than that through link 4–8 (Figure 2.73).

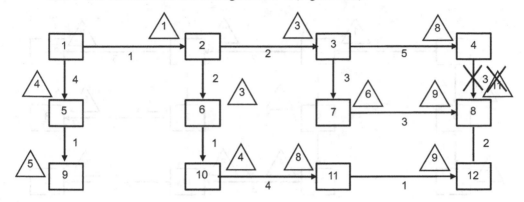

FIGURE 2.73   Elimination of link 4 to 8 for Problem 2.53.

Step 13: the travel time from the node with the shortest travel time (node 8) to the nearest nodes (node 12) (Figure 2.74).

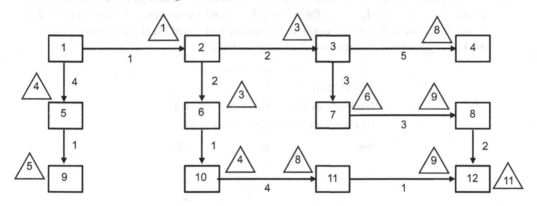

FIGURE 2.74   Travel times from the node with the shortest travel time (node 8) to the nearest nodes (node 12) for Problem 2.53.

Step 14: eliminate link 8 to 12 because the accumulative travel time to node 12 through link 11–12 is shorter than that through link 8–12 (Figure 2.75).

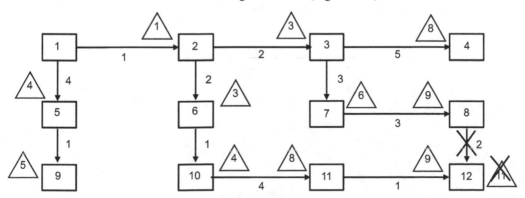

FIGURE 2.75   Elimination of link 8 to 12 for Problem 2.53.

Step 15: at the end, the final network for the minimum path tree (or what is so-called the skim tree) is composed (Figure 2.76).

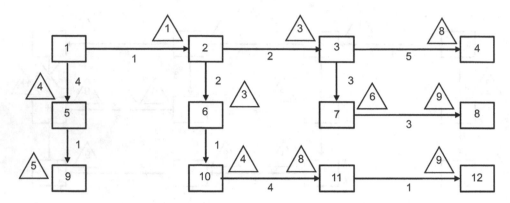

FIGURE 2.76   The final minimum path network (skim tree) for Problem 2.53.

2.54   For the problem above (Problem 2.53), if the number of auto trips between zone 1 and all other zones is as shown in Table 2.86, determine the number of trips on each link using the all-or-nothing trip assignment method and based on the minimum path network composed (this process is called network loading).

TABLE 2.86
Number of Auto Trips from Zone 1
to Other Zones for Problem 2.54

From	To	Trips
1	2	40
	3	55
	4	60
	5	90
	6	100
	7	50
	8	35
	9	80
	10	40
	11	70
	12	30

Solution:

The links on the minimum path tree (the skim tree) from zone 1 (node 1) to each of the other zones (nodes) are summarized in Table 2.87.

TABLE 2.87

The Minimum Path From Zone 1 to Each of the Other Zones for Problem 2.54

From	To	Trips	Links
1	2	40	1–2
	3	55	1–2, 2–3
	4	60	1–2, 2–3, 3–4
	5	90	1–5
	6	100	1–2, 2–6
	7	50	1–2, 2–3, 3–7
	8	35	1–2, 2–3, 3–7, 7–8
	9	80	1–5, 5–9
	10	40	1–2, 2–6, 6–10
	11	70	1–2, 2–6, 6–10, 10–11
	12	30	1–2, 2–6, 6–10, 10–11, 11–12

Based on the previous table that shows the links on the minimum path network for trips from zone 1 to all other zones, the number of trips on each link (network loading) is computed.

Sample Calculation:

Link 1–2 is used by trips from zone 1 to zones 2, 3, 4, 6, 7, 8, 10, 11, and 12. Therefore, the trips between zone 1 and each of these zones are assigned to link 1–2.

$$\text{Trips on link 1-2} = 40 + 55 + 60 + 100 + 50 + 35 + 40 + 70 + 30 = 480$$

In a similar manner, the number of trips on each of the other links is estimated. The results are shown in Table 2.88. This method is called the all-or-nothing algorithm because all the trips between zone pairs are either fully loaded or not loaded on the link.

TABLE 2.88

Number of Trips Assigned to Each Link of the Minimum Path Network for Problem 2.54

Link	Used by Trips from Zone 1 to Zones	Number of Trips
1–2	2, 3, 4, 6, 7, 8, 10, 11, 12	$40 + 55 + 60 + 100 + 50 + 35 + 40 + 70 + 30 = 480$
23	3, 4, 7, 8	$55 + 60 + 50 + 35 = 200$
3–4	4	60
1–5	5, 9	$90 + 80 = 170$
2–6	6, 10, 11, 12	$100 + 40 + 70 + 30 = 240$
3–7	7, 8	$50 + 35 = 85$
7–8	8	35
5–9	9	80
6–10	10, 11, 12	$40 + 70 + 30 = 140$
10–11	11, 12	$70 + 30 = 100$
11–12	12	30

2.55   For the nine-node network shown in the diagram below, determine the shortest travel path from node 5 to all other nodes using the minimum path method (the skim tree). The travel times are in minutes (Figure 2.77).

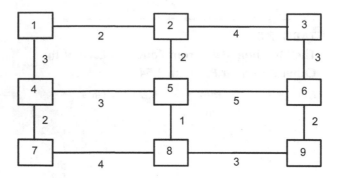

FIGURE 2.77   A nine-node network for Problem 2.54.

Solution:

Step 1: the travel times from node 5 to the nearest nodes (nodes 2, 4, 6, and 8) are determined (Figure 2.78).

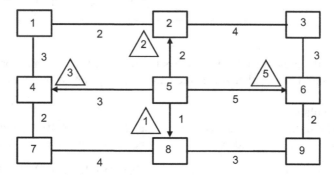

FIGURE 2.78   travel times from node 5 to the nearest nodes (nodes 2, 4, 6, and 8) for Problem 2.55.

Step 2: the travel times from the node with the shortest travel time (node 8) to the nearest nodes (nodes 7 and 9) are determined (Figure 2.79).

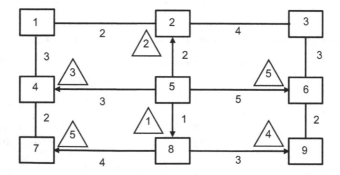

FIGURE 2.79   Travel times from the node with the shortest travel time (node 8) to the nearest nodes (nodes 7 and 9) for Problem 2.55.

Step 3: the travel times from the node with the shortest travel time (node 2) to the nearest nodes (nodes 1 and 3) are determined (Figure 2.80).

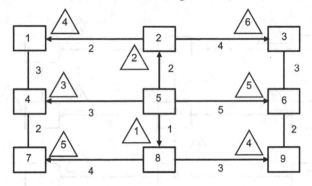

FIGURE 2.80 Travel times from the node with the shortest travel time (node 2) to the nearest nodes (nodes 1 and 3) for Problem 2.55.

Step 4: the travel times from the node with the shortest travel time (node 4) to the nearest nodes (nodes 1 and 7) are determined (Figure 2.81).

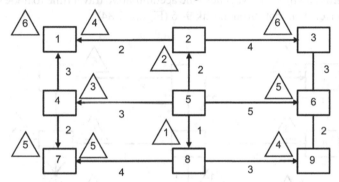

FIGURE 2.81 Travel times from the node with the shortest travel time (node 4) to the nearest nodes (nodes 1 and 7) for Problem 2.55.

Step 5: eliminate link 4 to 1 because the accumulative travel time to node 1 through link 2–1 is shorter than that through link 4–1. In addition, eliminate link 4 to 7. Although the accumulative travel time to node 7 through link 4–7 is the same as that through link 8–7, but the travel time to reach the preceding node (node 8 for link 8–7) through link 5–8 (1 minute) is shorter than the travel time to reach the preceding node (node 4 for link 4–7) through link 5–4 (3 minutes). Therefore, link 4–7 must be eliminated (Figure 2.82).

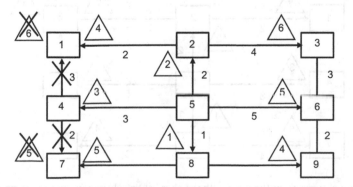

FIGURE 2.82 Elimination of link 4 to 1 and link 4 to 7 for Problem 2.55.

Step 6: the travel time from the node with the shortest travel time (node 9) to the nearest nodes (node 6) (Figure 2.83).

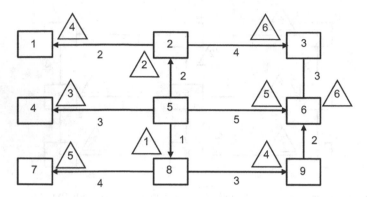

FIGURE 2.83 Travel times from the node with the shortest travel time (node 9) to the nearest nodes (node 6) for Problem 2.55.

Step 7: eliminate link 9 to 6 because the accumulative travel time to node 6 through link 5–6 is shorter than that through link 9–6 (Figure 2.84).

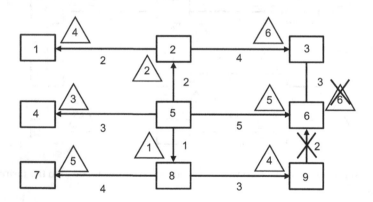

FIGURE 2.84 Elimination of link 9 to 6 for Problem 2.55.

Step 8: the travel time from the node with the shortest travel time (node 6) to the nearest nodes (node 3) (Figure 2.85).

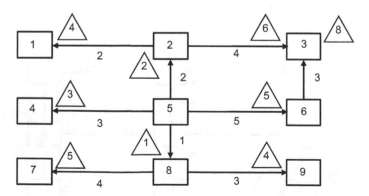

FIGURE 2.85 Travel times from the node with the shortest travel time (node 6) to the nearest nodes (node 3) for Problem 2.55.

Step 9: eliminate link 6 to 3 because the accumulative travel time to node 3 through link 2–3 is shorter than that through link 6–3 (Figure 2.86).

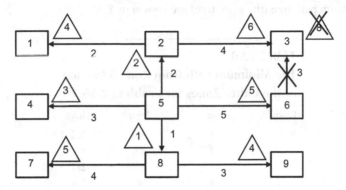

FIGURE 2.86   Elimination of link 6 to 3 for Problem 2.55.

Step 10: at the end, the final network for the minimum path tree (the skim tree) is composed, as shown in the diagram below (Figure 2.87).

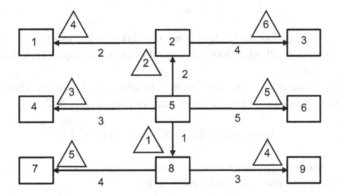

FIGURE 2.87   The final minimum path network (skim tree) for Problem 2.55.

2.56   Using Problem 2.55, if the number of auto trips between zone 5 and all other zones is as shown in Table 2.89, determine the number of trips on each link using the all-or-nothing trip assignment method and based on the minimum path network composed previously.

TABLE 2.89
Number of Auto Trips from Zone 5 to Other Zones for Problem 2.56

From	To	Trips
5	1	30
	2	100
	3	40
	4	80
	6	90
	7	45
	8	30
	9	70

Solution:

The links that are used from node 5 (zone 5) to each of the other nodes (zones) based on the minimum path tree (the skim tree) are shown in Table 2.90.

TABLE 2.90
The Minimum Path from Zone 5 to Each
of the Other Zones for Problem 2.56

From	To	Trips	Links
5	1	30	5–2, 2–1
	2	100	5–2
	3	40	5–2, 2–3
	4	80	5–4
	6	90	5–6
	7	45	5–8, 8–7
	8	30	5–8
	9	70	5–8, 8–9

Based on the previous table that shows the links on the minimum path network for trips from zone 5 to all other zones, the number of trips on each link is computed. Sample Calculation:

Link 5–2 is used by trips from zone 5 to zones 1, 2, and 3. Therefore, the trips between zone 5 and each of these zones are assigned to link 5–2.

$$\text{Trips on link 5-2} = 30 + 100 + 40 = 170$$

Link 5–8 is used by trips from zone 5 to zones 7, 8, and 9. Hence, the trips between zone 5 and each of these zones are assigned to link 5–8.

$$\text{Trips on link 5-8} = 45 + 30 + 70 = 145$$

In a similar manner, the number of trips on each of the other links is estimated. The results are shown in Table 2.91.

TABLE 2.91
Number of Trips Assigned to Each Link of the Minimum
Path Network for Problem 2.56

Link	Used by Trips from Zone 5 to Other Zones	Number of Trips
2–1	1	30
2–3	3	40
5–2	1, 2, 3	30 + 100 + 40 = 170
5–4	4	80
5–6	6	90
5–8	7, 8, 9	45 + 30 + 70 = 145
8–7	7	45
8–9	9	70

2.57 The following diagram shows the travel times on the links that connect six zones. Determine the minimum path from zone 6 to all other zones and construct the skim tree. The travel times between the zones are in minutes (Figure 2.88).

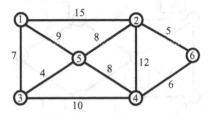

FIGURE 2.88 A six-zone network for Problem 2.57.

Solution:

Step 1: the travel times from zone 6 to the nearest zones (zones 2 and 4) are determined (Figure 2.89).

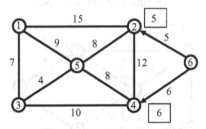

FIGURE 2.89 Travel times from zone 6 to the nearest zones (zones 2 and 4) for Problem 2.57.

Step 2: the travel times from the zone with the shortest travel time (zone 2) to the nearest zones (zones 1, 4, and 5) are determined (Figure 2.90).

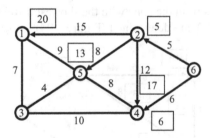

FIGURE 2.90 Travel times from the zone with the shortest travel time (zone 2) to the nearest zones (zones 1, 4, and 5) for Problem 2.57.

Step 3: eliminate link 2 to 4 because the accumulative travel time to zone 4 through link 6–4 is shorter than that through link 2–4 (Figure 2.91).

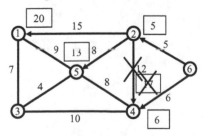

FIGURE 2.91 Elimination of link 2 to 4 for Problem 2.57.

Step 4: the travel times from the zone with the shortest travel time (zone 4) to the nearest zones (zones 3 and 5) are determined (Figure 2.92).

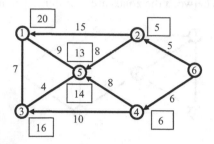

FIGURE 2.92   Travel times from the zone with the shortest travel time (zone 4) to the nearest zones (zones 3 and 5) for Problem 2.57.

Step 5: eliminate link 4 to 5 because the accumulative travel time to zone 5 through link 2–5 is shorter than that through link 4–5 (Figure 2.93).

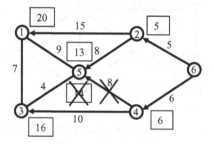

FIGURE 2.93   Elimination of link 4 to 5 for Problem 2.57.

Step 6: the travel times from the zone with the shortest travel times (zone 5) to the nearest zones (zones 1 and 3) are determined (Figure 2.94).

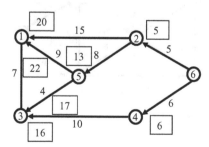

FIGURE 2.94   Travel times from the zone with the shortest travel time (zone 5) to the nearest zones (zones 1 and 3) for Problem 2.57.

Step 7: eliminate link 5 to 1 because the accumulative travel time to zone 1 through link 2–1 is shorter than that through link 5–1, and eliminate link 5–3 because the accumulative travel time to zone 3 through link 4–3 is shorter than that through link 5–3 (Figure 2.95).

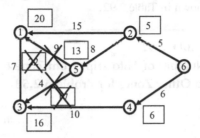

FIGURE 2.95 Elimination of link 5 to 1 and link 5 to 3 for Problem 2.57.

Step 8: the travel times from the zone with the shortest travel times (zone 3) to the nearest zones (zone 1) are determined (Figure 2.96).

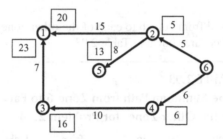

FIGURE 2.96 Travel times from the zone with the shortest travel time (zone 3) to the nearest zones (zone 1) for Problem 2.57.

Step 9: eliminate link 3–1 because the accumulative travel time to zone 1 through link 2–1 is shorter than that through link 3–1 (Figure 2.97).

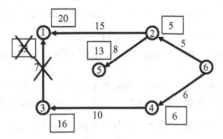

FIGURE 2.97 Elimination of link 3 to 1 for Problem 2.57.

Step 10: the final minimum path network (the skim tree) is composed, as shown in Figure 2.98.

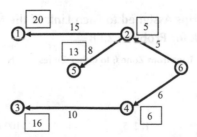

FIGURE 2.98 The final minimum path network (skim tree) for Problem 2.57.

2.58   In the problem above (Problem 2.57), load the number of trips on each link of the minimum path network using the all-or-nothing trip assignment method and based on the final skim tree (the minimum path network) if the number of trips between zone 6 and the other zones is as shown in Table 2.92.

TABLE 2.92
Number of Auto Trips from Zone 6
to Other Zones for Problem 2.58

T_{61}	100
T_{62}	80
T_{63}	90
T_{64}	120
T_{65}	140

Solution:

The links that are used from zone 6 to each of the other zones based on the minimum path tree (the skim tree) are shown in Table 2.93.

TABLE 2.93
The Minimum Path from Zone 6 to Each
of the Other Zones for Problem 2.58

From	To	Trips	Links
6	1	100	6–2, 2–1
	2	80	6–2
	3	90	6–4, 4–3
	4	120	6–4
	5	140	6–2, 2–5

Based on the previous table that shows the links on the minimum path network for trips from zone 6 to all other zones, the number of trips on each link is estimated. Sample Calculation:

Link 6–2 is used by trips from zone 6 to zones 1, 2, and 5. Therefore, the trips between zone 6 and each of these three zones are assigned to link 6–2.

$$\text{Trips on link } 6\text{-}2 = 100 + 80 + 140 = 320$$

Following the same procedure, the number of trips on each of the other links is estimated. The results are shown in Table 2.94.

TABLE 2.94
Number of Trips Assigned to Each Link of the Minimum
Path Network for Problem 2.58

Link	Used by Trips from Zone 6 to Other Zones	Number of Trips
2–1	1	100
2–5	5	140
4–3	3	90
6–2	1, 2, 5	100+80+ 140=320
6–4	3, 4	90+120=210

Therefore, the final skim tree (the minimum path network) loaded with trips is shown in Figure 2.99.

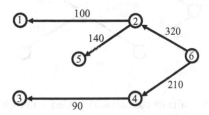

FIGURE 2.99   The Final minimum path network loaded with trips for Problem 2.58.

2.59 Given the minimum path network (the skim tree) in Figure 2.100, load the network using the all-or-nothing trip assignment method if the trips from zone 4 to the other zones are, as given in Table 2.95.

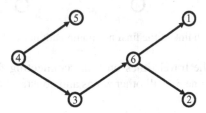

FIGURE 2.100   Given minimum path network for Problem 2.59.

TABLE 2.95
Number of Auto Trips from Zone 4
to Other Zones for Problem 2.59

T_{41}	200
T_{42}	150
T_{43}	180
T_{45}	250
T_{46}	300

Solution:

The links are loaded with trips based on the usage of each link when moving from zone 4 to the other zones on the skim tree (the minimum path network). For instance, link 4–3 is used when moving from zone 4 to 3, from zone 4 to 6, from zone 4 to 1, and from zone 4 to 2. Therefore, the summation of T_{43}, T_{46}, T_{41}, and T_{42} is the number of trips using link 4–3. Following the same procedure, the other links on the skim tree are loaded, as shown in Figures 2.101 and 2.102.

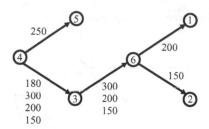

FIGURE 2.101   Links of the final minimum path network loaded with trips for Problem 2.59.

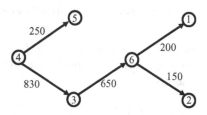

FIGURE 2.102   Total trips on each link of the final minimum path network for Problem 2.59.

2.60   Figure 2.103 shows the travel times on links connecting four zones. Determine the minimum path from zone 3 to all other zones and construct the skim tree (show a step-by-step procedure).

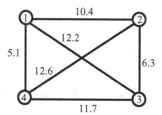

FIGURE 2.103   A four-zone network for Problem 2.60.

Solution:

Step 1: determine the travel times from zone 3 to the nearest zones (zones 2 and 4) (Figure 2.104).

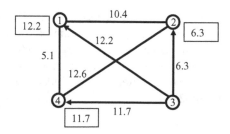

FIGURE 2.104   Travel times from the zone with the shortest travel time (zone 3) to the nearest zones (zones 2 and 4) for Problem 2.60.

Step 2: determine the travel times from the zone with the shortest travel time (zone 2) to the nearest zones (zones 1 and 4) (Figure 2.105).

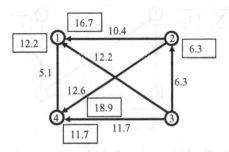

FIGURE 2.105 Travel times from the zone with the shortest travel time (zone 2) to the nearest zones (zones 1 and 4) for Problem 2.60.

Step 3: eliminate link 2–1 because the accumulative travel time to zone 1 through link 3–1 is shorter than that through link 2–1, and eliminate link 2–4 because accumulative travel time to zone 4 through link 3–4 is shorter than that through link 2–4 (Figure 2.106).

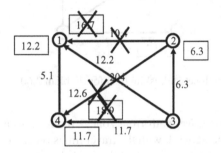

FIGURE 2.106 Elimination of link 2 to 1 and link 2 to 4 for Problem 2.60.

Step 4: determine the travel times from the zone with the shortest travel times (zone 4) to the nearest zones (zone 1) (Figure 2.107).

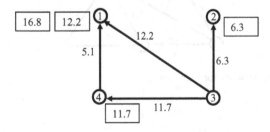

FIGURE 2.107 Travel times from the zone with the shortest travel time (zone 4) to the nearest zones (zone 1) for Problem 2.60.

Step 5: eliminate link 4–1 because the accumulative travel time to zone 4 through link 3–1 is shorter than that through link 4–1 (Figure 2.108).

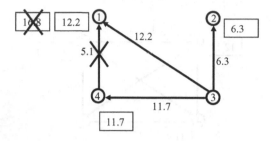

FIGURE 2.108 Elimination of link 4 to 1 for Problem 2.60.

Step 6: the final minimum path network (the skim tree) is composed, as shown in Figure 2.109.

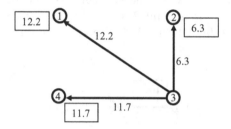

FIGURE 2.109 The minimum path network (skim tree) for Problem 2.60.

2.61   Determine the shortest travel path from zone 2 (node 2) to all other zones (nodes) in the following five-zone network with the travel times on each link shown for each zone pair. Show a step-by-step procedure and the final skim tree. The travel times are given in minutes (Figure 2.110).

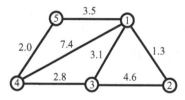

FIGURE 2.110 A five-zone network for Problem 2.61.

Solution:

Step 1: determine the travel times from zone 2 to the nearest zones (zones 1 and 3) (Figure 2.111).

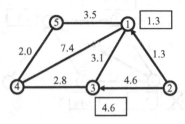

FIGURE 2.111   Travel times from zone 2 to the nearest zones (zones 1 and 3) for Problem 2.61.

Step 2: determine the travel times from the zone with the shortest travel time (zone 1) to the nearest zones (zones 3, 4, and 5) (Figure 2.112).

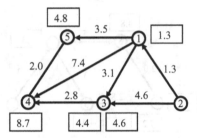

FIGURE 2.112   Travel times from the zone with the shortest travel time (zone 1) to the nearest zones (zones 3, 4, and 5) for Problem 2.61.

Step 3: eliminate link 2–3 because the accumulative travel time to zone 3 through link 1–3 is shorter than that through link 2–3 (Figure 2.113).

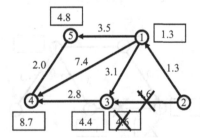

FIGURE 2.113   Elimination of link 2 to 3 for Problem 2.61.

Step 4: determine the travel times from the zone with the shortest travel time (zone 3) to the nearest zones (zone 4) (Figure 2.114).

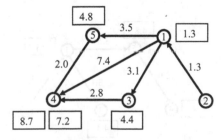

FIGURE 2.114   Travel times from the zone with the shortest travel time (zone 3) to the nearest zones (zone 4) for Problem 2.61.

Step 5: eliminate link 1–4 because the accumulative travel time to zone 4 through link 3–4 is shorter than that through link 1–4 (Figure 2.115).

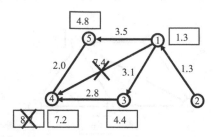

FIGURE 2.115 Elimination of link 1 to 4 for Problem 2.61.

Step 6: determine the travel times from the zone with the shortest travel time (zone 5) to the nearest zones (zone 4) (Figure 2.116).

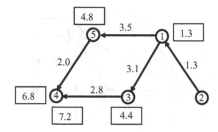

FIGURE 2.116 Travel times from the zone with the shortest travel time (zone 5) to the nearest zones (zone 4) for Problem 2.61.

Step 7: eliminate link 3–4 because the accumulative travel time to zone 4 through link 5–4 is shorter than that through link 3–4 (Figure 2.117).

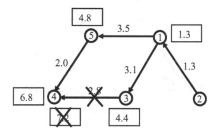

FIGURE 2.117 Elimination of link 3 to 4 for Problem 2.61.

Step 8: the final network for the minimum path (the skim tree) is composed as in Figure 2.118:

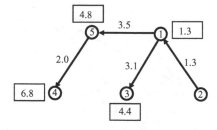

FIGURE 2.118 The final minimum path network (skim tree) for Problem 2.61.

2.62 Based on the final skim tree (the minimum path tree) composed in the problem above (Problem 2.61), load the network using the all-or-nothing trip assignment method if the number of trips from zone 2 to all other zones is shown in Table 2.96.

TABLE 2.96
Number of Auto Trips from Zone 2
to Other Zones for Problem 2.62

From Zone	To Zone	Number of Trips
2	1	110
	3	140
	4	180
	5	200

Solution:

Each link in the minimum path network is loaded with the trips that use the link when moving from zone 2 to the other zones on the skim tree. For instance, link 1–3 is used when moving from zone 2–3 only. Therefore, the T_{23} (the trips from zone 2 to zone 3) is the number of trips using link 1–3. Link 2–1 is used when moving from zone 2 to each of the zones 1, 3, 5, and 4. Therefore, the summation of T_{21}, T_{23}, T_{25}, and T_{24} (the trips from zone 2 to each of the zones 1, 3, 5, and 4, respectively) is the number of trips using link 2–1. Following the same procedure, the other links on the skim tree are loaded, as shown in Figure 2.119 and Figure 2.120:

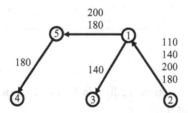

FIGURE 2.119 Links of the final minimum path network loaded with trips for Problem 2.62.

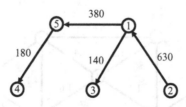

FIGURE 2.120 Total trips on each link of the final minimum path network for Problem 2.62.

2.63   If the travel times (in minutes) on each link between each node pair are given in Figure 2.121 for a seven-node network:

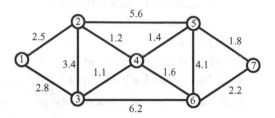

FIGURE 2.121   A seven-node network for Problem 2.63.

(1) Determine the shortest travel path from node 1 to all other nodes (zones) and show the final skim tree.

(2) Load the network based on the final skim tree using the all-or-nothing trip assignment if the zone-to-zone trips are as shown in Table 2.97.

TABLE 2.97

Number of Auto Trips from Zone 1 to Other Zones for Problem 2.63

T_{12}	100
T_{13}	120
T_{14}	150
T_{15}	200
T_{16}	240
T_{17}	80

Solution:

(1) The following step-by-step procedure is followed to determine the final minimum path network (the skim tree):

Step 1: determine the travel times from node 1 to the nearest nodes (nodes 2 and 3) (Figure 2.122).

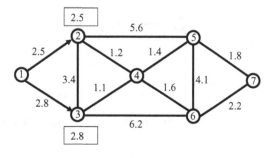

FIGURE 2.122   Travel times from node 1 to the nearest nodes (nodes 2 and 3) for Problem 2.63.

Step 2: determine the travel times from the node with the shortest travel time (node 2) to the nearest nodes (nodes 3, 4, and 5) (Figure 2.123).

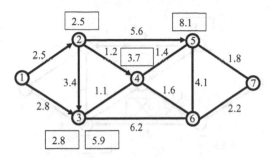

FIGURE 2.123   Travel times from the node with the shortest travel time (node 2) to the nearest nodes (nodes 3, 4, and 5) for Problem 2.63.

Step 3: eliminate link 2–3 because the accumulative travel time to node 3 through link 1–3 is shorter than that through link 2–3 (Figure 2.124).

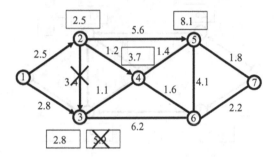

FIGURE 2.124   Elimination of link 2 to 3 for Problem 2.63.

Step 4: determine the travel times from the node with the shortest travel time (node 3) to the nearest nodes (nodes 4 and 6) (Figure 2.125).

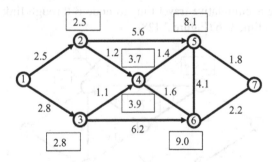

FIGURE 2.125   Travel times from the node with the shortest travel time (node 3) to the nearest nodes (nodes 4 and 6) for Problem 2.63.

Step 5: eliminate link 3–4 because the accumulative travel time to node 4 through link 2–4 is shorter than that through link 3–4 (Figure 2.126).

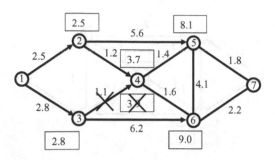

FIGURE 2.126   Elimination of link 3 to 4 for Problem 2.63.

Step 6: determine the travel times from the node with the shortest travel time (node 4) to the nearest nodes (nodes 5 and 6) (Figure 2.127).

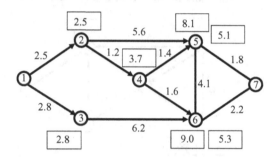

FIGURE 2.127   Travel times from the node with the shortest travel time (node 4) to the nearest nodes (nodes 5 and 6) for Problem 2.63.

Step 7: eliminate link 2–5 because the accumulative travel time to node 5 through link 4–5 is shorter than that through link 2–5. In addition, eliminate link 3 to 6 because the accumulative travel time to node 6 through link 4–6 is shorter than that through link 3–6 (Figure 2.128).

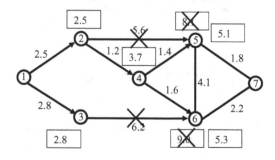

FIGURE 2.128   Elimination of link 2 to 5 and link 3 to 6 for Problem 2.63.

Step 8: determine the travel times from the node with the shortest travel time (node 5) to the nearest nodes (nodes 6 and 7) (Figure 2.129).

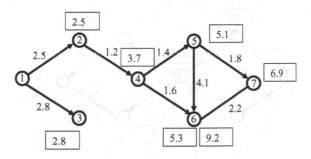

FIGURE 2.129   Travel times from the node with the shortest travel time (node 5) to the nearest nodes (nodes 6 and 7) for Problem 2.63.

Step 9: eliminate link 5–6 because the accumulative travel time to node 6 through link 4–6 is shorter than that through link 5–6 (Figure 2.130).

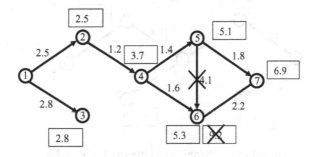

FIGURE 2.130   Elimination of link 5 to 6 for Problem 2.63.

Step 10: determine the travel times from the node with the shortest travel time (node 6) to the nearest nodes (node 7) (Figure 2.131).

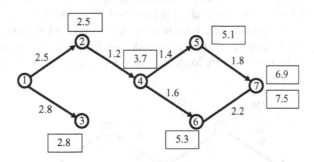

FIGURE 2.131   Travel times from the node with the shortest travel time (node 6) to the nearest nodes (node 7) for Problem 2.63.

Step 11: eliminate link 6–7 because the accumulative travel time to node 7 through link 5–7 is shorter than that through link 6–7 (Figure 2.132).

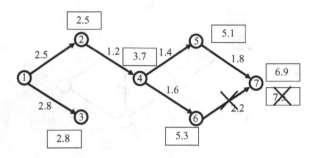

FIGURE 2.132   Elimination of link 6 to 7 for Problem 2.63.

Step 12: The final network for the minimum path (the skim tree) is composed, as shown in Figure 2.133.

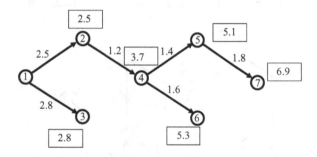

FIGURE 2.133   The final minimum path network (skim tree) for Problem 2.63.

(2)   The minimum path network (final skim tree) is loaded using the all-or-nothing trip assignment method:

Each link in the minimum path network is loaded with the trips that use the link when moving from node 1 to the other nodes on the skim tree. For instance, link 2–4 is used when moving from node 1 to each of the nodes 4, 5, 6, and 7. Therefore, the summation of T_{14}, T_{15}, T_{16}, and T_{17} (the trips from node 1 to each of the nodes 4, 5, 6, and 7, respectively) is the number of trips using link 2–4. Following the same procedure, the other links on the skim tree are loaded, as shown in Figure 2.134 and Figure 2.135.

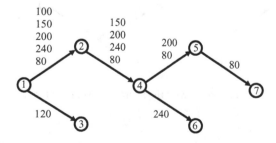

FIGURE 2.134   Links of the final minimum path network loaded with trips for Problem 2.63.

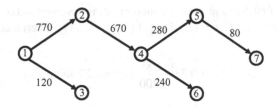

FIGURE 2.135 Total trips on each link of the final minimum path network for Problem 2.63.

2.64 The auto trips between zones 1 and 2 is 2000, and there are three available routes A, B, and C between the two (as shown in Figure 2.136). Determine the number of trips assigned to route A using the minimum time path with the capacity restraint method in the first iteration at a volume increment of 500 trips if the specific parameters for the highways between the two zones for capacity restraint are as given in Table 2.98.

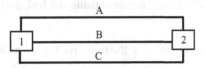

FIGURE 2.136 Three routs A, B, and C between zones 1 and 2 for Problem 2.64.

TABLE 2.98

Capacity Restraint Parameters for the Three Highways between the Two Zones for Problem 2.64

Link Performance Parameter	Route A	Route B	Route C
Free-Flow Travel Time (min)	15.0	13.0	12.0
Capacity (vph)	1800	2000	2200
α		0.83	
β		2.7	

Solution:

The capacity restraint formula is given by the following mathematical expression (it is also called the link performance function):

$$t = t_0 \left[1 + \alpha \left(\frac{q}{q_{max}} \right)^{\beta} \right] \quad (2.39)$$

Where:
t = the travel time on the link
t_0 = the free-flow travel time
q = the flow rate (volume per hour) on the link
q_{max} = the maximum flow rate (capacity) of the link
α and β = the specific parameters of the highway (highway-related parameters)

Iteration 1: The first 500 vph (volume increment) is assigned to the path with the minimum travel time (route C, $t_0 = 12$ minutes) (Figure 2.137). Therefore:

$$t_C = 12\left[1 + 0.83\left(\frac{500}{2200}\right)^{2.7}\right] = 12.2 \text{ minutes}$$

In this case, the number of trips assigned to rout A is zero.

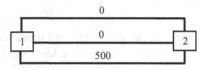

FIGURE 2.137   Iteration 1 of the minimum time path with capacity restraint method for Problem 2.64.

2.65   Using Problem 2.64, determine the travel time (minutes) and the number of trips assigned to route C in the fourth iteration using the link performance function.

Solution:

After iteration 1, the travel times on the three routes are as shown in Table 2.99.

TABLE 2.99			
Estimated Travel Times for Routes *A*, *B*, and *C* after Iteration #1 for Problem 6.65			
Link Performance Parameter	**Route *A***	**Route *B***	**Route *C***
Travel Time (min)	15.0	13.0	12.2

Iteration 2: The second 500 vph is assigned to the path with the minimum travel time in the first iteration (route C, $t_C = 12.2$ minutes) (Figure 2.138). Therefore:

$$t = t_0\left[1 + \alpha\left(\frac{q}{q_{max}}\right)^{\beta}\right]$$

$\Rightarrow$

$$t_C = 12\left[1 + 0.83\left(\frac{1000}{2200}\right)^{2.7}\right] = 13.2 \text{ minutes}$$

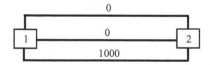

FIGURE 2.138   Iteration 2 of the minimum time path with capacity restraint method for Problem 2.65.

The travel times for the three routes A, B, and C after iteration 2 of the minimum time path with capacity restraint method are shown in Table 2.100.

TABLE 2.100

Estimated Travel Times for Routes *A*, *B*, and *C* after Iteration #2 for Problem 6.65

Link Performance Parameter	Route *A*	Route *B*	Route *C*
Travel Time (min)	15.0	13.0	13.2

Iteration 3: The third 500 vph is assigned to the path with the minimum travel time in the second iteration (route *B*, $t_B = 13$ minutes) (Figure 2.139; Table 2.101). Therefore:

$$t_B = 13\left[1 + 0.83\left(\frac{500}{2000}\right)^{2.7}\right] = 13.3 \text{ minutes}$$

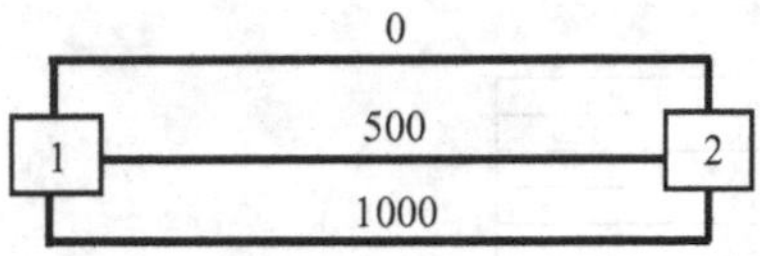

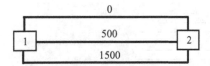

FIGURE 2.139 Iteration 3 of the minimum time path with capacity restraint method for Problem 2.65.

TABLE 2.101

Estimated Travel Times for Routes *A*, *B*, and *C* after Iteration #3 for Problem 6.65

Link Performance Parameter	Route *A*	Route *B*	Route *C*
Travel Time (min)	15.0	13.3	13.2

Iteration 4: The fourth 500 vph is assigned to the path with the minimum travel time in the third iteration (route *C*, $t_C = 13.2$ minutes) (Figure 2.140; Table 2.102). Therefore:

$$t_C = 12\left[1 + 0.83\left(\frac{1500}{2200}\right)^{2.7}\right] = 15.5 \text{ minutes}$$

FIGURE 2.140 Iteration 4 of the minimum time path with capacity restraint method for Problem 2.65.

TABLE 2.102

Estimated Travel Times for Routes *A*, *B*, and *C* after Iteration #4 for Problem 6.65

Link Performance Parameter	Route *A*	Route *B*	Route *C*
Travel Time (min)	15.0	13.3	15.5

Therefore, the number of trips assigned to route C after iteration 4 is equal to 1500 vph and the travel time is 15.5 minutes.

This problem is solved using the MS Excel program. The formulas and the "If" statements needed in the iterative process to determine the travel time and the number of trips on each route (link) are shown in the Excel worksheet shown below. Using the same function, it can be copied to the right to cover more routes (more columns in Excel) and down to include more iterations (more rows in Excel).

Figure 2.141 shows the MS Excel worksheet used for the iterative approach of the minimum time path with capacity restraint method.

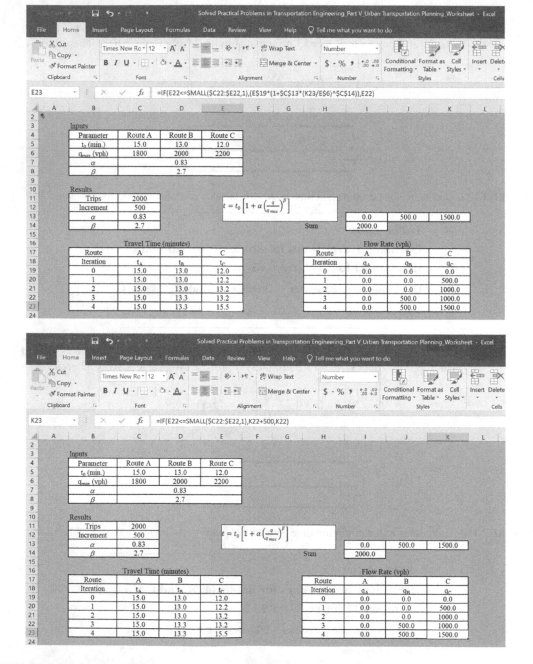

FIGURE 2.141 Image of MS Excel worksheet used for the iterative approach of the minimum time path with capacity restraint method for Problem 2.65.

2.66 Six available highways (A, B, C, D, E, and F) connect between two zones 1 and 2, as shown in the diagram below. The total number of trips during the peak hour between the two zones is 5000 vph. The auto trips between zones 1 and 2 is 5000 trips, and there are six available routes between the two zones (as shown in Figure 2.142). Determine the travel time and the number of trips assigned to each highway using the minimum time path with capacity restraint method until the total number of trips (5000 vph) is loaded or when an equilibrium occurs between the six highways (whichever comes first) using a traffic flow increment of 50 vph if the specific parameters for the highways between the two zones for capacity restraint are as given in Table 2.103.

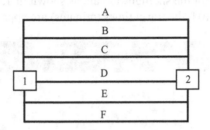

FIGURE 2.142 Six routs A, B, C, D, E, and F between zones 1 and 2 for Problem 2.66.

TABLE 2.103

Capacity Restraint Parameters for the Three Highways between the Two Zones for Problem 2.66

Link Performance Parameter	Route A	Route B	Route C	Route D	Route E	Route F
Free-Flow Travel Time (min)	23.0	24.0	25.0	26.0	26.5	27.0
Capacity (vph)	2800	2800	2800	1900	1900	1900
α			0.71			
β			2.1			

Solution:

The capacity restraint formula (the link performance function) given by the following mathematical expression is used to estimate the travel time on each highway:

$$t = t_0\left[1+\alpha\left(\frac{q}{q_{max}}\right)^{\beta}\right]$$

Sample Calculation:

Iteration 1: The first 50 vph is assigned to the path with the minimum travel time (highway A). Therefore:

$$t_A = 23\left[1+0.71\left(\frac{50}{2800}\right)^{2.1}\right] \cong 23.0 \text{ minutes}$$

Iteration 2: Since the travel time for highway A is still the lowest among the available six highways, the second 50 vph is also assigned to highway A. Therefore:

$$t_A = 23\left[1+0.71\left(\frac{100}{2800}\right)^{2.1}\right] \cong 23.0 \text{ minutes}$$

The iterative approach is continued to load the trips at an increment of 50 in each iteration on the highway with the minimum travel time compared to the travel times of the available highways until the total number of trips (5000 vph) is completely loaded on the six highways or when an equilibrium occurs between the six highways (whichever comes first).

Using the MS Excel program, the proper formulas and "If" statements needed in the iterative process to determine the travel time and the number of trips on each highway are used, as shown in the Excel worksheet shown in Figure 2.143. Finally, after 100 iterations when the total number of trips (5000 vph) is loaded on the six highways, the estimated travel times of the six highways are as shown in Table 2.104.

As seen in Table 2.104, the travel times (minutes) are approximately at equilibrium.

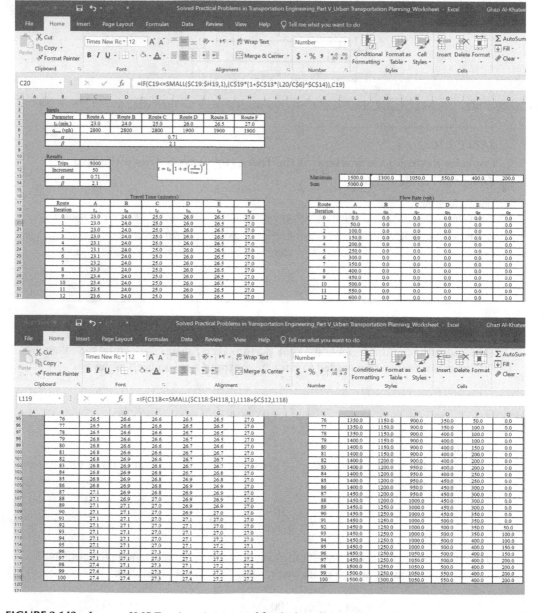

FIGURE 2.143 Image of MS Excel worksheet used for the iterative approach of the minimum time path with capacity restraint method for Problem 2.66.

TABLE 2.104

Estimated Travel Times for Routes *A*, *B*, *C*, *D*, *E*, and *F* after Iteration #100 for Problem 6.66

Iteration	Travel Time (minutes)					
	t_A	t_B	t_C	t_D	t_E	t_F
100	27.4	27.4	27.3	27.4	27.2	27.2

Multiple- Choice Questions and Answers for Chapter 1

1. The factors affecting travel demand include:
   a. Quality of available transportation services.
   b. Purpose of trip.
   c. Location of land use.
   d. Socio-economic characteristics of people living in the zone.
   e. **All of the above.**
2. The process of determining the number of trips that will begin or end in a traffic analysis zone within a study area is called:
   a. Trip distribution.
   b. **Trip generation.**
   c. Modal split.
   d. Traffic assignment.
   e. Trip end.
3. The trips that are either produced by a traffic zone or attracted to a traffic zone are called:
   a. Productions.
   b. Attractions.
   c. Origins.
   d. Destinations.
   e. **Trip ends.**
4. The process of developing a relationship between productions (or attractions) and land use is part of the:
   a. Trip distribution.
   b. **Trip generation.**
   c. Modal split.
   d. Traffic assignment.
   e. Trip end.
5. Trips that are defined in terms of the direction of a given inter-zonal trip are called:
   a. Origins.
   b. Destinations.
   c. Productions.
   d. Attractions.
   e. **a and b.**
6. Trips that are defined in terms of the land use associated with each trip end are called:
   a. Productions.
   b. Attractions.
   c. **a and b.**
   d. Origins.
   e. Destinations.

7. One of the following methods/models is *not* among the methods used for trip generation:
   a. Cross-classification method.
   b. **Gravity model.**
   c. Rates based on activity units model.
   d. Regression models.
   e. None of the above.

8. One of the following is *not* among the characteristics of disaggregate models for trip generation:
   a. They are household-based models.
   b. Rationale.
   c. The level of aggregation is based on the availability of data.
   d. **Zonal characteristics are considered.**
   e. All of the above.

9. One of the following is <u>not</u> among the characteristics of aggregate models for trip generation:
   a. They are zone-based models.
   b. Population characteristics are considered.
   c. Zonal characteristics are considered.
   d. **They account for zones with similar average characteristics.**
   e. None of the above.

10. The evaluation of regression models used for trip generation is accomplished through:
    a. Statistical significance.
    b. Logic of the model.
    c. Simplicity of the model.
    d. **All of the above.**
    e. None of the above.

11. The factors that impact which trip generation model to use include:
    a. Reliability of the model.
    b. Accuracy of the model.
    c. Availability of data.
    d. Availability of computer software and tools.
    e. **All of the above.**

12. The process by which the trips produced in one zone are assigned to other zones in the study zone is called:
    a. **Trip distribution.**
    b. Trip generation.
    c. Modal split.
    d. Traffic assignment.
    e. Trip end.

13. The purpose of trip distribution is to:
    a. Determine the transport mode split.
    b. Determine the shortest path between two zones.
    c. **Determine the volume of trips between zones.**
    d. Determine the number of trip ends for a study zone.
    e. All of the above.

14. The Factors affecting trip distribution include:
    a. Production.
    b. Attractiveness.
    c. Cost of travel.
    d. Time of travel.
    e. **All of the above.**

15. One of the following models/methods is <u>not</u> among the models used for trip distribution:
    a. Gravity model.
    b. Fratar model.
    c. **Cross-classification method.**
    d. Intervening opportunities method.
    e. Logit model.
16. The model which states that the number of trips between two zones is directly proportional to the attractions generated by the destination zone and inversely proportional to a function of travel time between the two zones is:
    a. **Gravity model.**
    b. Fratar model.
    c. Cross-classification method.
    d. Intervening opportunities method.
    e. Logit model.
17. One of the following is considered a limitation of the gravity model when used for trip distribution:
    a. The model requires extensive calibration.
    b. The absence of variables related to the characteristics of the individual or households who select the destination.
    c. The model cannot reflect changes in travel time between zones.
    d. **a and b.**
    e. The model cannot be used to predict traffic between zones where no traffic currently exists.
18. One of the following is considered a limitation of the growth factor models when used for trip distribution:
    a. The model requires extensive calibration.
    b. The model cannot reflect changes in travel time between zones.
    c. The model cannot be used to predict traffic between zones where no traffic currently exists.
    d. The absence of variables related to the characteristics of the individual or households who select the destination.
    e. **b and c.**
19. The step of the travel demand forecasting process that determines the number (or percentage) of trips between zones that are made by each transportation mode is called:
    a. Trip distribution.
    b. Trip generation.
    c. **Modal split.**
    d. Traffic assignment.
    e. Trip end.
20. The factors affecting mode choice include:
    a. Availability of transit service or auto ownership.
    b. Income of travelers.
    c. Relative advantages of each mode over the other available modes.
    d. Type of trip.
    e. **All of the above.**
21. One of the following models is <u>not</u> among the models used for mode choice:
    a. Direct Generation Models.
    b. Trip End Models.
    c. **Fratar model.**
    d. Trip Interchange Models.
    e. Probabilistic Models.

22. The models in which the mode choice is made after trip generation and before trip distribution are:
    a. Direct Generation Models.
    b. **Trip End Models.**
    c. Fratar model.
    d. Trip Interchange Models.
    e. Probabilistic Models.

23. In trip end models for modal choice, the relationship between the percentage of travel by transit and the urban travel factor (UTF) follows:
    a. A logarithmic model.
    b. An exponential function.
    c. A quadratic function.
    d. **An S curve.**
    e. A linear line.

24. In which of the following modal choice models do travelers choose their mode of transportation based on the competitiveness and the quality of service of the available modes:
    a. Direct Generation Models.
    b. Trip End Models.
    c. Trip Interchange Models.
    d. Probabilistic Models.
    e. **c and d.**

25. One of the following models is <u>not</u> among the modal choice probabilistic models:
    a. **Direct Generation Models.**
    b. Logit Models.
    c. Incremental logit models.
    d. Probit Models.
    e. Discriminant Analysis Models.

26. The utility function in probabilistic logit models for modal choice measures:
    a. The percentage of travel by transit.
    b. The travel time by a specific mode.
    c. **The degree of satisfaction for mode users based on their choice.**
    d. The total cost of the trip.
    e. The probability of choosing a transportation mode.

27. The step of the travel demand forecasting process in which the highway routes that will be used are determined and the number of autos and buses that are expected on each highway route are also determined is called:
    a. Trip distribution.
    b. Trip generation.
    c. Modal split.
    d. **Traffic assignment.**
    e. Trip end.

28. The basic approaches that can be used for traffic assignment purposes include:
    a. Diversion curves.
    b. Minimum time path assignment.
    c. Minimum time path with capacity restraint.
    d. **All of the above.**
    e. None of the above.

29. Using the diversion curves approach for traffic assignment, the traffic between two zones is determined as a function of:
    a. **Relative travel time or cost.**
    b. Minimum route time.
    c. Maximum route capacity.
    d. a and b.
    e. a and c.
30. In the minimum time path algorithm for traffic assignment, the traffic is assigned:
    a. Relative to the travel time on the route.
    b. **On the route that has the shortest travel time between the two zones.**
    c. Based on travel time and capacity of the route.
    d. Based on the minimum cost of the route.
    e. None of the above.
31. The relationship between travel time (speed) and volume on each link between two zones or what is called the link performance function is used in one of the following traffic assignment approaches:
    a. Diversion curves.
    b. Minimum time path assignment.
    c. **Minimum time path with capacity restraint.**
    d. All of the above.
    e. b and c.

Part II

Highway Survey

3 Terminology

Chapter 3 includes questions that cover the terminology used in highway surveys. A highway surveyor should be aware of the terms and concepts related to the type of survey that is used in the planning and pre-construction stage of roads and highways. It involves the measurement of horizontal and vertical angles, distances as well as elevations. Detailed plans and profiles of proposed highways are the outcome of this type of survey. Elevations along the highway centerline and across cross-sections established at regular intervals along the highway are used to compute construction earthwork cut and fill quantities. The questions in this chapter present a review of the basic terminology and concepts used for highway surveys. The questions are in a multiple-choice format and the answers are available at the end of Part II.

4 Errors

Chapter 4 includes practical problems and questions related to survey errors. The errors covered in this part are mainly associated with highway measurements of linear distances and angles. Errors can be blunders, constant errors, systematic errors, or random errors. The focus in this part will be on the last two types of error since blunders can be easily detected and constant errors are a special type of systematic errors. Systematic errors are detected when an environmental event takes place. These errors can typically be modeled using a mathematical formula and therefore the related measurements can be corrected accordingly. On the other hand, random errors theoretically reflect the limitations of the measuring instrument and the human operator, and can be minimized by using better instruments and proper procedures, and performing repeated measurements. The following practical problems and questions will focus on these two types of errors (systematic and random).

4.1   Determine the range of a distance measurement of 62 ± 0.5 m at 2σ.

Solution:

The standard error of the measurement at 1σ is ± 0.5 m. At 2σ, the standard error is determined as shown in in the procedure below:

$$\text{Standard Error at } 2\sigma = 2 \times \text{Standard Error at } 1\sigma \tag{4.1}$$

$\Rightarrow$

$$\text{Standard Error at } 2\sigma = 2 \times 0.5 = \pm 1.0 \text{ m} \left(\pm 3.3 \text{ ft} \right)$$

Therefore, the range at 2σ is determined as:

$$\text{Range at } 2\sigma = \text{Distance Measurement} \pm \text{Standard Error at } 2\sigma \tag{4.2}$$

$\Rightarrow$

$$\text{Range at } 2\sigma = 62 \pm 1.0 = 61.0 - 63.0$$

Figure 4.1 shows the MS Excel worksheet used to perform the computations of this problem.

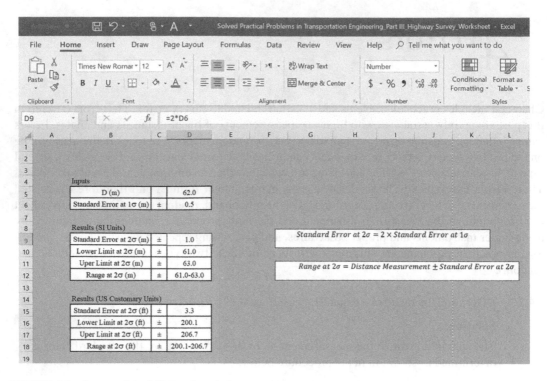

FIGURE 4.1   Image of the MS Excel worksheet used for the computations of Problem 4.1.

4.2   If the probability that the random error of a measured distance falls within the range of ±0.10 m is 95.4%, estimate the standard error (σ) of the measured distance.

Solution:

Table 4.1 shows the probability values for different error ranges. The limits for the error range are simply the z-value in a normal distribution of errors and the probability value is the area under the normal distribution curve within that range.

TABLE 4.1

Probability Values at Different Error Ranges (Normal Distribution Values)

Error Range	Probability
±0.6745σ	50.0
±1.00σ	68.3
±1.6449σ	90.0
±2.00σ	95.4
±3.00σ	99.7

Kavanagh, Barry; Mastin, Tom, Surveying: Principles and Applications: International Edition, 9th Ed., ©2014, Reprinted by permission of Pearson Education, Inc., New York, New York, USA.

At a probability of 95.4%, the error range is $\pm2.00\sigma$ (the z-value at 95.4% = 2). Therefore,

$$2\sigma = \pm0.10 \text{ m}$$

$\Rightarrow$

$$\sigma = \pm0.05 \text{ m} \left(\pm0.16 \text{ ft}\right)$$

Figure 4.2 shows the MS Excel worksheet used to perform the computations of this problem.

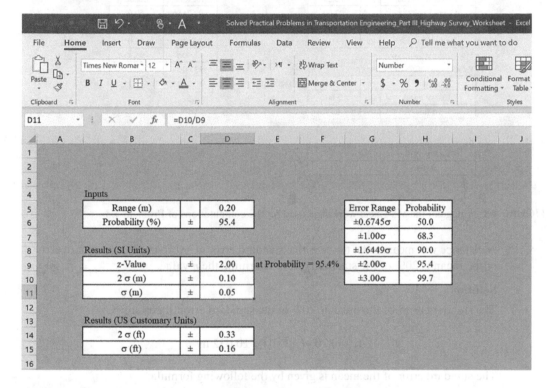

FIGURE 4.2 Image of the MS Excel worksheet used for the computations of Problem 4.2.

4.3 If the range of the length measurement of a passenger care is 18 to 22 ft at 90% probability, determine the standard error of the length measurement.

Solution:

The table in Problem 4.2 is used again in the solution of this problem. At a 90% probability level, the z-value of the error range is ±1.6449 and the error range is $\pm1.6449\sigma$. Therefore:

$$1.6449\sigma = \frac{22-18}{2} = 2 \text{ ft}$$

$\Rightarrow$

$$\sigma = \pm1.22 \text{ ft} \left(\pm0.37 \text{ m}\right)$$

Figure 4.3 shows the MS Excel worksheet used to perform the computations of this problem.

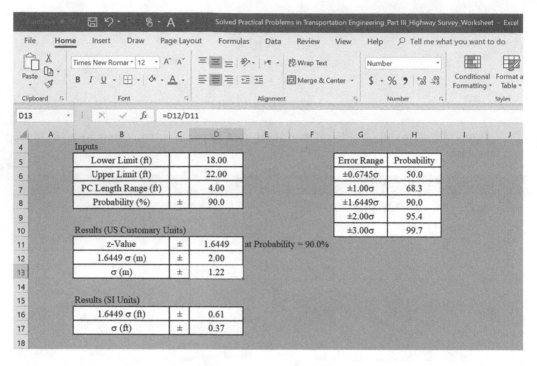

FIGURE 4.3 Image of the MS Excel worksheet used for the computations of Problem 4.3.

4.4   A distance was measured n times with a standard error of ±0.16 m. If the standard error of the mean is 20% of the standard error, determine the n value.

Solution:

The standard error of the mean is 20% of the standard error, therefore:

$$\sigma_{\bar{x}} = 0.20 \times (\pm0.16) = \pm0.032 \text{ m}$$

The standard error of the mean is given by the following formula:

$$\sigma_{\bar{x}} = \pm\frac{\sigma_x}{\sqrt{n}} \tag{4.3}$$

Where:
$\sigma_x =$ standard error (deviation) for a single measurement
$\sigma_{\bar{x}} =$ standard error of the mean of measurements
$n =$ number of measurements

$\Rightarrow$

$$0.032 = \pm\frac{0.16}{\sqrt{n}}$$

Solving for n provides:

$$n = 25$$

Figure 4.4 shows the MS Excel worksheet used to perform the computations of this problem.

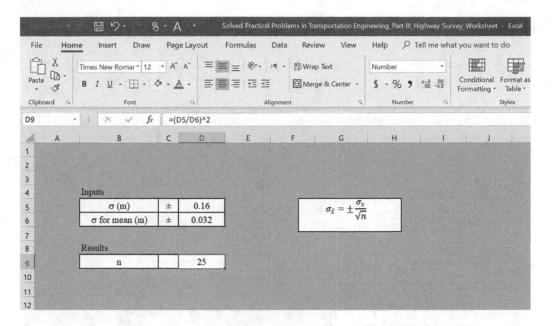

FIGURE 4.4 Image of the MS Excel worksheet used for the computations of Problem 4.4.

4.5 The maximum error of a distance measurement is ±0.222 m, determine the standard error and the probable error of the measurement.

Solution:

The maximum error is defined as shown in the following expression:

$$\text{Maximum Error} = \pm 3\sigma \tag{4.4}$$

Where:
σ = standard deviation
3 = the z-value at 99.7% probability (reliability) for a normal distribution of errors (shown in the table of Problem 4.2)

⇒

$$0.222 = \pm 3\sigma$$

⇒

$$\sigma = \pm 0.074 \text{ m} \left(\pm 0.243 \text{ ft}\right)$$

The probable error is defined as:

$$\text{Probable Error} = \pm 0.6745\sigma \tag{4.5}$$

Where:
σ = standard deviation
0.6745 = the z-value at 50% probability (reliability) for a normal distribution of errors (shown in the table of Problem 4.2).

⇒

$$\text{Probable Error} = \pm 0.6745 \left(0.07\right) = \pm 0.050 \text{ m} \left(\pm 0.164 \text{ ft}\right)$$

Figure 4.5 shows the MS Excel worksheet used to perform the computations of this problem.

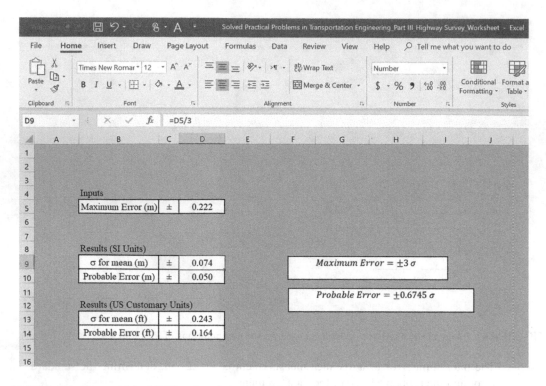

FIGURE 4.5 Image of the MS Excel worksheet used for the computations of Problem 4.5.

4.6   A distance is measured to be 82.222 m (26.975 ft) with a relative precision of 1/1000 at 2σ, determine the probable error and the maximum error of the distance measurement.

Solution:

The same table used in Problem 4.2 will be used in the solution of this problem. The table is shown below with the z-values in the error range corresponding to the probability levels:

Error Range	Probability
$\pm 0.6745\sigma$	50.0
$\pm 1.00\sigma$	68.3
$\pm 1.6449\sigma$	90.0
$\pm 2.00\sigma$	95.4
$\pm 3.00\sigma$	99.7

Kavanagh, Barry; Mastin, Tom, Surveying: Principles and Applications: International Edition, 9th Ed., ©2014, Reprinted by permission of Pearson Education, Inc., New York, New York, USA.

The formula used for relative precision is shown below:

$$\text{Relative Precision at } 1\sigma = \frac{1}{\left(\dfrac{D}{\sigma_D}\right)} \qquad (4.6)$$

Where:

D = distance

σ_D = standard deviation of distance measurement

In a similar manner, the relative precision at 2σ can be given by the following expression:

$$\text{Relative Precision at } 2\sigma = \frac{1}{\left(\dfrac{D}{2\sigma_D}\right)} \qquad (4.7)$$

$\Rightarrow$

$$0.001 = \frac{1}{\left(\dfrac{82.222}{2\sigma_D}\right)}$$

$\Rightarrow$

$$\sigma_D = 0.041 \text{ m} \left(0.135 \text{ ft}\right)$$

The probable error is defined as:

$$\text{Probable Error} = \pm 0.6745\sigma$$

$\Rightarrow$

$$\text{Probable Error} = \pm 0.6745\left(0.041\right) = \pm 0.028 \text{ m} \left(\pm 0.091 \text{ ft}\right)$$

The maximum error is defined as:

$$\text{Maximum Error} = \pm 3\sigma$$

$\Rightarrow$

$$\text{Maximum Error} = \pm 3\left(0.0411\right) = \pm 0.123 \text{ m} \left(\pm 0.405 \text{ ft}\right)$$

The MS Excel worksheet used to perform the computations in this problem is shown in Figure 4.6.

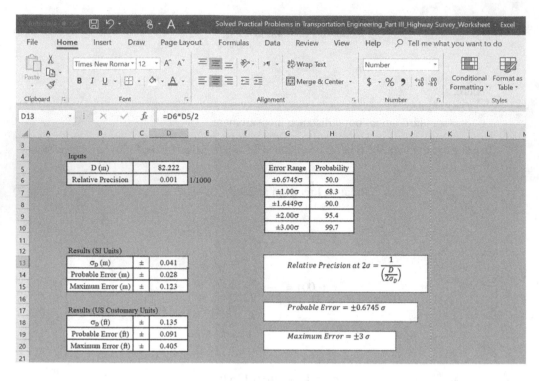

FIGURE 4.6 Image of the MS Excel worksheet used for the computations of Problem 4.6.

4.7   A 100-ft tape is to be used to measure a total distance of 2000 ft (609.6 m) in a parking
lot. How accurately should each 100-ft (30.5 m) distance be measured so that the 90%
error would not exceed ±0.20 ft (0.06 m) in the total distance of 2000 ft (609.6 m).

Solution:

The 90% error is defined as shown in the expression below:

$$90\% \text{ Error} = \pm1.6449\sigma \tag{4.8}$$

Where:
σ = standard deviation
1.6449 = the z-value at 90% probability (reliability) for a normal distribution of
errors (shown in the table of Problem 4.2)

In this case, since the 90% error is given for the total distance measurement, therefore:

$$\pm0.20 = \pm1.6449\sigma_{\text{total}}$$

$\Rightarrow$

$$\sigma_{\text{total}} = \pm0.122 \text{ ft}\left(0.037 \text{ m}\right)$$

This is the total standard deviation (σ_{total}) for the total distance measurement done in 20
measurements (because the length of the tape is 100 ft). But the total standard deviation
is given by the following formula:

$$\sigma_{total} = \pm\sqrt{n}\sigma \qquad (4.9)$$

Where:

σ_{total} = total standard deviation for total distance
n = number of measurements
σ = standard deviation for each measurement

$\Rightarrow$

$$0.122 = \pm\sqrt{20}\sigma$$

$\Rightarrow$

$$\sigma = \pm0.027 \text{ ft}\left(\pm0.008 \text{ m}\right)$$

Consequently, each 100-ft distance measurement should be made such that the standard error in the 100-ft measurement does not exceed ±0.027 ft.

Figure 4.7 shows the MS Excel worksheet used to perform the computations of this problem.

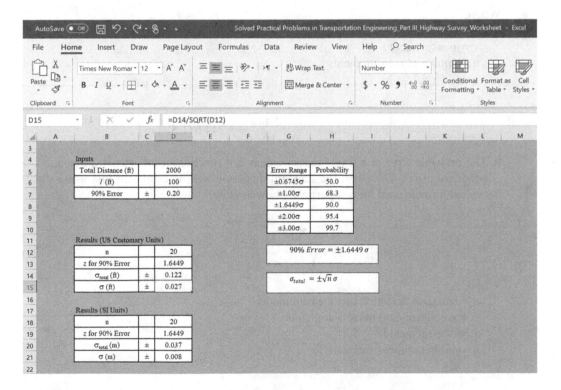

FIGURE 4.7 Image of the MS Excel worksheet used for the computations of Problem 4.7.

4.8   A median width of a highway is measured ten times with the following values:

Median Width (ft)	5.234	5.236	5.238	5.250	5.231	5.244	5.246	5.255	5.268	5.242

Determine the following:

a. The most probable value of the measured width.
b. The probable error of a single measurement.
c. The maximum error of a single measurement.
d. 90% error.

Solution:

a. The most probable value of the measured width is the average (mean) of the ten measurements. In other words, it is determined using the formula used to compute the mean shown in the following expression:

$$\bar{x} = \frac{\sum_{i=1}^{n} x_i}{n}$$

(4.10)

Where:
$\bar{x}$ =average (mean) of measurements
x_i=measurement i
n=number of measurements

$\Rightarrow$

$$\text{most probable value} = \bar{x} = \frac{52.444}{10} = 5.244 \text{ ft}(1.598 \text{ m})$$

b. The probable error of a single measurement is a function of the standard deviation (standard error). Therefore, the standard deviation will be estimated first using the typical formula used for this purpose (shown below):

$$\sigma_x = \sqrt{\frac{\sum_{i=1}^{n} (x_i - \bar{x})^2}{n-1}}$$

(4.11)

Where:
σ_x=standard deviation of x measurements
$\bar{x}$ =average (mean) of measurements
x_i=measurement i
n=number of measurements

$\Rightarrow$

$$\sigma_x = \sqrt{\frac{0.0011084}{10-1}} = \mp 0.011 \text{ ft}(\mp 0.003 \text{ m})$$

The probable error is given by the following formula (Equation 4.5 in Problem 4.5):

$$\text{Probable Error} = \pm 0.6745\sigma$$

$\Rightarrow$

$$\text{Probable Error} = \pm 0.6745(0.011) = 0.007 \text{ ft} (0.002 \text{ m})$$

c. The maximum error of a single measurement is determined using the same formula used earlier in Problem 4.5, which is shown below:

$$\text{Maximum Error} = \pm 3\sigma$$

$\Rightarrow$

$$\text{Maximum Error} = \pm 3(0.011) = 0.033 \text{ ft} (0.010 \text{ m})$$

d. The 90% error is also determined using the same formula used in Problem 4.7, which is shown below:

$$90\% \text{ Error} = \pm 1.6449\sigma$$

$\Rightarrow$

$$90\% \text{ Error} = \pm 1.6449(0.011) = 0.018 \text{ ft} (0.006 \text{ m})$$

Figure 4.8 illustrates the MS Excel worksheet used to do the computations for this problem.

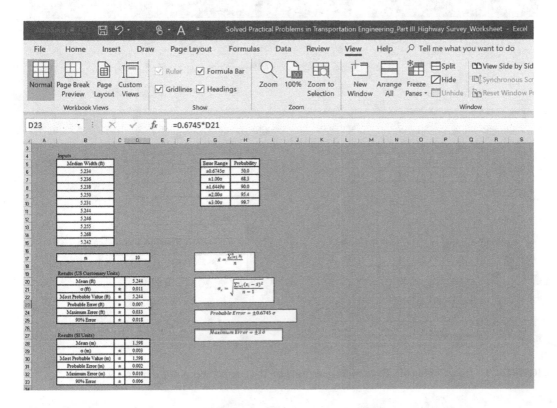

FIGURE 4.8 Image of the MS Excel worksheet used for the computations of Problem 4.8.

4.9   If the probability of the random error of measuring the width of a local street lane to fall
       within the range of ±0.08 m is 68.3%, determine the standard error (σ) of the measured
       lane width.

Solution:

The 68.3% error is defined as shown in the expression below:

$$68.3\% \text{ Error} = \pm 1.00\sigma \tag{4.12}$$

Where:
σ = standard deviation
1.00 = the z-value at 68.3% probability (reliability) for a normal distribution of errors
(shown in the table of Problem 4.2)

$\Rightarrow$

$$\pm 0.08 = \pm 1.00\sigma$$

$\Rightarrow$

$$\sigma = \pm 0.08 \text{ m} \left(\pm 0.26 \text{ ft} \right)$$

Figure 4.9 shows the MS Excel worksheet used to do the computations of this problem.

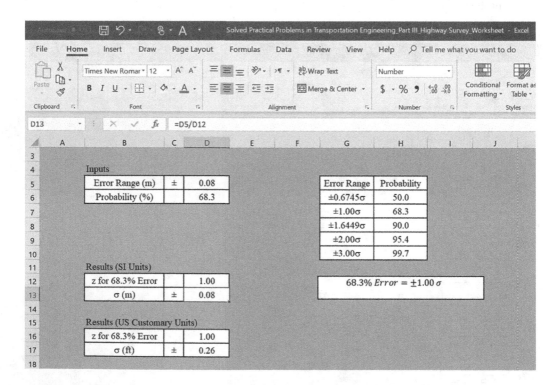

FIGURE 4.9   Image of the MS Excel worksheet used for the computations of Problem 4.9.

4.10   If the standard error of a measured lane width of a roadway is ±5 cm (±1.97 in), determine how many repeated measurements of the width should be performed so that the standard error of the mean is ±5 mm (±0.197 in).

Solution:

The number of repeated measurements is determined using the formula for the standard error of the mean shown below:

$$\sigma_{\bar{x}} = \pm \frac{\sigma_x}{\sqrt{n}}$$

$\Rightarrow$

$$0.005 = \pm \frac{0.05}{\sqrt{n}}$$

Solving for n provides:

$$n = 100 \text{ measurments}$$

Figure 4.10 illustrates the MS Excel worksheet used to do this computation.

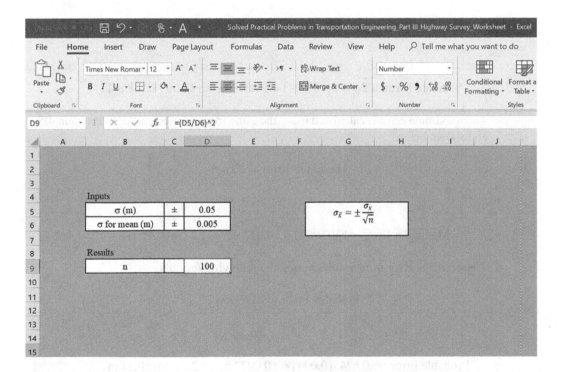

FIGURE 4.10   Image of the MS Excel worksheet used for the computations of Problem 4.10.

4.11   A width of a small parking lot is measured 100 times with a mean value of 95.46 m (313.19 ft). If the standard error of the measured width is ±4 cm (±1.57 in), determine the following:

a. The standard error of the mean.
b. The standard error at 2σ.
c. The maximum error.
d. The probable error.

Solution:

a. The standard error of the mean of the 100 measurements is determined using the following formula:

$$\sigma_{\bar{x}} = \pm \frac{\sigma_x}{\sqrt{n}}$$

⇒

$$\sigma_{\bar{x}} = \pm \frac{0.04}{\sqrt{100}} = \pm 0.004 \text{ m} = \pm 4 \text{ mm} \left(\pm 0.16 \text{ in} \right)$$

b. The standard error at 2σ is computed as follows:

$$\text{Standard Error at } 2\sigma = 2 \left(\text{Standard Error at } 1\sigma \right)$$

⇒

$$\text{Standard Error at } 2\sigma = 2 \left(\pm 0.004 \right) = \pm 0.008 \text{ m} = \pm 8 \text{ mm} \left(\pm 0.31 \text{ in} \right)$$

c. The maximum error is calculated using the same formula used earlier in Problem 4.5 and shown below:

$$\text{Maximum Error} = \pm 3\sigma$$

⇒

$$\text{Maximum Error} = \pm 3 \left(0.004 \right) = \pm 0.012 \text{ m} = \pm 1.2 \text{ cm} \left(0.47 \text{ in} \right)$$

d. The probable error is determined using the following formula:

$$\text{Probable Error} = \pm 0.6745\sigma$$

⇒

$$\text{Probable Error} = \pm 0.6745 \left(0.004 \right) = \pm 0.0027 \text{ m} = \pm 2.7 \text{ mm} \left(\pm 0.11 \text{ in} \right)$$

The MS worksheet used to perform the computations of this problem is illustrated in Figure 4.11.

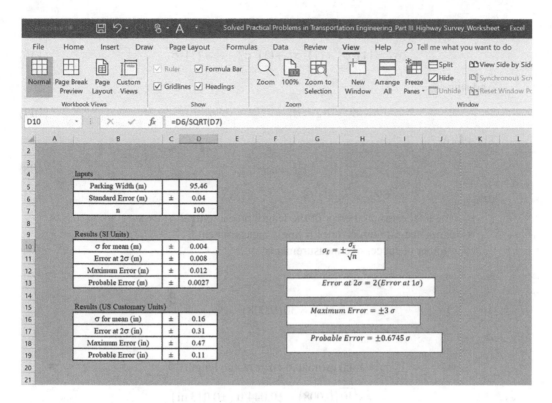

FIGURE 4.11 Image of the MS Excel worksheet used for the computations of Problem 4.11.

4.12 A series of measurements to determine the length of a parking lot of a shopping center is taken. The random error in taking each measurement is estimated to be ±0.008 ft (±0.24 cm). Determine the total estimated error of the length if 30 measurements are made.

Solution:

Using the law of the propagation of errors, an expression can be derived to determine the total estimated error as described in the procedure below:

$$\sigma_y^2 = \left(\frac{\partial F}{\partial x_1}\right)^2 \sigma_{x_1}^2 + \left(\frac{\partial F}{\partial x_2}\right)^2 \sigma_{x_2}^2 + \left(\frac{\partial F}{\partial x_3}\right)^2 \sigma_{x_3}^2 + \ldots + \left(\frac{\partial F}{\partial x_n}\right)^2 \sigma_{x_n}^2 \tag{4.13}$$

Where:

σ_y = estimated standard error of y (function, F of $x_1, x_2, x_3, \ldots x_n$)

$\sigma_{x1}, \sigma_{x2}, \sigma_{x3}, \ldots$ and σ_n = estimated standard error of $x_1, x_2, x_3, \ldots$ and x_n measurements, respectively

$\partial F/\partial x_1, \partial F/\partial x_2, \partial F/\partial x_3, \ldots$ and $\partial F/\partial x_n$ = partial derivatives of the function F with respect to $x_1, x_2, x_3, \ldots$ and x_n, respectively

In this case, the length of the parking lot is given as the total of all measurements; in other words, it is given by:

$$L = F(l_1 + l_2 + l_3 + \ldots + l_n) = l_1 + l_2 + l_3 + \ldots + l_n \tag{4.14}$$

$$\sigma_L^2 = \left(\frac{\partial F}{\partial l_1}\right)^2 \sigma_{l_1}^2 + \left(\frac{\partial F}{\partial l_2}\right)^2 \sigma_{l_2}^2 + \left(\frac{\partial F}{\partial l_3}\right)^2 \sigma_{l_3}^2 + \ldots + \left(\frac{\partial F}{\partial l_n}\right)^2 \sigma_{l_n}^2 \tag{4.15}$$

$\Rightarrow$

$$\sigma_L^2 = \left(1\right)^2 \sigma_{l_1}^2 + \left(1\right)^2 \sigma_{l_2}^2 + \left(1\right)^2 \sigma_{l_3}^2 + \ldots + \left(1\right)^2 \sigma_{l_n}^2 \qquad (4.16)$$

And since:

$$\sigma_{l_1} = \sigma_{l_2} = \sigma_{l_3} = \ldots = \sigma_{l_n} = \sigma_l \qquad (4.17)$$

Therefore:

$$\sigma_L^2 = n\sigma_l^2 \qquad (4.18)$$

Where:

σ_L=the total estimated error of the length measurement
σ_l=estimated random error of a single measurement l
n=number of successive measurements

$\Rightarrow$

$$\sigma_L^2 = 30\left(0.008\right)^2$$

$\Rightarrow$

$$\sigma_L = \text{total estimated error of the length}$$

$$= \sqrt{30}\left(0.008\right) = \pm0.044 \text{ ft }\left(\pm0.013 \text{ m}\right)$$

The MS Excel worksheet used to perform the computations of this problem is illustrated in Figure 4.12.

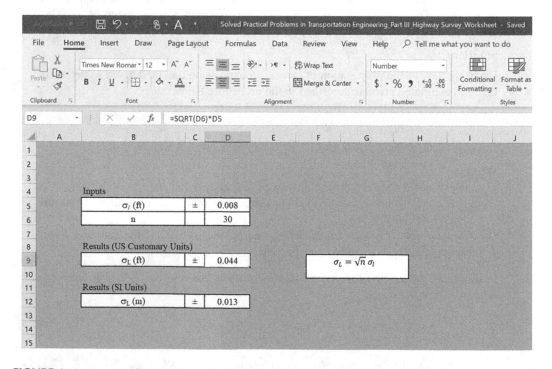

FIGURE 4.12   Image of the MS Excel worksheet used for the computations of Problem 4.12.

4.13   The length and the width of a rectangular parking lot are measured to be 242.26 m and
       125.38 m, respectively. If the standard errors for measuring the length and the width are
       ±5 cm and ±4 cm, respectively, determine the standard error and the probable error in
       computing the area of the parking lot.

Solution:

Using the law of the propagation of errors, an expression can be derived to determine
the standard error of the area as described below:

$$\sigma_y^2 = \left(\frac{\partial F}{\partial x_1}\right)^2 \sigma_{x_1}^2 + \left(\frac{\partial F}{\partial x_2}\right)^2 \sigma_{x_2}^2 + \left(\frac{\partial F}{\partial x_3}\right)^2 \sigma_{x_3}^2 + \ldots + \left(\frac{\partial F}{\partial x_n}\right)^2 \sigma_{x_n}^2$$

Since the area is given by:

$$A = L \times W$$

$\Rightarrow$

$$\sigma_A^2 = \left(\frac{\partial A}{\partial L}\right)^2 \sigma_L^2 + \left(\frac{\partial A}{\partial W}\right)^2 \sigma_W^2 \tag{4.20}$$

Where:

A, L, and W = area, length, and width of the parking lot, respectively
$\sigma_L = \sigma_L = \sigma_L$ = standard errors of area, length, and width of the parking lot, respectively
∂A/∂L and ∂A/∂W = partial derivatives of the area A with respect to L and W,
respectively

Since:

$$\frac{\partial A}{\partial L} = W \tag{4.21}$$

$$\frac{\partial A}{\partial W} = L \tag{4.22}$$

Therefore, the standard error of the area is given by the expression below:

$$\sigma_A^2 = \left(W\right)^2 \sigma_L^2 + \left(L\right)^2 \sigma_W^2 \tag{4.23}$$

$\Rightarrow$

$$\sigma_A^2 = \left(125.38\right)^2 \left(0.05\right)^2 + \left(242.26\right)^2 \left(0.04\right)^2$$

$\Rightarrow$

$$\sigma_A = \pm 11.54 \text{ m}^2$$

This means that; although the standard errors for measuring the length and the width
of the parking lot are ±5 cm and ±4 cm, which are relatively small, but the estimated

standard error of computing the area of the parking lot is large (± 11.54 m^2). Therefore, high accuracy should be considered when measuring linear measurements to compute other parameters such as area.

The probable error in computing the area is calculated using the known formula used previously and shown in the expression below:

$$\text{Probable Error} = \pm 0.6745 \sigma$$

$\Rightarrow$

$\text{Probable Error} = \pm 0.6745 (11.54) = 7.78 \text{ m}^2 \left(83.8 \text{ ft}^2\right)$, which is also high.

Figure 4.13 shows the MS Excel worksheet used to perform the computations of this problem.

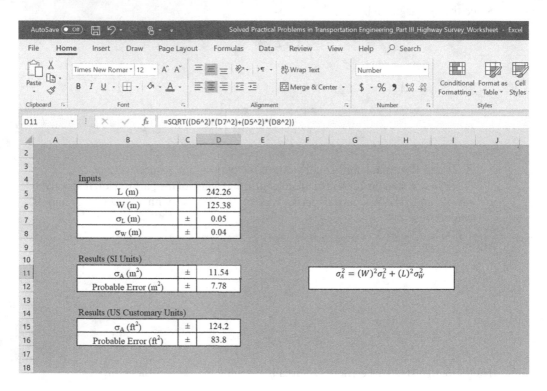

FIGURE 4.13   Image of the MS Excel worksheet used for the computations of Problem 4.13.

4.14   A simple horizontal curve of a highway is to be staked out by taking twelve successive measurements at twelve 100-m stations with a total standard error of not more than ± 0.10 m. How accurately should each 100-m distance be measured so that the total standard error (permissible value) is not exceeded?

Solution:

Using the law of the propagation of errors, the same expression derived in Problem 4.12 for the total standard error is used as follows:

$$\sigma_L^2 = n \sigma_l^2$$

$$\Rightarrow$$

$$(0.10)^2 = 12\sigma_l^2$$

$$\Rightarrow$$

$$\sigma_l = \pm 0.029 \text{ m} = +2.9 \text{ cm} \left(\pm 1.1 \text{ in} \right)$$

Therefore, the measurement of each station (each 100-m distance) should be made at a maximum accuracy of ±2.9 cm so that the standard error of the total length will not exceed ±0.10 m (±10 cm).

Figure 4.14 shows the MS Excel worksheet used to perform the computations of this problem.

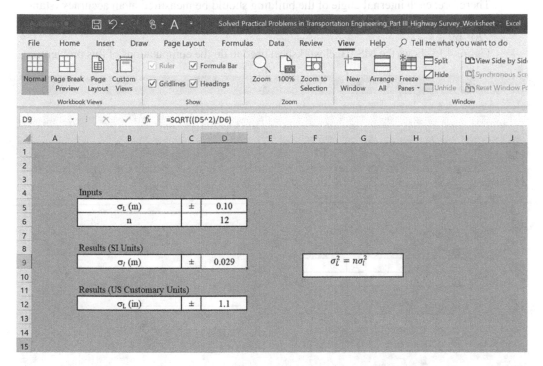

FIGURE 4.14 Image of the MS Excel worksheet used for the computations of Problem 4.14.

4.15 A terminal building at an airport has the shape of a six-sided closed traverse. The sum of the internal angles of the traverse is exactly 720°. In a surveying job to lay out the building, it is required to measure the internal angles of the building in the field such that the sum should not deviate from the 720° by more than ±3′. How accurately should each angle be measured?

Solution:

The law of the propagation of errors again is used, as shown below:

Since the total value of the measured internal angles is equal to the sum of the six measured angles, based on Equation 4.18 in Problem 4.12, the following expression using the law of the propagation of errors can be written:

$$\sigma_{\text{Total}}^2 = n\sigma^2$$

Where:

σ_{Total} = total standard error for the sum of angles
n = number of internal angles
σ = standard error for each angle measurement

⇒

$$(3)^2 = 6(\sigma)^2$$

⇒

$$\sigma = \pm 1.2'$$

Therefore, each internal angle of the building should be measured at an accuracy (standard error) of $\pm 1.2'$ so that the sum of the internal angles would not deviate from the 720° by $\pm 3'$.

Figure 4.15 shows the MS Excel worksheet used to do the computations of this problem.

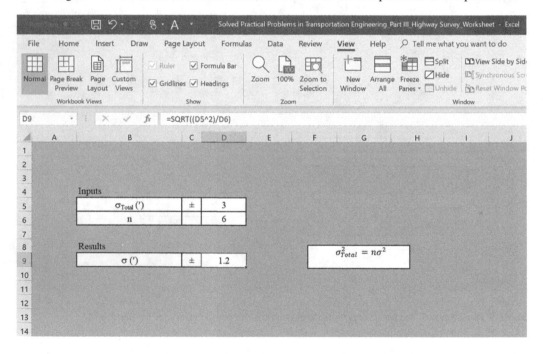

FIGURE 4.15   Image of the MS Excel worksheet used for the computations of Problem 4.15.

4.16   If the two sides a and b of a triangular corner island for a rural intersection shown in Figure 4.16 are measured to be 4.35 m and 4.15 m, respectively, with an estimated standard error of ± 0.05 m for each, compute the estimated standard error of the area of the island using the law of propagation. *Assume that the island has a perfect triangular shape.*

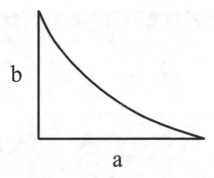

FIGURE 4.16 An intersection triangular island for Problem 4.16.

Solution:

The area of the triangular island is given by the following formula:

$$A = \frac{1}{2}(a)(b) \tag{4.24}$$

Using the law of propagation, the following expression is derived:

$$\sigma_A^2 = \left(\frac{\partial A}{\partial a}\right)^2 \sigma_a^2 + \left(\frac{\partial A}{\partial b}\right)^2 \sigma_b^2 \tag{4.25}$$

$$\Rightarrow$$

$$\sigma_A^2 = \left(\frac{1}{2}b\right)^2 \sigma_a^2 + \left(\frac{1}{2}a\right)^2 \sigma_b^2 \tag{4.26}$$

Where:

A = the area of the triangular corner island

a and b = the two sides of the triangular corner island

$$\Rightarrow$$

$$\sigma^2{}_A = \left(\frac{1}{2} \times 4.15\right)^2 (0.05)^2 + \left(\frac{1}{2} \times 4.35\right)^2 (0.05)^2$$

$$\Rightarrow$$

$$\sigma_A = \sqrt{\left(\frac{1}{2} \times 4.15\right)^2 (0.05)^2 + \left(\frac{1}{2} \times 4.35\right)^2 (0.05)^2} = \pm 0.15 \text{ m}^2 \left(\pm 1.62 \text{ ft}^2\right)$$

Figure 4.17 shows the MS Excel worksheet used to do the computations of this problem.

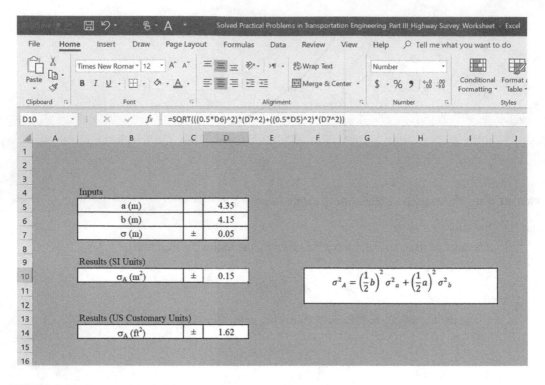

FIGURE 4.17 Image of the MS Excel worksheet used for the computations of Problem 4.16.

4.17   The area of a rectangular parking lot for a shopping mall is to be measured with a maximum allowable error of ±3.0 m². The length and width of the parking lot are measured to be 550.80 m and 320.90 m, respectively. Assuming that both the length and the width are to be measured with the same relative precision, determine:

a. The allowable relative precision in making the distance (length or width) measurement at 1σ.

b. The estimated standard error of the length measurement.

c. The estimated standard error of the width measurement.

Solution:

The allowable relative precision in making the distance (length or width) measurement at 1σ is determined using the formula below:

$$\frac{1}{N} = \frac{1}{\left(\dfrac{D}{\sigma_D}\right)}$$

(4.27)

Where:

$1/N$ = relative precision represented as a ratio of 1 to a whole number, N

D and σ_b = measured distance and its estimated standard error, respectively

Since the length and the width of the parking lot will have the same precision, it provides that:

$$\frac{1}{N} = \frac{1}{\left(\dfrac{L}{\sigma_L}\right)} = \frac{1}{\left(\dfrac{W}{\sigma_W}\right)}$$

$\Rightarrow$

$$\sigma_L = \frac{L}{N}$$

And:

$$\sigma_W = \frac{W}{N}$$

Using the law of the propagation of errors for the area of the parking lot, the formula derived earlier in Equation 4.23 in Problem 4.13 can be used as shown below:

$$\sigma_A^2 = (W)^2 \sigma_L^2 + (L)^2 \sigma_W^2$$

The maximum allowable error for measuring the area of the parking lot is given in this problem as ± 3.0 m^2, but the maximum error is given by the following formula:

$$\text{Maximum Error} = \pm 3\sigma$$

$\Rightarrow$

$$3.0 = \pm 3\sigma_A$$

$\Rightarrow$

$$\sigma_A = \pm 1.0 \text{ m}^2$$

Substituting the values of the different terms in the above equation (Equation 4.23 of Problem 4.13) provides the following:

$$(1.0)^2 = (320.90)^2 \left(\frac{550.80}{N} \right)^2 + (550.80)^2 \left(\frac{320.90}{N} \right)^2$$

Solving the above equation for N provides:

$$N = 249965 \cong 250000$$

Therefore, the relative precision in making the length or the width measurement at 1 σ is 1/250000.

b. The estimated standard error of the length measurement is calculated as follows:

$$\sigma_L = \frac{L}{N}$$

$\Rightarrow$

$$\sigma_L = \frac{550.80}{250000} = \pm 0.002 \text{ m} \left(\pm 0.024 \text{ ft} \right)$$

c. The estimated standard error of the width measurement is determined in a similar manner:

$$\sigma_W = \frac{W}{N}$$

$\Rightarrow$

$$\sigma_W = \frac{320.90}{250000} = \pm 0.001 \text{ m} \left(\pm 0.014 \text{ ft} \right)$$

Note that measuring the length and the width at estimated standard errors of ± 0.002 m and ± 0.001 m, respectively, results in an estimated standard error for the area of ± 1.0 m². It indicates that measuring linear distances is required to be performed at very high accuracy to avoid large errors in computing the area.

Figure 4.18 shows the MS Excel worksheet used to perform the computations of this problem.

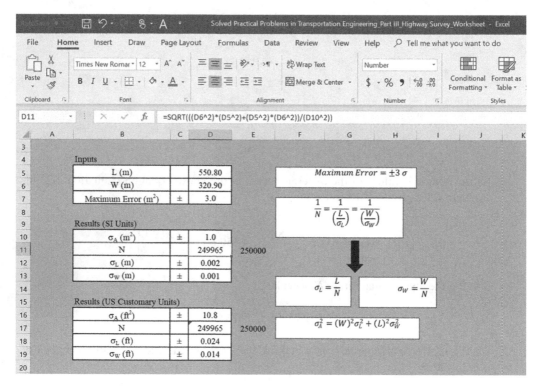

FIGURE 4.18   Image of the MS Excel worksheet used for the computations of Problem 4.17.

4.18   A circular playground is to be laid out in the suburb of a city. The radius of the play-
       ground is measured as 35.40 m (116.14 ft) to estimate the area of the playground. If the
       maximum allowable error of the area is ±9 m², determine the estimated standard error
       at which the radius measurement should be performed.

Solution:

Using the law of the propagation of errors, an expression can be derived to determine
the standard error of the area as described below:

$$\sigma_y^2 = \left(\frac{\partial F}{\partial x_1}\right)^2 \sigma_{x_1}^2 + \left(\frac{\partial F}{\partial x_2}\right)^2 \sigma_{x_2}^2 + \left(\frac{\partial F}{\partial x_3}\right)^2 \sigma_{x_3}^2 + \ldots + \left(\frac{\partial F}{\partial x_n}\right)^2 \sigma_{x_n}^2$$

Since the circular playground area is given by:

$$A = \pi r^2 \tag{4.28}$$

Where:
    A = the area of the circular playground
    r = the radius of the circular playground

⇒

$$\sigma_A^2 = \left(\frac{\partial A}{\partial r}\right)^2 \sigma_r^2 \tag{4.29}$$

Where:
    σ_A = the standard error of the area (A)
    $\partial A/\partial r$ = partial derivative of A with respect to r
    σ_r = the standard error of the radius (r)
    A = the area of the circular playground
    r = the radius of the circular playground

⇒

$$\sigma_A^2 = \left(2\pi r\right)^2 \sigma_r^2 \tag{4.30}$$

The maximum allowable error for measuring the area of the playground is given in this
problem as ±9 m², but the maximum error is given by the following formula:

$$\text{Maximum Error} = \pm 3\sigma$$

⇒

$$9 = \pm 3\sigma_A$$

⇒

$$\sigma_A = \pm 3 \text{ m}^2$$

Since:

$$\sigma_A^2 = \left(2\pi r\right)^2 \sigma_r^2$$

$$\Rightarrow$$

$$(3)^2 = (2\pi \times 35.40)^2 \sigma_r^2$$

Solving the above equation for σ_r provides:

$$\sigma_r = \pm 0.013 \, \text{m} \left(\pm 0.044 \, \text{ft} \right)$$

This result means that to estimate the area of the circular playground with a maximum allowable error of $\pm 9 \, \text{m}^2$, the radius of the circular playground should be measured with a standard error of ± 1.3 cm.

Figure 4.19 shows the MS Excel worksheet used to conduct the computations of this problem.

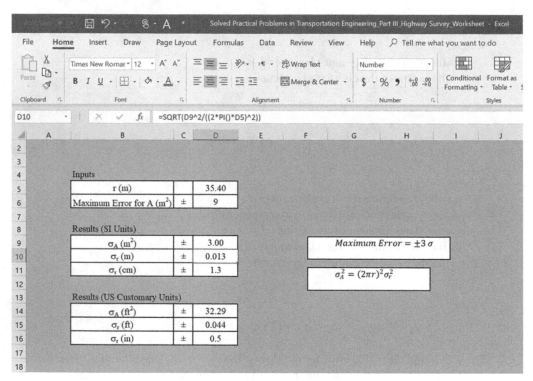

FIGURE 4.19   Image of the MS Excel worksheet used for the computations of Problem 4.18.

4.19   The area of an elliptical stadium is to be estimated. The semi-major axis (a) and the semi-minor axis (b) of the elliptical stadium are measured as 200 ± 0.10 m and 100 ± 0.10 m, respectively. Estimate the standard error of the area of the stadium.

Solution:

Again, the law of the propagation of errors with the expression shown below will be used to derive an expression for the standard error of the area:

$$\sigma_y^2 = \left(\frac{\partial F}{\partial x_1} \right)^2 \sigma_{x_1}^2 + \left(\frac{\partial F}{\partial x_2} \right)^2 \sigma_{x_2}^2 + \left(\frac{\partial F}{\partial x_3} \right)^2 \sigma_{x_3}^2 + \ldots + \left(\frac{\partial F}{\partial x_n} \right)^2 \sigma_{x_n}^2$$

the formula $A = \pi ab$, where: a and b are the semi-major axis and the semi-minor axis of the ellipse, respectively.

Since the elliptical stadium area is given by:

$$A = \pi ab \qquad (4.31)$$

Where:
    A = the area of the elliptical stadium
    a = semi-major axis of the elliptical stadium
    b = semi-minor axis of the elliptical stadium

$\Rightarrow$

$$\sigma_A^2 = \left(\frac{\partial A}{\partial a}\right)^2 \sigma_a^2 + \left(\frac{\partial A}{\partial b}\right)^2 \sigma_b^2 \qquad (4.32)$$

Where:
    σ_A = the standard error of the area (A)
    $\partial A/\partial a$ = partial derivative of A with respect to a
    $\partial A/\partial b$ = partial derivative of A with respect to b
    σ_a = the standard error of the semi-major axis (a)
    σ_b = the standard error of the semi-minor axis (b)
    A = the area of the elliptical stadium
    a = semi-major axis of the elliptical stadium
    b = semi-minor axis of the elliptical stadium

$\Rightarrow$

$$\sigma_A^2 = \left(\pi b\right)^2 \sigma_a^2 + \left(\pi a\right)^2 \sigma_b^2 \qquad (4.33)$$

$\Rightarrow$

$$\sigma_A^2 = \left(\pi \times 100\right)^2 \left(0.10\right)^2 + \left(\pi \times 200\right)^2 \left(0.10\right)^2$$

$\Rightarrow$

$$\sigma_A = \pm 70.2 \text{ m}^2 \left(\pm 756.1 \text{ ft}^2\right)$$

This result means that if the two axes of the elliptical stadium are measured at a standard error of ± 0.10 m (± 10 cm), the standard error for estimating the area will be ± 70.2 m^2.

Figure 4.20 shows the MS Excel worksheet used to do the computations of this problem.

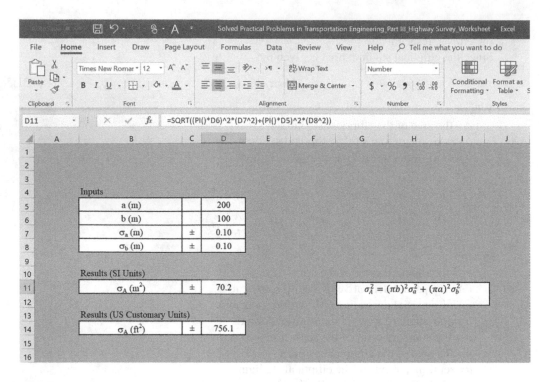

FIGURE 4.20 Image of the MS Excel worksheet used for the computations of Problem 4.19.

4.20 Three independent measurements of the length of a simple circular curve of a highway are obtained as shown in the following table:

L_1	300.50 m	σ_1	±0.05 m
L_2	300.00 m	σ_2	±0.01 m
L_3	301.10 m	σ_3	±0.10 m

Determine the weighted mean of the length measurement (L) and its estimated standard error.

Solution:

To determine the weight of each length measurement, a standard error of unit weight has to be chosen among the three standard errors given in this problem. The weight of each measurement is given by the formula below:

$$w_i = \left(\frac{\sigma_0}{\sigma_i} \right)^2 \tag{4.34}$$

Where:

w_i = weight of measurement x_i

σ_0 = standard error of unit weight (any of the given standard errors can be selected as the standard error of unit weight)

The standard error of unit weight, $\sigma_0 = \sigma_2$, based on this selected value, the weight of each measurement is determined as follows:

$\Rightarrow$

$$w_1 = \left(\frac{\sigma_2}{\sigma_1}\right)^2 = \left(\frac{0.01}{0.05}\right)^2 = 0.04$$

$$w_2 = \left(\frac{\sigma_2}{\sigma_2}\right)^2 = 1$$

$$w_3 = \left(\frac{\sigma_2}{\sigma_3}\right)^2 = \left(\frac{0.01}{0.10}\right)^2 = 0.01$$

The results are shown in Table 4.2.

TABLE 4.2
The Three Curve Length Measurements and Their Weights for Problem 4.20

L_1	300.50 m	w_1	0.04
L_2	300.00 m	w_2	1
L_3	301.10 m	w_3	0.01

In this case, the most probable value of the length measurement is given by the weighted mean of the n measurements, as shown in the following expression:

$$\bar{x} = \frac{\sum_{i=1}^{n} w_i x_i}{\sum_{i=1}^{n} w_i} \tag{4.35}$$

Where:
  $\bar{x}$ = weighted mean of n measurements
  w_i = weight of measurement x_i
  x_i = measurement $_i$

$\Rightarrow$

$$\bar{x} = \frac{(300.50)(0.04) + (300.00)(1) + (301.10)(0.01)}{0.04 + 1 + 0.01}$$

$$= 300.03 \text{ m} (984.35 \text{ ft})$$

The estimated standard error of the weighted mean is calculated using the formula shown below:

$$\sigma_{\bar{x}} = \frac{\sigma_0}{\sqrt{\sum_{i=1}^{n} w_i}} \qquad (4.36)$$

$\sigma_{\bar{x}}$ = estimated standard error of the mean
σ_0 = standard error of unit weight (in this case, σ_2 has been selected as the standard error of unit weight)
w_i = weight of measurement x_i

$\Rightarrow$

$$\sigma_{\bar{x}} = \frac{0.01}{\sqrt{0.04 + 1 + 0.01}} = \pm 0.01 \, \text{m} \left(\pm 0.03 \, \text{ft} \right)$$

Figure 4.21 shows the MS Excel worksheet used to perform the computations of this problem.

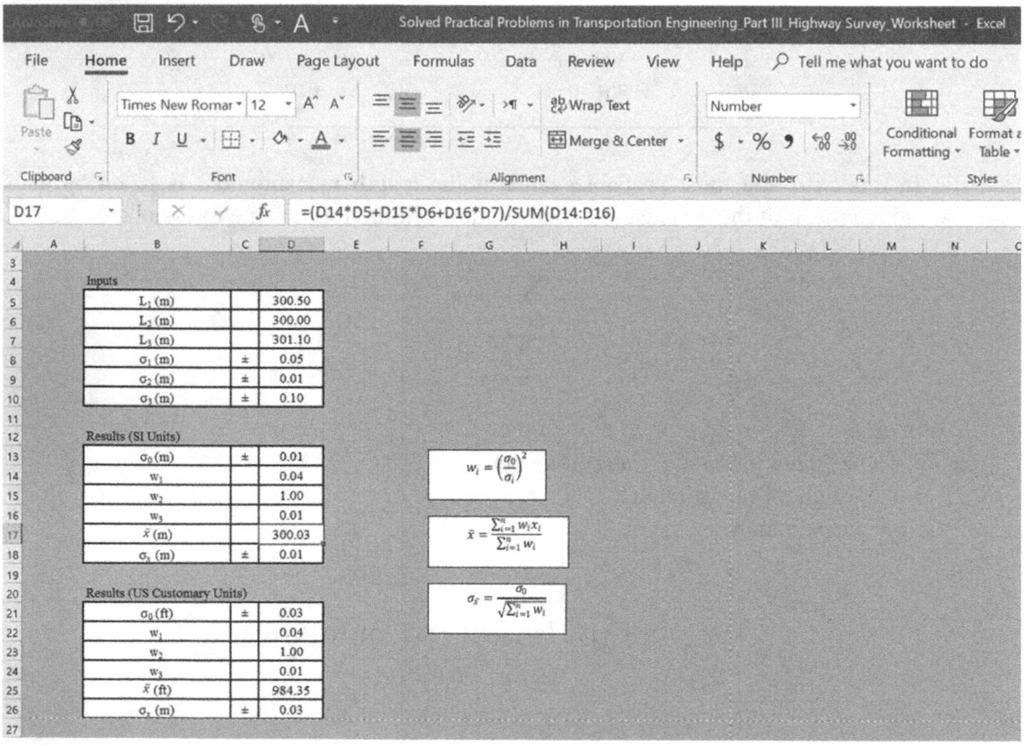

FIGURE 4.21   Image of the MS Excel worksheet used for the computations of Problem 4.20.

4.21 A chord length of a simple horizontal curve of 30.24 m (99.21 ft) is to be laid out in the field at a temperature of −10°C (14°F) using a 20-m steel tape calibrated at a temperature of 25°C (77°F). Determine the actual length of the tape that should be used to establish the desired chord length if the tape is fully supported during the measurement.

Solution:

The length correction, as well as the temperature correction, are applied here, as shown in the following procedure:

The length correction is given by the following formula:

$$C_l = \left(\frac{l-l'}{l'}\right)L \tag{4.37}$$

Where:

C_l = length correction
l = actual tape length
l' = nominal tape length
L = measured distance

⇒

$$C_l = \left(\frac{l-20}{20}\right)(30.24) = 1.512l - 30.24$$

On the other hand, the temperature correction is provided by the following expression:

$$C_t = 0.0000116(T_1 - T_0)L \tag{4.38}$$

Where:

0.0000116 = coefficient of thermal expansion of steel per 1°C
T_1 = field temperature (°C)
T_0 = temperature under which the tape is calibrated (°C)
L = measured length (m)

⇒

$$C_t = 0.0000116(-10-25)(30.24) = -0.01228 \text{ m}$$

In order to establish the desired chord length, the total correction should be equal to zero. Therefore, the length correction plus the temperature correction should be equal to zero as in the following expression:

$$1.512l - 30.24 - 0.01228 = 0$$

⇒

$$l = 20.01 \text{ m} \left(65.64 \text{ ft}\right)$$

In other words, the actual tape length should measure 1 cm more than 20 m so that at the end, when measuring the chord length, it will be exactly equal to 30.24 m.

Figure 4.22 shows the MS Excel worksheet used to conduct the computations of this problem.

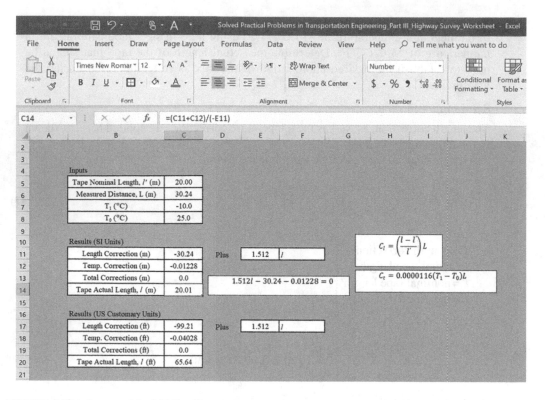

FIGURE 4.22 Image of the MS Excel worksheet used for the computations of Problem 4.21.

4.22 A 100-m longitudinal crack in a local street segment that requires maintenance is measured using a 50-m steel tape under a tension of 40 N (9 lbf). The tape has an actual length of 50.025 m (164.12 ft) when calibrated at a temperature of 25°C (77°F) and using a tension of 50 N (11.24 lbf) with the tape fully supported on the ground. Determine the temperature at which the measurement should be made so that the desired length is established knowing that the tape cross-sectional area is 0.026 cm² (0.004 in²) and the modulus of the elasticity of steel is 200 GPa (29×10⁶ psi). The measurement is to be made while the tape is fully supported on the ground.

Solution:

The length correction is given by the following formula:

$$C_l = \left(\frac{l-l'}{l'}\right)L$$

⇒

$$C_l = \left(\frac{50.025-50}{50}\right)100 = 0.05 \text{ m}\left(0.16 \text{ ft}\right)$$

The temperature correction is provided by the following expression:

$$C_t = 0.0000116\left(T_1-T_0\right)L$$

$$\Rightarrow$$

$$C_t = 0.0000116(T_1 - 25)100$$

$$\Rightarrow$$

$$C_t = 0.00116T_1 - 0.029$$

On the other hand, the tension correction is given by the following formula:

$$C_P = \left(\frac{P_1 - P_0}{AE}\right)L \qquad (4.39)$$

Where:
C_P = the elongation of the tape
L = the length of the tape
P_1 = the applied tension
P_0 = the calibration tension
A = the cross-sectional area of the tape
E = the modulus of elasticity of the tape material (typical E value for steel = 200 GPa)

The units must be consistent, therefore:

$E = 200 \text{ GPa} = 200 \times 10^9 \text{ N/m}^2$
$A = 0.026 \text{ cm}^2 = 0.026 \times 10^{-4} \text{ m}^2 = 2.6 \times 10^{-6} \text{ m}^2$

$$\Rightarrow$$

$$C_P = \left(\frac{40 - 50}{2.6 \times 200 \times 10^3}\right)100 = -0.00192 \text{ m} (-0.0063 \text{ ft})$$

In order to establish the desired length, the total correction should be equal to zero. In other words:

$$0.05 - 0.00192 + 0.00116T_1 - 0.029 = 0$$

$$\Rightarrow$$

$$T_1 = -16.4°C (2.4°F)$$

The cracking length measurement should be performed at a temperature of −16.4°C (2.4°F) so that at the end the 100-m length is established.

Figure 4.23 shows the MS Excel worksheet used to conduct the computations of this problem.

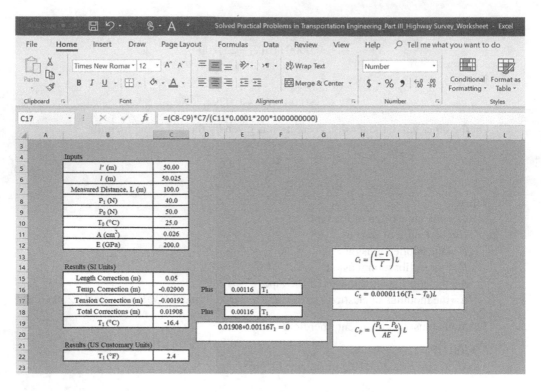

FIGURE 4.23 Image of the MS Excel worksheet used for the computations of Problem 4.22.

4.23 A highway surveyor intends to perform a surveying job at a field temperature of 35°C (95°F) using a 30-m tape. The surveying job includes measuring a distance between two points in a rough terrain in six stages; in each stage, the tape is raised above the ground and supported at the two ends. At the end of the six stages, the distance measures 165.70 m (543.64 ft). The steel tape has a length of 30.02 m (98.49 ft) when it is calibrated at a temperature of 25°C (77°F) and supported on the ground. At the same temperature (25°C = 77°F), the tape measures 30.00 m (98.42 ft) exactly when it is supported at the two ends under a tension force of 44 N (9.89 lbf). If the weight of the tape is 5.5 N (1.24 lbf), determine the tension that the surveyor applies to the tape to obtain the 165.70-m (543.64-ft) distance measurement. *Note: Assume that the tension error is neglected.*

Solution:

Note that at the calibration conditions, the length correction for a distance equal to the tape length (30 m) is equal to:

$$C_l = \left(\frac{l - l'}{l'} \right) L$$

⇒

$$C_l = \left(\frac{30.02 - 30}{30} \right) 30.00 = 0.02 \text{ m} \left(0.06 \text{ ft} \right)$$

Since the tape length measures 30.00 m exactly when it is supported at two ends at the calibration temperature (25°C), the length correction (or error) when the tape is fully supported on the ground at the same calibration temperature should be equal in magnitude to the sag correction (or error) but with an opposite sign for the same tape length. In other words, it is equal to:

$$C_s = \frac{W^2 L}{24 P^2} \qquad (4.40)$$

$\Rightarrow$

$$C_s = \frac{(5.5)^2 (30)}{24 (44)^2} = 0.02 \text{ m} (0.06 \text{ ft})$$

This is the case, as seen from the given data in this problem. Now for the entire measured distance, the corrections are calculated as shown in the following procedure:

The length correction is given by the following formula:

$$C_l = \left(\frac{l - l'}{l'}\right) L$$

$\Rightarrow$

$$C_l = \left(\frac{30.02 - 30}{30}\right) 165.70 = 0.11 \text{ m} (0.36 \text{ ft})$$

The temperature correction is determined using the following expression:

$$C_t = 0.0000116 (T_1 - T_0) L$$

$\Rightarrow$

$$C_t = 0.0000116 (35 - 25) 165.70 = 0.019 \text{ m} (0.06 \text{ ft})$$

On the other hand, the sag correction is calculated using the following formula:

$$C_s = \frac{W^2 L}{24 P^2}$$

Where:

C_s = the sag correction
W = the weight of the tape
L = the length of the tape
P = the applied tension

$\Rightarrow$

In each of the first five stages, since the distance in each stage is 30 m, the sag correction would be equal to:

$$C_{s\text{-stages } 1-5} = \frac{(5.5)^2 (30)}{24 P^2}$$

In the sixth (last) stage, since the distance would be 15.70 m, the weight of the tape over the 15.70-m distance would be equal to $5.5 \times \dfrac{15.70}{30}$ N, and the sag correction would be equal to:

$$C_{s\text{-stage }6} = \frac{\left(5.5 \times \dfrac{15.70}{30}\right)^2 (15.70)}{24P^2}$$

The total sag correction is therefore equal to:

$$C_s = 5 \times \frac{(5.5)^2 (30)}{24P^2} + \frac{\left(5.5 \times \dfrac{15.70}{30}\right)^2 (15.70)}{24P^2}$$

$\Rightarrow$

$$C_s = \frac{194.48}{P^2}$$

In order to obtain a final distance measurement of 165.70, the total corrections should be equal to zero. Knowing that the sag correction is always −ve; consequently, the following formula should apply:

$$C_{\text{Total}} = C_l + C_t - C_s = 0 \tag{4.41}$$

Where:
 C_{Total} = the total corrections
 C_l = the length correction
 C_t = the temperature correction
 C_s = the sag correction

$\Rightarrow$

$$C_{\text{Total}} = 0.11 + 0.019 - \frac{194.48}{P^2} = 0$$

Solving the above equation provides the following solution for the applied tension (P):

$$P = 38.8 \text{ N} \left(8.7 \text{ lbf}\right)$$

Therefore, a tension force of 38.8 N (8.7 lbf) should be applied to the tape so that the measured distance at the end would be 165.70 m (543.64 ft).

Figure 4.24 shows the MS Excel worksheet used to perform the computations of this problem.

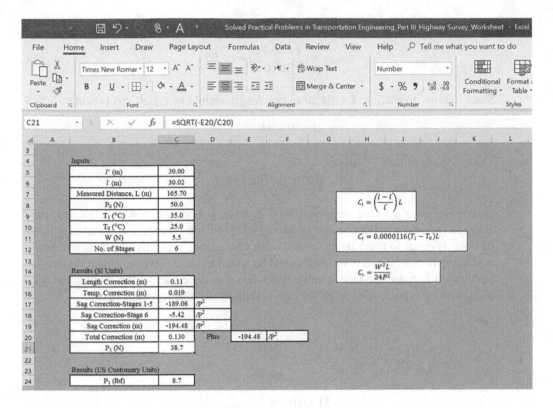

FIGURE 4.24 Image of the MS Excel worksheet used for the computations of Problem 4.23.

4.24 A linear distance of a highway segment on level terrain is measured by a highway surveyor using a 50-m tape, when it is fully supported on the ground at a field temperature
of 45°C (113°F), to be 140.25 m (460.14 ft). However, the tape measures exactly 50.00 m
(164.04 ft) when it is raised above the ground at a temperature of 25°C (77°F) and under
a tension of 45 N (10.12 lbf). Determine the corrected value of the measured distance if
the weight of the tape is 5 N (1.12 lbf). *Assume that the tension error is neglected.*

Solution:

Since the tape measures exactly 50.00 m (164.04 ft) when it is raised above the ground
at a temperature of 25°C and under a tension of 45 N (10.12 lbf), the length error (or
correction) for the 50-m tape length is equal to the sag error (or correction) for the 50-m
tape length with an opposite sign. Consequently:

$$C_s = \frac{W^2 L}{24 P^2}$$

⇒

$$C_{s\text{-}50\,m} - \frac{(5)^2 (50)}{24 (45)^2} - 0.026\ \text{m} \left(0.085\ \text{ft}\right)$$

$$C_l = \left(\frac{l - l'}{l'}\right) L$$

$\Rightarrow$

$$C_{l\text{-}50\,m} = \left(\frac{l-50}{50}\right)50 = l - 50$$

Since the two corrections for the 50-m tape length should be equal, this provides:

$$l - 50 = 0.026$$

$\Rightarrow$

$$l = 50.026 \text{ m}\left(164.13 \text{ ft}\right)$$

Or simply, since the tape measures 50.00 m when it is raised above the ground, the 50.00 m includes a deduction due to the sag error. In other words, the actual length of the tape is equal to the 50.00 m plus the sag correction. Typically, the sag correction is −ve, but in this case, it is +ve and should be added to the 50.00 m to obtain the actual tape length because the tape is calibrated when it is raised above the ground and the measurement is performed when the tape is supported on the ground. Hence:

$$C_s = \frac{W^2 L}{24 P^2}$$

$\Rightarrow$

$$C_{s\text{-}50\,m} = \frac{(5)^2 (50)}{24(45)^2} = 0.026 \text{ m}\left(0.085 \text{ ft}\right)$$

$\Rightarrow$

Actual tape length $= l = 50.00 + 0.026 = 50.026 \text{ m}\left(164.13 \text{ ft}\right)$

Now, the corrections that will be made to the measured distance are based on the length error in the three stages (50 m + 50 m + 40.25 m) as shown below:

$$C_l = \left(\frac{l-l'}{l'}\right)L$$

$\Rightarrow$

$$C_l = \left(\frac{50.026 - 50}{50}\right)140.25 = 0.073 \text{ m}\left(0.239 \text{ ft}\right)$$

The temperature correction is determined using the following expression:

$$C_t = 0.0000116\left(T_1 - T_0\right)L$$

$\Rightarrow$

$$C_t = 0.0000116\left(45 - 25\right)140.25 = 0.033 \text{ m}\left(0.108 \text{ ft}\right)$$

$$\text{Corrected Distance} = \text{Measured Distance} + \text{Total Corrections} \quad (4.42)$$

$\Rightarrow$

$$\text{The corrected distance} = 140.25 + 0.073 + 0.033$$

$$= 140.36 \, \text{m} \left(460.50 \, \text{ft} \right)$$

Figure 4.25 shows the MS Excel worksheet used to do the computations of this problem.

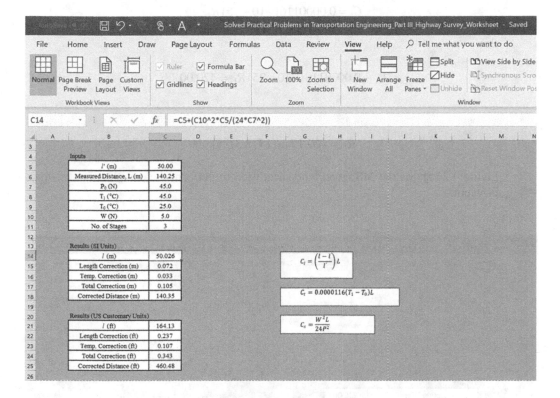

FIGURE 4.25 Image of the MS Excel worksheet used for the computations of Problem 4.24.

4.25 A parking lot with actual dimensions of 240.00 m (787.40 ft) by 420.00 m (1377.95 ft) is to be laid out using a steel tape that is calibrated at a temperature of 25°C (77°F) to have a length of 50.00 m (164.04 ft). Determine the readings of the tape to establish these dimensions if the measurements are conducted at a field temperature of −10°C (14°F).

Solution:

Since the measurements are performed at a field temperature that is different from the calibration temperature, a temperature error is involved. The temperature correction is calculated as follows:

$$C_t = 0.0000116 \left(T_1 - T_0 \right) L$$

$\Rightarrow$

For the width of the parking lot (240.00 m), if the tape reading of the width is R_w, then:

$$C_t = 0.0000116 \left(-10 - 25 \right) R_w$$

But:

$$R_W + 0.0000116(-10 - 25)R_W = 240.00$$

$$\Rightarrow$$

$$R_W = 240.10 \text{ m}\left(787.72 \text{ ft}\right)$$

For the length of the parking lot (420. m), if the tape reading of the width is R_L, then:

$$C_t = 0.0000116(-10 - 25)R_L$$

But:

$$R_L + 0.0000116(-10 - 25)R_L = 420.00$$

$$\Rightarrow$$

$$R_L = 420.17 \text{ m}\left(1378.51 \text{ ft}\right)$$

Figure 4.26 shows the MS Excel worksheet used to perform the computations of this problem.

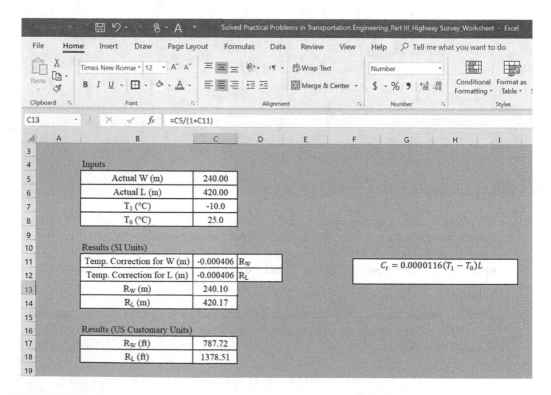

FIGURE 4.26   Image of the MS Excel worksheet used for the computations of Problem 4.25.

4.26   The actual distance between two stations along a highway segment is 100.00 m
       (324.08 ft). A 100-m fiberglass tape having a weight of 5 N (1.12 lbf) is used between
       the two stations and measures 100.05 m (328.25 ft) when it is supported on the ground.
       Calculate the tension that should be applied to the tape when it is raised above the
       ground so that it would measure exactly 100.00 m (324.08 ft). *Assume that the tension
       error is negligible.*

Solution:

Two types of corrections (or errors) are involved in this surveying job: length correction
and sag correction. The length correction is calculated using the typical formula shown
below:

$$C_l = \left(\frac{l - l'}{l'} \right) L$$

⇒

$$C_l = \left(\frac{100.05 - 100}{100} \right) 100 = 0.05 \text{ m} \left(0.16 \text{ ft} \right)$$

$$C_s = \frac{W^2 L}{24 P^2}$$

⇒

$$C_s = \frac{(5)^2 100}{24 P^2}$$

The sag correction is always −ve. In order for the tape to measure exactly
100.00 m between the two stations, the total corrections (or errors) will have to be
zero. Therefore:

$$0.05 - \frac{(5)^2 100}{24 P^2} = 0$$

Solving the above equation for P (the tension force) provides the following value:

$$P = 45.6 \text{ N} \left(10.3 \text{ lbf} \right)$$

In conclusion, a tension force of 45.6 N (10.3 lbf) should be applied to the tape when it is
raised above the ground in order to measure exactly 100.00 m between the two stations.
In this case, the two errors (the length error and the sag error) will cancel each other and
the final result will be 100.00 m.

Figure 4.27 shows the MS Excel worksheet used to perform the computations of this
problem.

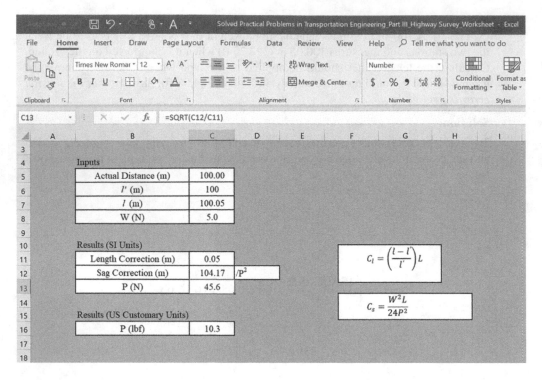

FIGURE 4.27 Image of the MS Excel worksheet used for the computations of Problem 4.26.

4.27 The length of an urban street block is measured as 274.62 m (900.98 ft) using a steel tape having a cross-sectional area of 0.020 cm² (0.003 in²) and a modulus of elasticity of 200 GPa (29,000,000 psi) at a tension of 25 N (5.62 lbf). If the same distance is measured a second time with the same tape at a tension of 50 N (11.24 lbf), compute the measured distance taken the second time. *Consider only the tension error and assume all other errors are neglected.*

Solution:

The tension correction is calculated using the following formula:

$$C_P = \left(\frac{P_1 - P_0}{AE} \right) L$$

⇒

In the first measurement:

$$C_{P1} = \left(\frac{25 - P_0}{\left(0.020 \times 10^{-4}\right)\left(200 \times 10^9\right)} \right) 274.62$$

And in the second measurement:

$$C_{P2} = \left(\frac{50 - P_0}{\left(0.020 \times 10^{-4}\right)\left(200 \times 10^9\right)} \right) 274.62$$

The difference between the two corrections is determined as:

$$C_{P1-2} = \left(\frac{25 - P_0}{(0.020 \times 10^{-4})(200 \times 10^9)} \right) 274.62$$

$$- \left(\frac{50 - P_0}{(0.020 \times 10^{-4})(200 \times 10^9)} \right) 274.62 = -0.017 \text{ m}$$

The measured distance in the second case is determined by adding the difference in the two corrections to the measured distance in the first case. The difference is –ve because the tension force in the second measurement is higher than the tension force in the first measurement. Therefore, it is expected that the measured distance for the second time to be lower than the measured distance for the first time.

$$\text{Measured Distance } 2 = 274.62 - 0.017 = 274.60 \text{ m} (900.92 \text{ ft})$$

Figure 4.28 shows the MS Excel worksheet used to perform the computations of this problem.

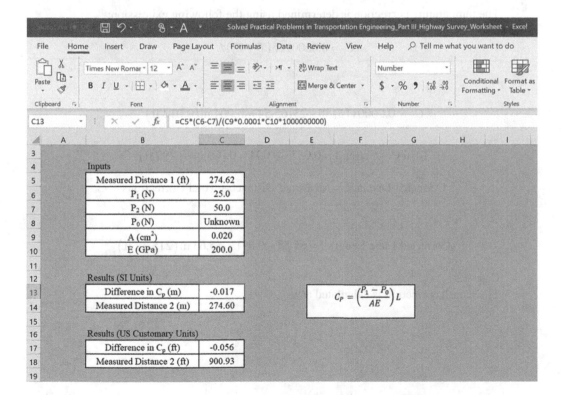

FIGURE 4.28   Image of the MS Excel worksheet used for the computations of Problem 4.27.

4.28   A highway surveyor measures the length of a right-turn lane at an intersection using two tapes with the following results and calibration conditions:

(1) Tape #1: nominal length = 30 m, calibrated tape length = 30.04 m at a temperature = 25°C (77°F), measured length = 65.54 m, field temperature = 30°C (86°F).

(2) Tape #2: nominal length = 30 m, measured length = 65.46 m, field temperature = 40°C (104°F).

Determine the actual length of the second tape at the calibration temperature (25°C = 77°F).

Solution:

Case #1:
The length correction is determined as follows:

$$C_l = \left(\frac{l - l'}{l'} \right) L$$

⇒

$$C_{l1} = \left(\frac{30.04 - 30}{30} \right) 65.54 = 0.087 \text{ m} \left(0.287 \text{ ft} \right)$$

The temperature correction is determined using the following expression:

$$C_t = 0.0000116 \left(T_1 - T_0 \right) L$$

⇒

$$C_{t1} = 0.0000116 \left(30 - 25 \right) 65.54 = 0.004 \text{ m} \left(0.012 \text{ ft} \right)$$

$$\text{Total Correction 1} = 0.087 + 0.004 = 0.091 \text{ m} \left(0.299 \text{ ft} \right)$$

$$\text{Corrected Distance} = \text{Measured Distance} + \text{Total Corrections}$$

⇒

$$\text{Corrected Lane Length 1} = 65.54 + 0.091 = 65.63 \text{ m} \left(215.32 \text{ ft} \right)$$

Case #2:
The length correction is calculated as:

$$C_l = \left(\frac{l - l'}{l'} \right) L$$

⇒

$$C_{l2} = \left(\frac{l - 30}{30} \right) 65.46 = 2.182l - 65.46$$

The temperature correction is determined as below:

$$C_t = 0.0000116 \left(T_1 - T_0 \right) L$$

$\Rightarrow$

$$C_{t2} = 0.0000116(40-25)65.46 = 0.011\,\text{m}\,(0.037\,\text{ft})$$

$$\text{Total Correction 2} = 2.182l - 65.46 + 0.011$$

$$\text{Corrected Distance} = \text{Measured Distance} + \text{Total Corrections}$$

$\Rightarrow$

$$\text{Corrected Lane Length 2} = 65.46 + 2.182l - 65.46 + 0.011$$

$$= 2.182l + 0.011$$

But:

$$\text{Corrected Lane Length 2} = \text{Corrected Lane Length 1} \qquad (4.43)$$

$\Rightarrow$

$$\text{Corrected Lane Length 2} = 65.63\ \text{m} = 2.182l + 0.011$$

Solving the above equation for l (the actual tape length in case #2) provides the following value for the tape length:

$$l = 30.07\ \text{m}\ (98.66\ \text{ft})$$

Figures 4.29 and 4.30 show the MS Excel worksheet used to perform the computations of this problem.

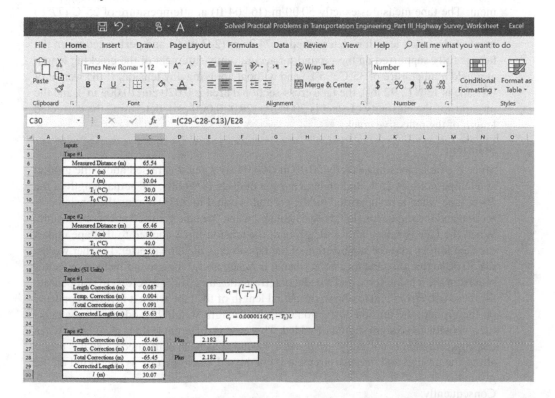

FIGURE 4.29 Image of the MS Excel worksheet used for the computations (SI units) of Problem 4.28.

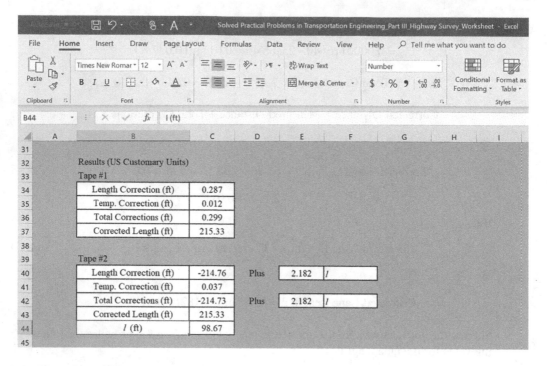

FIGURE 4.30 Image of the MS Excel worksheet used for the computations (US Customary units) of Problem 4.28.

4.29   A 50-m steel tape is used to measure the length of a reflective crack in a composite pave-
       ment. The tape measures exactly 50.00 m (164.04 ft) at a temperature of 25°C (77°F)
       and under a tension of 84 N (18.9 lbf) when it is raised above the ground. The measure-
       ment of the crack length is determined as 74.36 m (243.96 ft) and performed when the
       tape is fully supported on the ground for accurate measurement of the cracking length
       at a field temperature of 35°C (95°F) and without applying any tension. If the weight of
       the tape is 13 N (2.92 lbf), the tape cross-sectional area is 0.020 cm² (0.003 in²), and the
       modulus of the elasticity of steel is 200 GPa (29,000,000 psi), determine the corrected
       length of the reflective crack.

Solution:

Since the tape measures exactly 50.00 m (164.04 ft) when it is raised above the ground
at a temperature of 25°C and under a tension of 40 N (8.99 lbf), the actual length of the
50-m tape when it is supported on the ground is equal to 50.00 m plus the tension correction and the sag correction; both calculated when the tape is raised above the ground.
When the tape is raised above the ground, the tension force applied to the tape makes
the tape longer, and the sag position of the tape makes the tape shorter. In other words,
the 50.00 m that the tape measures when it is raised above the ground includes an addition from the tension force and a deduction from the sag error. Therefore, to compensate
for these two errors and to obtain the actual length of the tape when it is fully supported
on the ground, the tension correction (with a −ve sign) should be added and a sag correction (with a +ve sign) should be added. This case is different from typical cases when
the sag correction is normally −ve because the tape has been calibrated when it is raised
above the ground and the measurement is conducted when the tape is supported on the
ground.
Consequently:

$$\Rightarrow$$

The sag correction for the 50-m tape length is determined as:

$$C_s = \frac{W^2 L}{24 P^2}$$

$\Rightarrow$

$$C_{s\text{-}50\,m} = \frac{(13)^2 (50)}{24(84)^2} = 0.05 \text{ m} (0.16 \text{ ft})$$

The tension correction for the 50-m tape length is computed using the following formula:

$$C_P = \left(\frac{P_1 - P_0}{AE} \right) L$$

$\Rightarrow$

$$C_{P\text{-}50\,m} = \left(\frac{0 - 84}{(0.020 \times 10^{-4})(200 \times 10^9)} \right) 50 = -0.01 \text{ m} (-0.034 \text{ ft})$$

$\Rightarrow$

$$\text{Actual tape length} = l = 50.00 + (-0.01) + 0.05$$

$$= 50.04 \text{ m} (164.17 \text{ ft})$$

The actual length of the tape used to perform the measurement of the crack length is equal to 50.04 m (164.17 ft). Hence, the corrections that will be made to the measured cracking length are related to the length error and temperature error in the two stages (50 m + 24.36 m) as shown below:

$$C_l = \left(\frac{l - l'}{l'} \right) L$$

$\Rightarrow$

$$C_l = \left(\frac{50.04 - 50}{50} \right) 74.36 = 0.059 \text{ m} (0.195 \text{ ft})$$

The temperature correction is determined using the following expression:

$$C_t = 0.0000116 (T_1 - T_0) L$$

$\Rightarrow$

$$C_t = 0.0000116 (35 - 25) 74.36 = 0.009 \text{ m} (0.028 \text{ ft})$$

$$\text{Corrected Distance} = \text{Measured Distance} + \text{Total Corrections}$$

$\Rightarrow$

$$\text{The corrected crack length} = 74.36 + 0.059 + 0.009$$

$$= 74.43 \text{ m} (244.19 \text{ ft})$$

Figure 4.31 shows the MS Excel worksheet used to perform the computations of this problem.

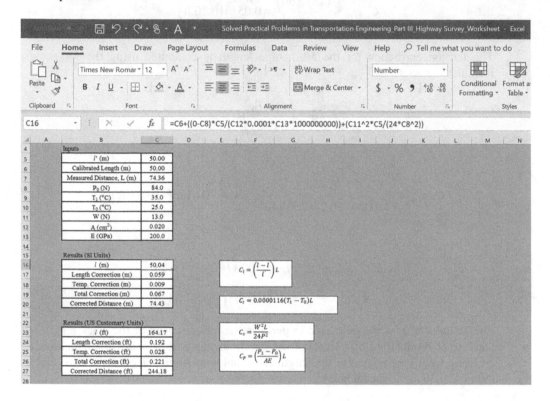

FIGURE 4.31   Image of the MS Excel worksheet used for the computations of Problem 4.29.

4.30   The distance between two stations on a roadway is measured using a 50-m steel tape that weighs 11 N (2.47 lbf) when it is supported at the two ends at 0, 50, and 100 m under a tension of 80 N (17.98 lbf). Determine the corrected distance between the two stations.

Solution:

A sag error is involved in the distance measurement in this case from 0 to 50 m and from 50 to 100 m. To determine the sag correction, the following formula is used:

$$C_s = \frac{W^2 L}{24 P^2}$$

$\Rightarrow$

$$C_{s\,0-50\,m} = \frac{(11)^2 (50)}{24(80)^2} = 0.04 \text{ m} (0.13 \text{ ft})$$

$$C_{s\,50-100\,m} = \frac{(10)^2\,(50)}{24(80)^2} = 0.04\ \text{m}\,(0.13\ \text{ft})$$

Since the sag correction is −ve, the corrected distance is therefore determined as:

The corrected distance = $100.00 - 0.04 - 0.04 = 99.92\ \text{m}\,(327.83\ \text{ft})$

Figure 4.32 shows the MS Excel worksheet used to perform the computations of this problem.

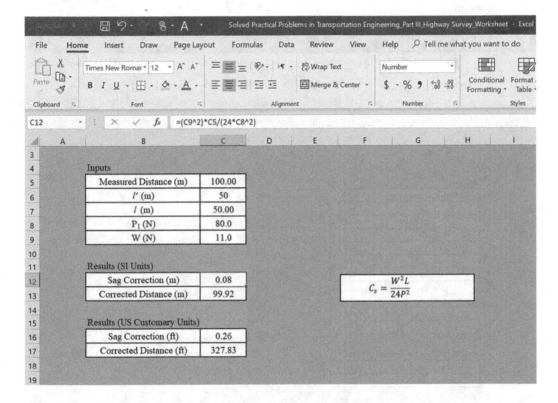

FIGURE 4.32 Image of the MS Excel worksheet used for the computations of Problem 4.30.

4.31 A 30-m steel tape with a cross-sectional area of 0.026 cm² has a correct length under a tension of 45 N (10.1 lbf). If the distance between two points is measured using this tape to be 30 m when the tension is doubled (90 N = 20.2 lbf), compute the corrected distance between the two points.

Solution:

The tension correction is given by the following formula:

$$C_P = \left(\frac{P_1 - P_0}{AE}\right)L$$

A typical value for the modulus of the elasticity of steel = 200 GPa is used.

⇒

$$C_P = \left(\frac{90 - 45}{\left(0.026 \times 10^{-4} \right) \left(200 \times 10^9 \right)} \right) 30 = 0.003 \text{ m}$$

The corrected distance is therefore determined as:

$$\text{The corrected distance} = 30.00 + 0.003 = 30.003 \text{ m} \left(98.43 \text{ ft} \right)$$

Figure 4.33 shows the MS Excel worksheet used to perform the computations of this problem.

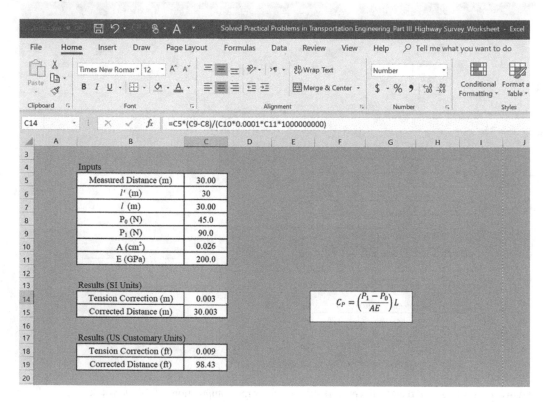

FIGURE 4.33   Image of the MS Excel worksheet used for the computations of Problem 4.31.

4.32   Determine the tension that should be applied to a steel tape at a field temperature of 40°C (104°F) so that its length will be exactly 50.00 m (164.04 ft) if the tape has a calibrated length of 49.98 m at a temperature of 25°C (77°F). The cross-sectional area of the tape is 0.024 cm² and the modulus of the elasticity of steel is 200 GPa (29,000,000 psi).

Solution:

Three errors are involved in the measurement in this case: the length error, the temperature error, and the tension error. Therefore, three corrections must be made. The length correction is calculated using the following formula:

$$C_l = \left(\frac{l - l'}{l'} \right) L$$

$\Rightarrow$

$$C_l = \left(\frac{49.98 - 50}{50} \right) 50.00 = -0.02 \text{ m} \left(-0.07 \text{ ft} \right)$$

The temperature correction is provided by the following expression:

$$C_t = 0.0000116 \left(T_1 - T_0 \right) L$$

$\Rightarrow$

$$C_t = 0.0000116 \left(40 - 25 \right) 50.00 = 0.0087 \text{ m} \left(0.0285 \text{ ft} \right)$$

On the other hand, the tension correction is given by the following formula:

$$C_P = \left(\frac{P_1 - P_0}{AE} \right) L$$

$\Rightarrow$

$$C_P = \left(\frac{P_1 - 0}{\left(0.024 \times 10^{-4} \right) \left(200 \times 10^9 \right)} \right) 50.00$$

To have an exact tape length of 50.00 m, the total corrections of the tape must be equal to zero. Hence:

$$C_{\text{Total}} = C_l + C_t + C_P = 0 \tag{4.44}$$

Where:
   C_{Total} = the total corrections
   C_l = the length correction
   C_t = the temperature correction
   C_s = the tension correction

$\Rightarrow$

$$C_{\text{Total}} = -0.02 + 0.0087 + \left(\frac{P_1}{\left(0.024 \times 10^{-4} \right) \left(200 \times 10^9 \right)} \right) 50.00 = 0$$

Solving the above equation for P_1 provides the following value for the tension:

$$P_1 = 108.5 \text{ N} \left(24.4 \text{ lbf} \right)$$

Figure 4.34 shows the MS Excel worksheet used to perform the computations of this problem.

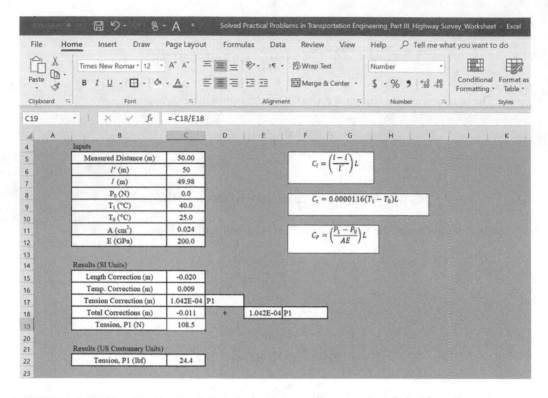

FIGURE 4.34 Image of the MS Excel worksheet used for the computations of Problem 4.32.

4.33   A distance is measured using a 50-m steel tape at a temperature of 25°C (77°F) to be 142.76 m (468.37 ft). On a different day, the distance is measured again using the same tape to be 142.82 m (468.57 ft). If the actual length of the tape at the temperature (25°C = 77°F) is 50.06 m, determine the second temperature.

Solution:

There are two cases in this problem. The corrections in each case are determined, as shown below:

Case #1:

Length Correction:

$$C_l = \left(\frac{l - l'}{l'}\right) L$$

⇒

$$C_{l1} = \left(\frac{50.06 - 50}{50}\right) 142.76 = 0.17 \text{ m} \left(0.56 \text{ ft}\right)$$

Corrected Distance:

$$\text{Corrected Distance} = 142.76 + 0.17 = 142.93 \text{ m} \left(468.93 \text{ ft}\right)$$

Case #2:

Length Correction:

$$C_l = \left(\frac{l - l'}{l'}\right)L$$

$\Rightarrow$

$$C_{l2} = \left(\frac{50.06 - 50}{50}\right)142.82 = 0.17\,\mathrm{m}\,(0.56\,\mathrm{ft})$$

Temperature Correction:

$$C_t = 0.0000116(T_1 - T_0)L$$

$\Rightarrow$

$$C_{t2} = 0.0000116(T_2 - 25)142.82 = 1.6567 \times 10^{-3}T_2 - 4.1418 \times 10^{-2}$$

Total Corrections:

$$C_{\mathrm{Total}} = 0.17 + 1.6567 \times 10^{-3}T_2 - 4.1418 \times 10^{-2}$$

Corrected Distance:

Corrected Distance = Measured Distance + Total Corrections

$\Rightarrow$

Corrected Distance 2 = 142.82 + 0.17 + 1.6567

$$\times 10^{-3}T_2 - 4.1418 \times 10^{-2}$$

But:

Corrected Distance 2 = Corrected Distance 1

$\Rightarrow$

$$142.82 + 0.17 + 1.6567 \times 10^{-3}T_2 - 4.1418 \times 10^{-2} = 142.93$$

Solving the above equation for T_2 provides the following (Figure 4.35):

$$T_2 = -11.3\,°\mathrm{C}\,(11.7\,°\mathrm{F})$$

Figure 4.35 shows the MS Excel worksheet used to perform the computations of this problem.

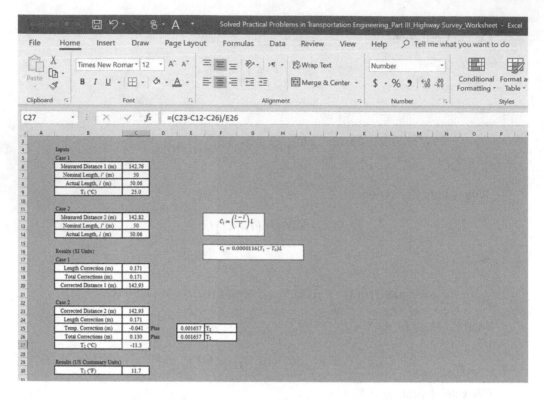

FIGURE 4.35 Image of the MS Excel worksheet used for the computations of Problem 4.33.

4.34 An on-street parking distance is measured using two 50-m tapes (tape #1 and tape #2) at the same temperature to be 125.26 m and 125.45 m, respectively. If the actual length of tape #1 is 50.04 m, determine the actual length of tape #2.

Solution:

Tape #1:
The length correction is determined as follows:

$$C_l = \left(\frac{l - l'}{l'}\right)L$$

⇒

$$C_{l1} = \left(\frac{50.04 - 50}{50}\right)125.26 = 0.10 \text{ m}\left(0.33 \text{ ft}\right)$$

Corrected Distance = Measured Distance + Total Corrections

⇒

Corrected Distance 1 = 125.26 + 0.10 = 125.36 m$\left(411.29 \text{ ft}\right)$

Tape #2:
The length correction is calculated as:

$$C_l = \left(\frac{l - l'}{l'}\right)L$$

$\Rightarrow$

$$C_{12} = \left(\frac{l-50}{50}\right)125.45 = 2.509l - 125.45$$

Corrected Distance = Measured Distance + Total Corrections

$\Rightarrow$

Corrected Distance $2 = 125.45 + 2.509l - 125.45 = 2.509l$

But:

Corrected Distance 2 = Corrected Distance 1

$\Rightarrow$

Corrected Distance $2 = 125.36$ m $= 2.509l$

$\Rightarrow$

$$l = 49.96 \text{ m} \left(163.92 \text{ ft}\right)$$

Figure 4.36 shows the MS Excel worksheet used to conduct the computations of this problem.

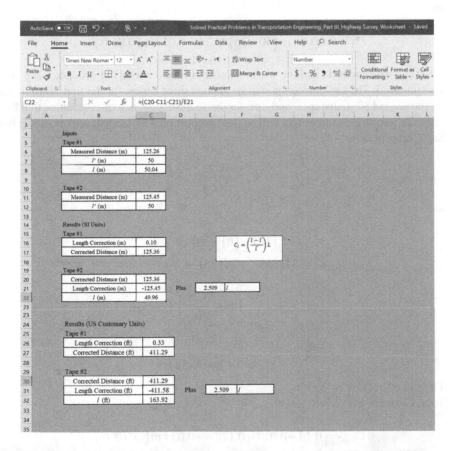

FIGURE 4.36 Image of the MS Excel worksheet used for the computations of Problem 4.34.

4.35 Two 30-m steel tapes are used to measure the same distance at a temperature of 40°C (104°F) to be 83.28 m (273.23 ft) and 83.48 m (273.88 ft), respectively. The first measurement is performed using tape #1 when it is raised above the ground and under a tension of 25 N (5.62 lbf), while the second measurement is done using tape #2 when it is fully supported on the ground. If the weight of tape #1 is 6 N (1.35 lbf) and its actual length is 30.05 m (98.59 ft) at 25°C (77°F), determine the actual length of tape #2 and the corrected distance. *Assume that the difference in the tension error between the two measurements is neglected.*

Solution:

Measurement #1 (Tape #1):
Length Correction:

$$C_l = \left(\frac{l - l'}{l'} \right) L$$

$\Rightarrow$

$$C_{l1} = \left(\frac{30.05 - 30}{30} \right) 83.28 = 0.14 \text{ m} \left(0.46 \text{ ft} \right)$$

Temperature Correction:

$$C_t = 0.0000116 \left(T_1 - T_0 \right) L$$

$\Rightarrow$

$$C_{t1} = 0.0000116 \left(40 - 25 \right) 83.28 = 0.014 \text{ m} \left(0.05 \text{ ft} \right)$$

Since the measurement is done when the tape is raised above the ground, a sag error is also involved. The sag correction will be applied three times, for the first 30-m distance, for the second 30-m distance, and then for the 23.28-m distance (please notice that for this distance, part of the tape weight will be used proportional to the distance) as shown below.
Sag Correction:

$$C_s = \frac{W^2 L}{24 P^2}$$

$\Rightarrow$

$$C_{s1} = \frac{(6)^2 (30)}{24 (25)^2} + \frac{(6)^2 (30)}{24 (25)^2}$$

$$+ \frac{\left(6 \times \dfrac{23.28}{30} \right)^2 (23.28)}{24 (25)^2}$$

$$= 0.178 m \left(0.58 \, ft \right)$$

The sag correction is −ve.
Total Corrections:

$$C_{\text{Total1}} = 0.14 + 0.014 - 0.178 = -0.024 \text{ m} \left(-0.08 \text{ ft} \right)$$

Corrected Distance:

$$\text{Corrected Distance} = \text{Measured Distance} + \text{Total Corrections}$$

$\Rightarrow$

$$\text{Corrected Distance } 1 = 83.28 + (-0.024) = 83.26 \text{ m} (273.15 \text{ ft})$$

Measurement #2 (Tape #2):
Length Correction:

$$C_l = \left(\frac{l-l'}{l'}\right) L$$

$\Rightarrow$

$$C_{l2} = \left(\frac{l-30}{30}\right) 83.48 = 2.783l - 83.48$$

Temperature Correction:

$$C_t = 0.0000116 (T_1 - T_0) L$$

$\Rightarrow$

$$C_{t2} = 0.0000116 (40 - 25) 83.48 = 0.015 \text{ m} (0.05 \text{ ft})$$

Since the measurement is done when the tape is fully supported on the ground, a sag error is not involved.
Total Corrections:

$$C_{\text{Total2}} = 2.783l - 83.48 + 0.015$$

Corrected Distance:

$$\text{Corrected Distance} = \text{Measured Distance} + \text{Total Corrections}$$

$\Rightarrow$

$$\text{Corrected Distance } 2 = 83.48 + 2.783l - 83.48 + 0.015$$

$$= 2.783l + 0.015$$

But:

$$\text{Corrected Distance } 2 = \text{Corrected Distance } 1$$

$\Rightarrow$

$$2.783l + 0.015 = 83.26$$

$\Rightarrow$

$$l = 29.91 \text{ m} (98.14 \text{ ft})$$

Figures 4.37 and 4.38 show the MS Excel worksheet used to perform the computations of this problem.

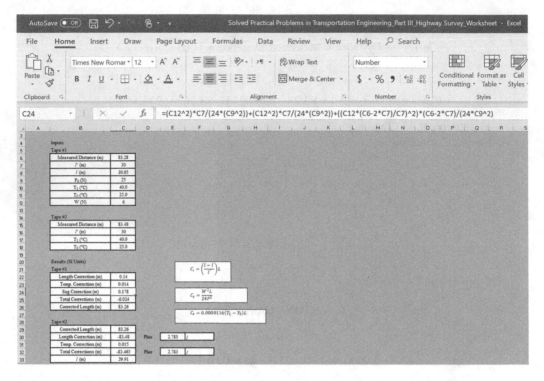

FIGURE 4.37 Image of the MS Excel worksheet used for the computations of Problem 4.35 (SI units).

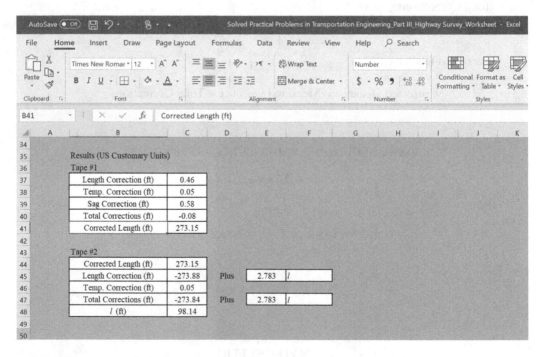

FIGURE 4.38 Image of the MS Excel worksheet used for the computations of Problem 4.35 (US Customary units).

4.36 An inclined distance on an upgrade segment of a roadway (see Figure 4.39) is measured using a 100-m tape to be 1150.52 m (3774.67 ft). If the upgrade slope is 5%, determine the correct horizontal distance obtained using the slope correction formula.

Solution:

The slope correction is determined using the following formula:

$$C_g = \frac{\Delta h^2}{2s} \tag{4.45}$$

Where:

C_g = the slope or grade correction
Δh = the vertical distance
s = the sloped (inclined) distance

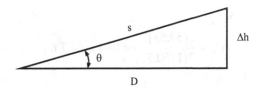

FIGURE 4.39 A schematic diagram of an upgraded segment of a highway for Problem 4.36.

$$\Delta h = s \times \sin \theta \tag{4.46}$$

Where:

Δh = the vertical distance
θ = the angle of the sloped (inclined distance) with the horizontal line (as shown in the above diagram)

But:

$$\theta = \tan^{-1}\left(\frac{G}{100}\right) \tag{4.47}$$

Where:

G = the grade (%)

$\Rightarrow$

$$\theta = \tan^{-1}\left(\frac{5}{100}\right) = 0.05 \text{ rad} = 2.862°$$

$$\Delta h = 1150.52 \times \sin(2.862°) = 57.45 \text{ m}(188.5 \text{ ft})$$

$\Rightarrow$

$$C_g = \frac{(57.45)^2}{2(1150.52)} = 1.43 \text{ m}(4.7 \text{ ft})$$

The slope correction is −ve. Therefore:

$$\text{The Corrected Distance} = 1150.52 - 1.43$$

$$= 1149.09 \text{ m} \left(3769.97 \text{ ft}\right)$$

Another simpler procedure is presented below:
Δh can be simply determined as:

$$\Delta h \cong s \times \frac{G}{100} \tag{4.48}$$

⇒

$$\Delta h \cong 1150.52 \times \frac{5}{100} = 57.53 \text{ m} \left(188.73 \text{ ft}\right)$$

⇒

$$C_g = \frac{\left(57.53\right)^2}{2\left(1150.52\right)} = 1.44 \text{ m} \left(4.7 \text{ ft}\right)$$

⇒

$$\text{The Corrected Distance} = 1150.52 - 1.44 = 1149.08 \text{ m} \left(3769.95 \text{ ft}\right)$$

Figure 4.40 shows the MS Excel worksheet used to perform the computations of this problem.

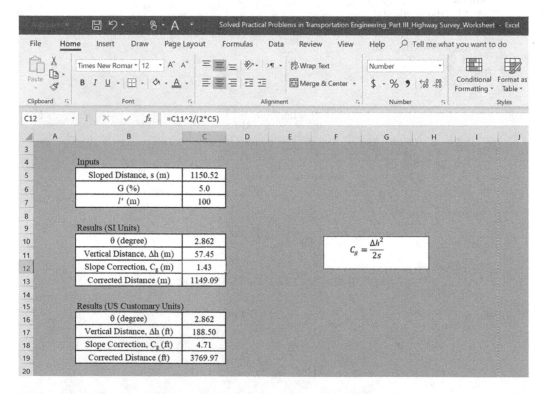

FIGURE 4.40   Image of the MS Excel worksheet used for the computations of Problem 4.36.

5 Leveling

Chapter 5 emphasizes leveling as one of the most important aspects of a highway survey. Leveling in a survey is considered an initial stage that is normally conducted prior to the construction of any highway project. The process of leveling involves determining the heights of different points along the highway project and in the transverse direction across regular stations. Establishing the heights (elevations) of the points is done in the leveling process with respect to a reference line or datum with a known elevation. This section focuses on practical problems related to direct leveling, including simple, differential, and fly leveling. Profile leveling is covered along with highway cross-sections, earthwork volumes, and the mass haul diagrams of highway projects. In addition, some problems include special cases of leveling such as reciprocal leveling and stadia leveling in addition to cases where the leveling staff is inverted.

5.1 A leveling device is used to take a reading on a leveling staff at a point 500 m (1640.4 ft) from the leveling device. Determine the combined error due to earth curvature and refraction.

Solution:

Due to the atmospheric refraction, the actual line of sight is refracted (concave downward), as shown in Figure 5.1:

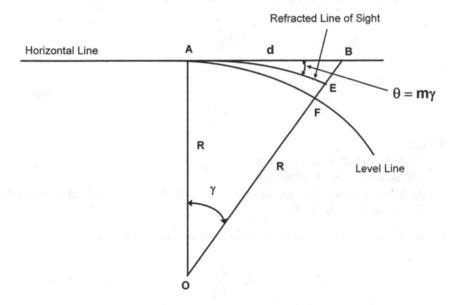

FIGURE 5.1 The combined effect of earth curvature and refraction on leveling for Problem 5.1.

In this figure, θ is given by the following formula:

$$\theta = m\gamma \tag{5.1}$$

Where:

θ=the angle between the horizontal line and the refracted line

m=the coefficient of refraction

γ=the central angle subtended at the earth's center by the distance between the leveling device and the object

Since θ is very small, $\theta \cong \sin\theta \cong \tan\theta$; therefore, the distance BE in the figure can be represented as:

$$\overline{BE} = \overline{AB}\theta \tag{5.2}$$

Or:

$$\overline{BE} = \overline{AB}(m\gamma) \tag{5.3}$$

Where:

BE=the magnitude of refraction

AB=the tangent distance (d) between the leveling device and the leveling staff (object)

m=the coefficient of refraction

γ=the central angle subtended at the earth's center by the distance between the leveling device and the object

Because γ is very small, $\gamma \cong \sin\gamma \cong \tan\gamma$; therefore, γ can be written as:

$$\gamma \cong \frac{d}{R} \tag{5.4}$$

$\Rightarrow$

$$\overline{BE} = m\frac{d^2}{R} \tag{5.5}$$

Where:

BE=the magnitude of refraction

m=the coefficient of refraction

$d=AB$=the tangent distance between the leveling device and the leveling staff (object)

R=the radius of the earth

But the magnitude of curvature (C) is given by the following expression:

$$C = \frac{d^2}{2R} \tag{5.6}$$

Where:

C=magnitude of curvature (BF in the figure above)

$d=AB$=the tangent distance between the leveling device and the leveling staff (object)

R=the radius of the earth

In other words, $2C = d^2/R$; consequently, the magnitude of refraction (BE) can be expressed as:

$$\overline{BE} = m(2C) = 2mC \tag{5.7}$$

The m (the coefficient of refraction) has an average value of 0.071. Therefore, the magnitude of refraction (BE) is given by the following formula:

$$\overline{BE} = 0.14C \tag{5.8}$$

In other words, the magnitude of atmospheric refraction is typically approximated to be 14% of the effect of curvature C, although m varies with atmospheric conditions.

The combined effect of the earth curvature and atmospheric refraction (C_r) is the distance EF in Figure 5.1 above and can be expressed as:

$$\overline{EF} = \overline{BF} - \overline{BE} \tag{5.9}$$

$\Rightarrow$

$$C_r = C - 0.14C = 0.86C \tag{5.10}$$

$\Rightarrow$

$$C_r = 0.86 \frac{d^2}{2R} \tag{5.11}$$

Knowing that the radius of the earth (R) is equal to 6378 km, the above formula can be written as:

$$C_r = 6.742 \times 10^{-5} d^2 \tag{5.12}$$

In this case, d, R, and C_r must be in units of km. To have the result (C_r) in units of m, the result must be multiplied by 1000 (conversion factor). In other words, the formula can be rewritten as:

$$C_r = 0.0674 d^2 \tag{5.13}$$

Where:

C_r = the combined effect of the earth curvature and atmospheric refraction (in units of m)

d = the tangent distance between the leveling device and the leveling staff or the object (in units of km)

In this problem, the distance is given as 500 m = 0.5 km, therefore:

$$C_r = 0.0674(0.5)^2 = 0.0169 \text{ m} = 16.9 \text{ mm} (0.66 \text{ in})$$

The MS Excel worksheet used to perform the computations in this problem is shown in Figure 5.2.

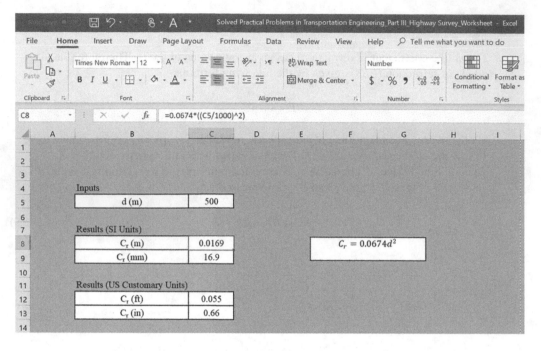

FIGURE 5.2   Image of MS Excel worksheet used for the computations of Problem 5.1.

5.2   Compute the curvature error due to earth curvature in measuring a reading on a leveling staff using a leveling device that is 850 m (2788.7 ft) from the staff.

Solution:

The curvature error is shown in Figure 5.3 as distance *BF*. The curvature error (*C*) is given by the following formula:

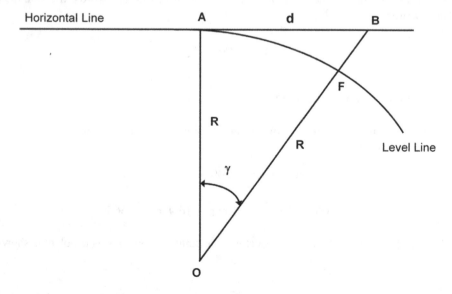

FIGURE 5.3   The effect of earth curvature on leveling for Problem 5.2.

$$C = \frac{d^2}{2R}$$

The radius of the earth, $R = 6378$ km. A "1000" is a conversion factor used to convert km into m.

$\Rightarrow$

$$C = \frac{(850)^2}{2(6378 \times 1000)} = 0.0566 \text{ m} = 56.6 \text{ mm} (2.23 \text{ in})$$

The MS Excel worksheet used to conduct the computations in this problem is shown in Figure 5.4.

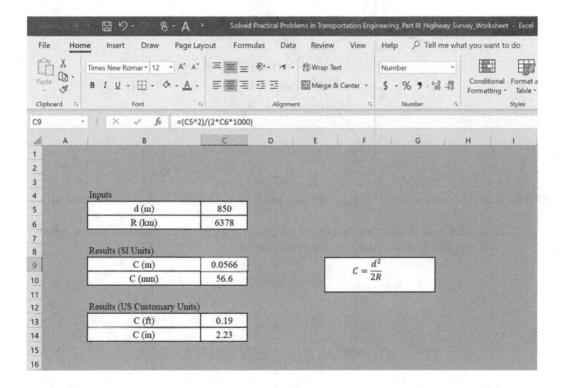

FIGURE 5.4 Image of MS Excel worksheet used for the computations of Problem 5.2.

5.3   A leveling device is used to take a reading of 2.863 m on a leveling staff at an object
       1500 m (4921.3 ft) from the leveling device. Determine the corrected reading taking
       into consideration the combined error due to earth curvature and refraction.

Solution:

The combined effect of curvature and refraction errors is represented in Figure 5.5 by
the distance *CD*:

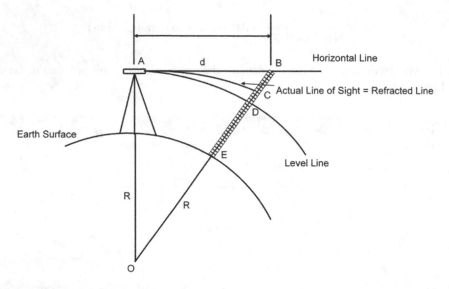

FIGURE 5.5   The combined effect of earth curvature and refraction on level readings for Problem 5.3.

The combined effect of curvature and refraction errors is determined, as shown earlier
in a previous problem using the following formula:

$$C_r = 0.0674d^2$$

⇒

$$C_r = 0.0674(1.500)^2 = 0.152 \text{ m}(0.50 \text{ ft})$$

⇒

$$\text{The Corrected Level Reading} = \text{Level Reading} - C_r \qquad (5.14)$$

⇒

$$\text{The Corrected Level Reading} = 2.863 - 0.152 = 2.711 \text{ m}(8.896 \text{ ft})$$

The MS Excel worksheet used to conduct the computations in this problem is shown in
Figure 5.6.

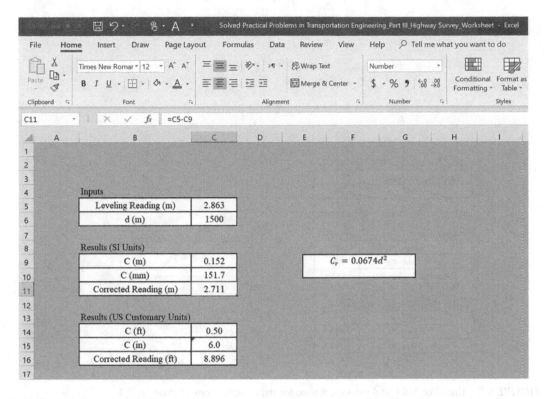

FIGURE 5.6 Image of MS Excel worksheet used for the computations of Problem 5.3.

5.4   A level device is used to take a reading of 3.518 m on a leveling staff at a target point 300 m from the level device. Estimate the corrected level reading for the earth curvature.

Solution:

The curvature error (C) is given by the following formula:

$$C = \frac{d^2}{2R}$$

⇒

$$C = \frac{(300)^2}{2(6378 \times 1000)} = 0.007 \text{ m} = 7 \text{ mm} (0.28 \text{ in})$$

The corrected level reading for the earth curvature is computed using the formula shown below :

$$\text{The Corretced Level Reading} = \text{Level Reading} - C \tag{5.15}$$

⇒

$$\text{The Corretced Level Reading} = 3.518 - 0.007 = 3.511 \text{ m} (11.519 \text{ ft})$$

The MS Excel worksheet used to perform the computations in this problem is shown in Figure 5.7.

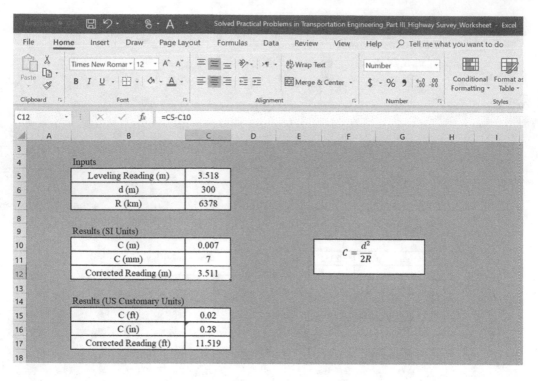

FIGURE 5.7   Image of MS Excel worksheet used for the computations of Problem 5.4.

5.5   The reading on a leveling staff taken by a level device is 2.918 m. If the corrected reading for the combined error due to earth curvature and refraction is 2.895 m, compute the distance between the level device and the staff.

Solution:

The corrected level reading is given by the following formula:

$$\text{The Corretced Level Reading} = \text{Level Reading} - C_r$$

The combined error due to earth curvature and refraction is the difference between the original level reading and the corrected reading. Therefore:

$$C_r = 2.918 - 2.895 = 0.023 \text{ m} \left(0.075 \text{ ft}\right)$$

But the combined error due to curvature and refraction is determined using the following formula:

$$C_r = 0.0674d^2$$

$$\Rightarrow$$

$$0.023 = 0.0674d^2$$

Solving the above equation for d provides the following value:

$$d = 0.584 \text{ km} = 584 \text{ m} \left(1917 \text{ ft}\right)$$

The MS Excel worksheet used to do the computations in this problem is shown in Figure 5.8.

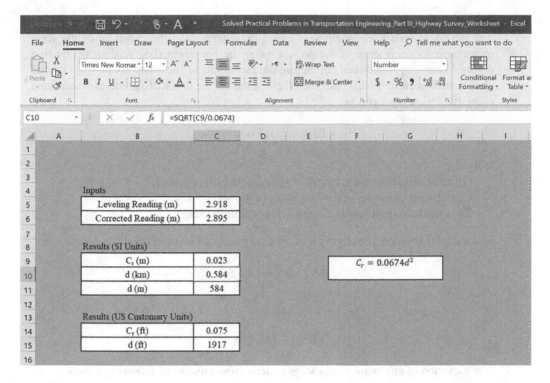

FIGURE 5.8   Image of MS Excel worksheet used for the computations of Problem 5.5.

5.6   Use the *height of instrument (HI) method* for differential leveling to do the necessary
computations and complete Table 5.1. Perform the three verification checks in the HI
method.
*Note: Fill in the table for each white-colored blank. The gray-colored blanks do not
require any computations.*

TABLE 5.1
HI Differential Leveling Record for Problem 5.6

Station	BS (m)	IS (m)	FS (m)	HI (m)	Elevation (m)
1 + 00	4.0			115.1	
2 + 00		2.6			
3 + 00					113.1
TP1			3.5		
	2.8				
4 + 00					112.2
5 + 00		4.1			
TP2			2.8		
	2.2				
6 + 00		1.5			
7 + 00		3.8			
TP3					110.0
	3.4				
8 + 00					111.1

Solution:

The height of instrument method is based on determining the elevation of the level instrument for each stage of the project using the backsight (BS) reading of the first point of each stage, and then computing the elevations of the other points from the HI elevation. Therefore, in this case, the HI elevation for the first three points is determined using the following formula:

$$HI_1 = \text{Elev. of 1st Point} + BS_1 \tag{5.16}$$

Where:

HI_1 = the elevation of the height of instrument in location #1

Elev. 1 = the elevation of the first point in the stage

BS_1 = the backsight reading at the first point in the stage

Since the HI is given, the elevation of the first point can be determined as below:

$$115.1 = \text{Elev.1} + 4.0$$

⇒

$$\text{Elev.1} = 111.1 \text{ m}$$

The elevation of any point in any stage is determined using the following formula:

$$\text{Elevation of any Point} = HI - IS \text{ or } FS \tag{5.17}$$

Where:

HI = the elevation of the height of instrument

IS = intermediate sight reading of the point

FS = foresight reading of the point

⇒

$$\text{Elevation of point } \#2 = 115.1 - 2.6 = 112.5 \text{ m}$$

The elevation of point #3 is given, the intermediate sight reading at this point is required. Therefore, the formula in Equation 5.17 above is used, as shown below:

$$\text{Elevation of any Point} = HI - IS \text{ or } FS$$

⇒

$$113.1 = 115.1 - IS \text{ of Point } \#3$$

⇒

$$IS \text{ of Point } \#3 = 2.0 \text{ m}$$

The elevation of the turning point (TP1) is determined using the same formula:

$$\text{Elevation of any Point} = HI - IS \text{ or } FS$$

⇒

$$\text{Elevation of TP1} = 115.1 - 3.5 = 111.6 \text{ m}$$

Turning point #1 (TP1) is the last point at which a backsight reading is taken from the first location of the level instrument and the first point to take a foresight reading from the next location of the level instrument. In other words, the first stage is finished and the second stage has started with the TP1. A new height of instrument is determined and used to determine the elevations of the points in the second stage.

The height of instrument elevation in the second stage is determined using the following formula:

$$\text{HI}_{i+1} = \text{Elev. of TP}_i + \text{BS}_{\text{TPi}} \tag{5.18}$$

Where:

$\text{HI}_{i+1} =$ the elevation of the height of instrument in location #$i + 1$

Elev. of $\text{TP}_i =$ the elevation of the turning point #i (TP_i)

$\text{BS}_{\text{TPi}} =$ the backsight reading at the TPi

⇒

$$\text{HI}_2 = \text{Elev. of TP1} + \text{BS}_{\text{TP1}}$$

⇒

$$\text{HI}_2 = 111.6 + 2.8 = 114.4 \text{ m}$$

In a similar manner, the elevations or the readings of all the other points in the project are determined. Table 5.2 summarizes the results:

TABLE 5.2

Elevations for a Differential Leveling Project Using the HI Method for Problem 5.6

Station	BS (m)	IS (m)	FS (m)	HI (m)	Elevation (m)
1 + 00	4.0			115.1	111.1
2 + 00		2.6			112.5
3 + 00		2.0			113.1
TP1			3.5		111.6
	2.8			114.4	
4 + 00		2.2			112.2
5 + 00		4.1			110.3
TP2			2.8		111.6
	2.2			113.8	
6 + 00		1.5			112.3
7 + 00		3.8			110.0
TP3			3.8		110.0
	3.4			113.4	
8 + 00			2.3		111.1
Total	12.4	16.2	12.4		

To verify that the computations in the HI method are correct, three verification checks are done:

1. Number of backsight (BS) readings = number of foresight (FS) readings.

$$\text{No. of BS Readings} = \text{No. of FS Readings} \tag{5.19}$$

$\Rightarrow$

$$4 = 4$$

$\Rightarrow$ OK

2. The reduced level (RL) or elevation of the last point – the reduced level or elevation of the first point (BM) = summation of BS readings – summation of FS readings.

$$\text{RL of Last Point} - \text{RL of 1st Point} = \sum \text{BS Readings} - \sum \text{FS Readings} \tag{5.20}$$

$\Rightarrow$

$$111.1 - 111.1 = 12.4 - 12.4$$

$$0 = 0$$

$\Rightarrow$ OK

3. Summation of reduced levels (or elevations) of all points except the first point (BM) = summation of the HI elevation times number of IS and FS readings at each location – summation of the IS readings – summation of the FS readings.

$$\sum \text{RL's except 1st Point} = \sum_{i=1}^{n} \left(\text{HI}_i \right) \left(\text{No. of IS and FS Readings}_i \right)$$
$$- \sum \text{IS Readings} - \sum \text{FS Readings} \tag{5.21}$$

Where:
i = location i for the level instrument
n = number of locations for the level instrument (or number of stages)

$\Rightarrow$

$$112.5 + 113.1 + 111.6 + 112.2 + 110.3 + 111.6 + 112.3$$
$$+ 110.0 + 110.0 + 111.1$$
$$= \left[(115.1 \times 3) + (114.4 \times 3) + (113.8 \times 3) + (113.4 \times 1) \right] - 16.2 - 12.4$$

$\Rightarrow$

$$1114.7 = 1143.3 - 16.2 - 12.4 = 1114.7$$

⇒ OK

Figure 5.9 shows the MS Excel worksheet used to determine the elevations by the HI method in this problem.

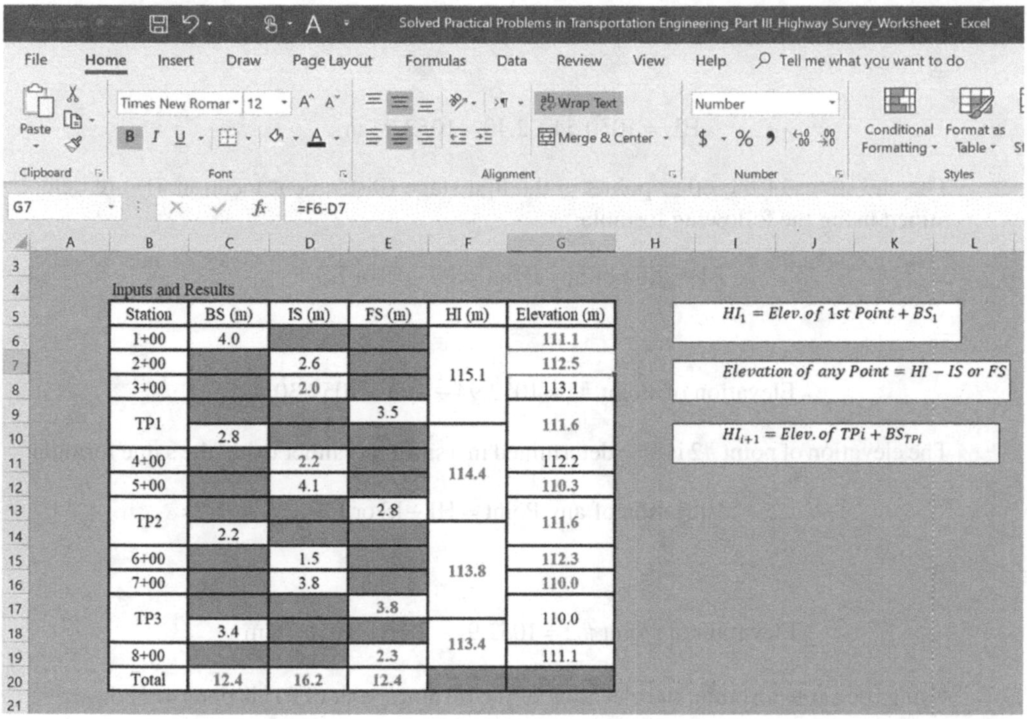

Station	BS (m)	IS (m)	FS (m)	HI (m)	Elevation (m)
1+00	4.0			115.1	111.1
2+00		2.6			112.5
3+00		2.0			113.1
TP1	2.8		3.5	114.4	111.6
4+00		2.2			112.2
5+00		4.1			110.3
TP2	2.2		2.8	113.8	111.6
6+00		1.5			112.3
7+00		3.8			110.0
TP3	3.4		3.8	113.4	110.0
8+00			2.3		111.1
Total	12.4	16.2	12.4		

FIGURE 5.9 Image of MS Excel worksheet used for determining the elevations by the HI method for Problem 5.6.

5.7 Complete the following record of differential levels and determine the elevations of all survey points in Table 5.3 using the height of instrument (HI) method. Verify that the computations are correct using the three verification checks in the HI method.

TABLE 5.3
HI Differential Leveling Record for Problem 5.7

Point	BS (m)	IS (m)	FS (m)	HI (m)	Elevation (m)
BM	2.40				1050.54
1		1.64			
2		2.16			
TP1			4.26		
	6.12				
TP2			2.50		
	8.20				
3		3.84			
4			1.62		

Solution:

Using the height of instrument method, the elevations and the other required values are determined as shown in the procedure described below:

$$HI_1 = \text{Elev. of 1st Point} + BS_1$$

$\Rightarrow$

$$HI_1 = 1050.54 + 2.40 = 1052.94 \text{ m}$$

The elevations of the other points in the first stage (instrument location #1) are determined using the following formula:

$$\text{Elevation of any Point} = HI - IS \text{ or } FS$$

$\Rightarrow$

$$\text{Elevation of Point \#1} = 1052.94 - 1.64 = 1051.30 \text{ m}$$

The elevation of point #2 is also determined in a similar manner using the same formula:

$$\text{Elevation of any Point} = HI - IS \text{ or } FS$$

$\Rightarrow$

$$\text{Elevation of Point \# 2} = 1052.94 - 2.16 = 1050.78 \text{ m}$$

Using the same formula, the elevation of the turning point (TP1) is determined:

$$\text{Elevation of any Point} = HI - IS \text{ or } FS$$

$\Rightarrow$

$$\text{Elevation of TP1} = 1052.94 - 4.26 = 1048.68 \text{ m}$$

At the turning point (TP1), a new height of instrument is determined and used to compute the elevations of the points in the second stage.

The height of instrument elevation in the second stage is determined using the following formula:

$$HI_{i+1} = \text{Elev. of } TP_i + BS_{TPi}$$

$\Rightarrow$

$$HI_2 = \text{Elev. of } TP1 + BS_{TP1}$$

$\Rightarrow$

$$HI_2 = 1048.68 + 6.12 = 1054.80 \text{ m}$$

The elevation of the turning point (TP2) is determined using the HI_2 and the following formula:

$$\text{Elevation of any Point} = HI - IS \text{ or } FS$$

$$\Rightarrow$$

$$\text{Elevation of TP2} = 1054.80 - 2.50 = 1052.30 \text{ m}$$

At the turning point (TP2), a new height of instrument is determined and used to compute the elevations of the points in the third stage.

The height of instrument elevation in the third stage is determined using the following formula:

$$HI_{i+1} = \text{Elev. of } TP_i + BS_{TPi}$$

$$\Rightarrow$$

$$HI_3 = \text{Elev. of } TP2 + BS_{TP2}$$

$$\Rightarrow$$

$$HI_3 = 1052.30 + 8.20 = 1060.50 \text{ m}$$

The elevations of points #3 and #4 are determined using the following formula and the height of instrument in location #3 (HI_3):

$$\text{Elevation of any Point} = HI - IS \text{ or } FS$$

$$\Rightarrow$$

$$\text{Elevation of Point } \#3 = 1060.50 - 3.84 = 1056.66 \text{ m}$$

In a similar manner:

$$\text{Elevation of Point } \#4 = 1060.50 - 1.62 = 1058.88 \text{ m}$$

Table 5.4 summarizes all the results:

TABLE 5.4

Elevations for a Differential Leveling Project Using the HI Method for Problem 5.7

Point	BS (m)	IS (m)	FS (m)	HI (m)	Elevation (m)
BM	2.40			1052.94	1050.54
1		1.64			1051.30
2		2.16			1050.78
TP1			4.26		1048.68
	6.12			1054.80	
TP2			2.50		1052.30
	8.20			1060.50	
3		3.84			1056.66
4			1.62		1058.88
Total	16.72	7.64	8.38		

The following three verification checks are performed to verify the computations in the HI method:

1. Number of backsight (BS) readings = number of foresight (FS) readings.

$$\text{No. of BS Readings} = \text{No. of FS Readings}$$

$\Rightarrow$

$$3 = 3$$

$\Rightarrow$ OK

2. The reduced level (RL) or elevation of the last point – the reduced level or elevation of the first point (BM) = summation of BS readings – summation of FS readings.

$$\text{RL of Last Point} - \text{RL of 1st Point} = \sum \text{BS Readings} - \sum \text{FS Readings}$$

$\Rightarrow$

$$1058.88 - 1050.54 = 16.72 - 8.38$$

$$8.34 = 8.34$$

$\Rightarrow$ OK

3. Summation of reduced levels (or elevations) of all points except the first point (BM) = summation of the HI elevation times the number of IS and FS readings at each location – summation of the IS readings – summation of the FS readings.

$$\sum \text{RL's except 1st Point} = \sum_{i=1}^{n} \left(\text{HI}_i \right) \left(\text{No. of IS and FS Readings}_i \right)$$

$$- \sum \text{IS Readings} - \sum \text{FS Readings}$$

$\Rightarrow$

$$1050.54 + 1051.30 + 1050.78 + 1048.68 + 1052.30$$

$$+ 1056.66 + 1058.88$$

$$= \left[\left(1052.94 \times 3 \right) + \left(1054.80 \times 1 \right) + \left(1060.50 \times 2 \right) \right] - 7.64 - 8.38$$

$\Rightarrow$

$$6318.60 = 6334.62 - 7.64 - 8.38 = 6318.60$$

$\Rightarrow$ OK

Figure 5.10 shows the MS Excel worksheet used to determine the elevations by the HI method in this problem.

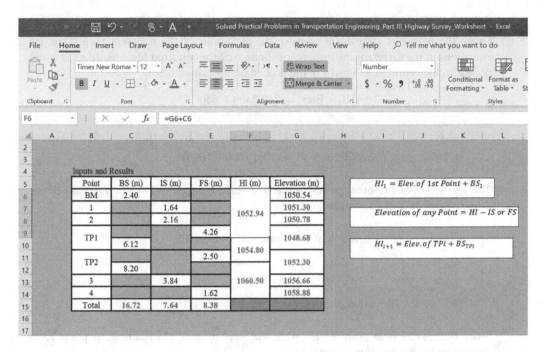

FIGURE 5.10 Image of MS Excel worksheet used for determining the elevations by the HI method for Problem 5.7.

5.8   Use the *height of instrument (HI) method* for differential leveling to complete Table 5.5. Use the three verification checks in the HI method to double-check the computations. *Note: Fill in the table for each white-colored blank. The gray-colored blanks do not require any computations.*

TABLE 5.5
HI Differential Leveling Record for Problem 5.8

Station	BS (m)	IS (m)	FS (m)	HI (m)	Elevation (m)
1 + 00	3.0			112.4	
2 + 00		1.6			
3 + 00					112.2
TP1			3.0		
	1.4				
4 + 00					110.1
5 + 00		4.0			
TP2			2.6		
	2.2				
6 + 00		1.3			
7 + 00		3.4			
TP3					109.3
	2.7				
8 + 00			2.6		

Solution:

Using the height of instrument method, the elevations of the different points, the HI elevations at the different instrument locations, and the required level readings at some points are determined as shown in the procedure described below:

$$HI_1 = \text{Elev. of 1st Point} + BS_1$$

$\Rightarrow$

$$\text{Elev. of Point \#1} = 112.4 - 3.0 = 109.4 \text{ m}$$

The elevations of the other points in the first stage (instrument location #1) are determined using the following formula:

$$\text{Elevation of any Point} = HI - IS \text{ or } FS$$

$\Rightarrow$

$$\text{Elevation of Point \#2} = 112.4 - 1.6 = 110.8 \text{ m}$$

Since the elevation of point #3 is given, the level reading (IS reading) at point #3 is determined using the same formula:

$$\text{Elevation of any Point} = HI - IS \text{ or } FS$$

$\Rightarrow$

$$112.2 = 112.4 - IS \text{ at Point \#3}$$

$\Rightarrow$

$$IS \text{ at Point \#3} = 0.2 \text{ m}$$

The last point in the first stage (instrument location #1) is the turning point (TP1). Using the same formula, the elevation of the turning point (TP1) is determined:

$$\text{Elevation of any Point} = HI - IS \text{ or } FS$$

$\Rightarrow$

$$\text{Elevation of TP1} = 112.4 - 3.0 = 109.4 \text{ m}$$

At the turning point (TP1), a new height of instrument is determined and used to compute the elevations of the points in the second stage.

The height of instrument elevation in the second stage (HI_2) is determined using the following formula:

$$HI_{i+1} = \text{Elev. of } TP_i + BS_{TPi}$$

$\Rightarrow$

$$HI_2 = \text{Elev. of TP1} + BS_{TP1}$$

$\Rightarrow$

$$HI_2 = 109.4 + 1.4 = 110.8 \text{ m}$$

Now the height of instrument in location #2 (HI_2) is used to compute the elevations of the points in the second stage, as shown below:

$$\text{Elevation of any Point} = HI - IS \text{ or } FS$$

Since the elevation of point #4 is given, the level reading (IS reading) at point #3 is determined using the same formula:

$\Rightarrow$

$$110.1 = 110.8 - IS \text{ at Point } \#4$$

$\Rightarrow$

$$IS \text{ at Point } \#4 = 0.7 \text{ m}$$

The elevation of point #5 is also determined using the same formula:

$$\text{Elevation of any Point} = HI - IS \text{ or } FS$$

$\Rightarrow$

$$\text{Elevation of Point } \#5 = 110.8 - 4.0 = 106.8 \text{ m}$$

Also, the elevation of the turning point (TP2) is computed using the same formula:

$$\text{Elevation of TP2} = 110.8 - 2.6 = 108.2 \text{ m}$$

A new height of instrument is determined at the turning point (TP2) and used to compute the elevations of the points in the third stage.

The height of instrument elevation in the third stage (HI_3) is determined using the following formula:

$$HI_{i+1} = \text{Elev. of TP}_i + BS_{TPi}$$

$\Rightarrow$

$$HI_3 = \text{Elev. of TP2} + BS_{TP2}$$

$\Rightarrow$

$$HI_3 = 108.2 + 2.2 = 110.4 \text{ m}$$

Using the height of instrument at location #3 (HI_3), the elevations of the points in this stage (the third stage) are calculated:

$$\text{Elevation of any Point} = HI - IS \text{ or } FS$$

$\Rightarrow$

$$\text{Elevation of Point } \#6 = 110.4 - 1.3 = 109.1 \text{ m}$$

$$\text{Elevation of Point } \#7 = 110.4 - 3.4 = 107.0 \text{ m}$$

From the elevation of the turning point (TP3) that is given, the FS reading at this point can be determined:

$$\text{Elevation of TP3} = \text{HI} - \text{FS at TP3}$$

$\Rightarrow$

$$109.3 = 110.4 - \text{FS at TP3}$$

$\Rightarrow$

$$\text{FS at TP3} = 1.1 \text{ m}$$

In a similar manner, a new height of instrument is determined at the turning point (TP3) and used to compute the elevations of the points in the fourth stage.

The height of instrument elevation in the fourth stage (HI_4) is determined using the following formula:

$$\text{HI}_{i+1} = \text{Elev. of TP}_i + \text{BS}_{\text{TPi}}$$

$\Rightarrow$

$$\text{HI}_4 = \text{Elev. of TP3} + \text{BS}_{\text{TP3}}$$

$\Rightarrow$

$$\text{HI}_4 = 109.3 + 2.7 = 112.0 \text{ m}$$

The elevations of the points in this stage (the fourth stage) are determined using the HI_4 (from location #4), as shown below:

$$\text{Elevation of any Point} = \text{HI} - \text{IS or FS}$$

$\Rightarrow$

$$\text{Elevation of Point } \#8 = 112.0 - 2.6 = 109.4 \text{ m}$$

The results determined above are summarized in Table 5.6:

TABLE 5.6
Elevations for a Differential Leveling Project Using the HI Method for Problem 5.8

Station	BS (m)	IS (m)	FS (m)	HI (m)	Elevation (m)
1 + 00	3.0			112.4	109.4
2 + 00		1.6			110.8
3 + 00		0.2			112.2
TP1			3.0		109.4
	1.4			110.8	
4 + 00		0.7			110.1
5 + 00		4.0			106.8
TP2			2.6		108.2
	2.2			110.4	
6 + 00		1.3			109.1
7 + 00		3.4			107.0
TP3			1.1		109.3
	2.7			112.0	
8 + 00			2.6		109.4
Total	9.3	11.2	9.3		

The following three verification checks are performed to verify the computations in the HI method:

$$\text{No. of BS Readings} = \text{No. of FS Readings}$$

$\Rightarrow$

$$4 = 4$$

$\Rightarrow$ OK

$$\text{RL of Last Point} - \text{RL of 1st Point} = \sum \text{BS Readings} - \sum \text{FS Readings}$$

$\Rightarrow$

$$109.4 - 109.4 = 9.3 - 9.3$$

$$0 = 0$$

$\Rightarrow$ OK

$$\sum \text{RL's except 1st Point} = \sum_{i=1}^{n} \left(\text{HI}_i \right) \left(\text{No. of IS and FS Readings}_i \right)$$

$$- \sum \text{IS Readings} - \sum \text{FS Readings}$$

$\Rightarrow$

$$110.8 + 112.2 + 109.4 + 110.1 + 106.8 + 108.2 + 109.1$$

$$+ 107.0 + 109.3 + 109.4$$

$$= \left[(112.4 \times 3) + (110.8 \times 3) + (110.4 \times 3) + (112.0 \times 1) \right] - 11.2 - 9.3$$

$\Rightarrow$

$$1092.3 = 1112.8 - 11.2 - 9.3 = 1092.3$$

$\Rightarrow$ OK

Figure 5.11 shows the MS Excel worksheet used to determine the elevations by the HI method in this problem.

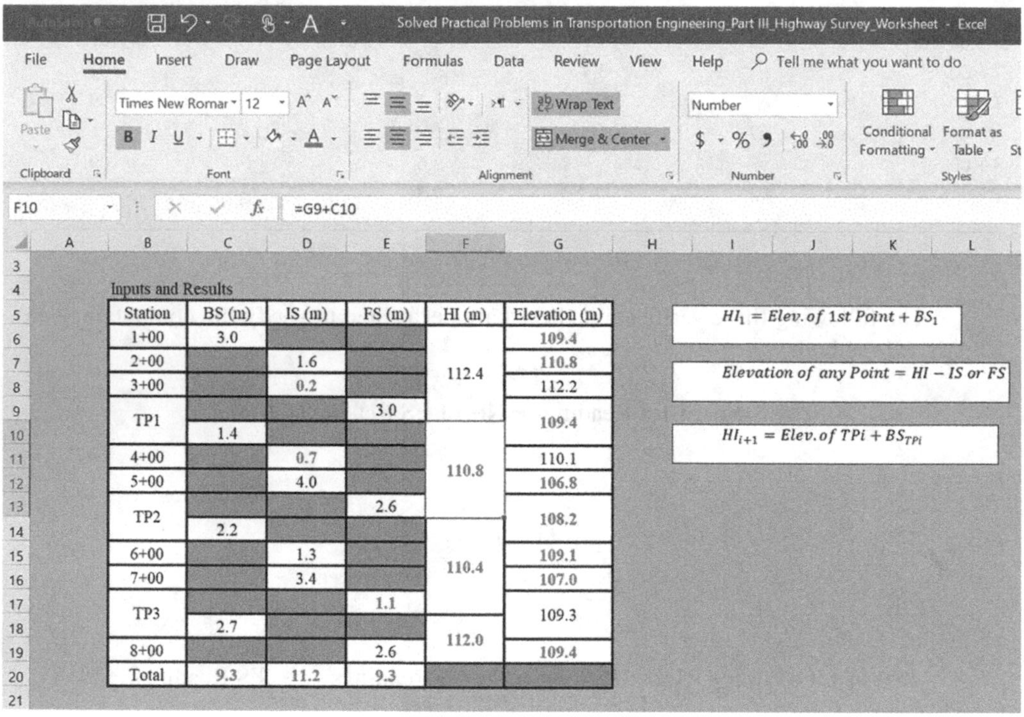

FIGURE 5.11 Image of MS Excel worksheet used for determining the elevations by the HI method for Problem 5.8.

5.9 Use the rise and fall (R&F) method to compute the elevations of all the points in the record of differential levels shown in Table 5.7. Perform the two verification checks to confirm the computations and the results in the R&F method.

TABLE 5.7
R&F Differential Leveling Record for Problem 5.9

Point	BS (m)	IS (m)	FS (m)	Rise (m)	Fall (m)	Elevation (m)
BM	1.524					150.000
1		2.341				
2		1.280				
TP1			3.112			
	2.106					
TP2			3.114			
	5.612					
3		1.224				
4			2.462			

Solution:

The amount of rise or fall between two successive points is determined using the following formula:

$$R \text{ or } F = R_i - R_{i+1} \tag{5.22}$$

Where:
R = rise magnitude
F = fall magnitude
R_i = level reading at point i
R_{i+1} = level reading at point $i + 1$ (next point)

If the sign is +ve, it is considered a rise; and if the sign is −ve, it is considered a fall. Therefore, between points BM and #1, the amount of rise or fall is computed as:

$$R \text{ or } F = 1.524 - 2.341 = -0.817 \text{ m}$$

Since the sign is −ve, this is considered a fall, hence:

$$F = 0.817 \text{ m}$$

The elevation of any point is determined using the formula below:

$$\text{Elev. of Point}_{i+1} = \text{Elev. of Point}_i + R \text{ or } F \tag{5.23}$$

Where:
Elev. of Point $i + 1$ = the elevation of point $i + 1$ (next point)
Elev. of Point i = the elevation of point i (previous point)
R or F = the amount of rise or fall (a +ve sign is used for the rise and a −ve sign is used for the fall in this formula)

$\Rightarrow$

$$\text{Elev. of Point \#1} = \text{Elev. of BM} + R \text{ or } F$$

$\Rightarrow$

$$\text{Elev. of Point \#1} = 150.000 - 0.817 = 149.183 \text{ m}$$

Between points #1 and #2, the magnitude of rise or fall is determined using the same formula:

$$R \text{ or } F = R_i - R_{i+1}$$

$\Rightarrow$

$$R \text{ or } F = 2.341 - 1.280 = 1.061 \text{ m}$$

Since the sign is +ve, this is considered a rise, hence:

$$R = 1.061 \text{ m}$$

The elevation of point #2 is calculated as:

$$\text{Elev. of Point}_{i+1} = \text{Elev. of Point}_i + R \text{ or } F$$

$\Rightarrow$

$$\text{Elev. of Point \#2} = \text{Elev. of Point \#1} + R \text{ or } F$$

$\Rightarrow$

$$\text{Elev. of Point \#2} = 149.183 + 1.061 = 150.244 \text{ m}$$

Between points #2 and TP1, the magnitude of rise or fall is determined using the same formula:

$$R \text{ or } F = R_i - R_{i+1}$$

$\Rightarrow$

$$R \text{ or } F = 1.280 - 3.112 = -1.832 \text{ m}$$

Since the sign is −ve, this is considered a fall, hence:

$$F = 1.832 \text{ m}$$

Therefore, the elevation of TP1 is calculated as:

$$\text{Elev. of Point}_{i+1} = \text{Elev. of Point}_i + R \text{ or } F$$

$\Rightarrow$

$$\text{Elev. of TP1} = \text{Elev. of Point \#2} + R \text{ or } F$$

$\Rightarrow$

$$\text{Elev. of TP1} = 150.244 - 1.832 = 148.412 \text{ m}$$

In a similar manner, the elevations of the remaining points are determined using the two formulas described above. The results of all the points in the record of differential levels are summarized in Table 5.8:

TABLE 5.8
Elevations for a Differential Leveling Project Using the R&F Method for Problem 5.9

Point	BS (m)	IS (m)	FS (m)	Rise (m)	Fall (m)	Elevation (m)
BM	1.524					150.000
1		2.341			0.817	149.183
2		1.280		1.061		150.244
TP1			3.112		1.832	148.412
	2.106					
TP2			3.114		1.008	147.404
	5.612					
3		1.224		4.388		151.792
4			2.462		1.238	150.554
Total				5.449	4.895	

The following two verification checks are performed to double-check the computations in the R&F method:

1. The number of backsight (BS) readings should be equal to the number of foresight (FS) readings.

$$\text{No. of BS Readings} = \text{No. of FS Readings} \tag{5.24}$$

$\Rightarrow$

$$3 = 3$$

$\Rightarrow$ OK

2. The reduced level (RL) or elevation of the last point – the reduced level or elevation of the first point (BM) = summation of rise values – summation of fall values.

$$\text{RL of Last Point} - \text{RL of 1st Point} = \sum \text{Rises} - \sum \text{Falls} \tag{5.25}$$

$\Rightarrow$

$$150.554 - 150.000 = 5.449 - 4.895$$

$$0.554 = 0.554$$

$\Rightarrow$ OK

Figure 5.12 shows the MS Excel worksheet used to determine the elevations by the R&F method in this problem.

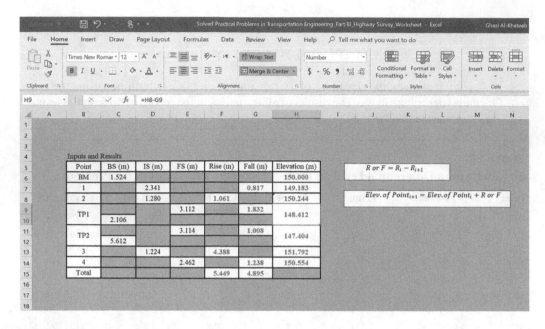

FIGURE 5.12 Image of MS Excel worksheet used for determining the elevations by the R&F method for Problem 5.9

5.10   A leveling task was accomplished inside a building, as shown in Figure 5.13. Complete the differential leveling record in Table 5.9 using the *height of instrument (HI) method*. *Note: the level staff at points A and B is inverted.*

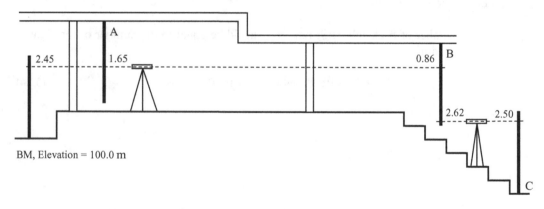

FIGURE 5.13 HI differential leveling project for Problem 5.10.

Solution:

The readings at the different points are recorded in Table 5.9. The HI method will be used to determine the elevations of points A, B, and C shown in the diagram above.

In the HI method, the HI elevation is determined using the following formula:

TABLE 5.9

HI Differential Leveling Record for Problem 5.10

Point	BS (m)	IS (m)	FS (m)	HI (m)	Elevation (m)
BM	2.45				100.00
A		−1.65			
B			−0.86		
	−2.62				
C			2.50		

$$HI_1 = \text{Elev. of 1st Point} + BS_1$$

$\Rightarrow$

$$HI_1 = 100.00 + 2.45 = 102.45 \text{ m}$$

The elevations of the different points in this stage are determined using the formula shown below:

$$\text{Elevation of any Point} = HI - IS \text{ or } FS$$

$\Rightarrow$

$$\text{Elevation of Point } A = 102.45 - (-1.65) = 104.10 \text{ m}$$

In a similar manner, the elevations of points B and C are calculated as follows:

$$\text{Elevation of any Point} = HI - IS \text{ or } FS$$

$\Rightarrow$

$$\text{Elevation of Point } B = 102.45 - (-0.86) = 103.31 \text{ m}$$

Point B is a turning point since a foresight reading from instrument location #1 is taken and a backsight reading from instrument location #2 is also taken. Therefore, a new HI elevation should be determined to be used in the next stage to calculate the elevation of point C.

$\Rightarrow$

$$HI_{i+1} = \text{Elev. of } TP_i + BS_{TPi}$$

$\Rightarrow$

$$HI_2 = \text{Elev. of } TP1 + BS_{TP1}$$

$\Rightarrow$

$$HI_2 = 103.31 + (-2.62) = 100.69 \text{ m}$$

Elevation of any Point = HI − IS or FS

$\Rightarrow$

Elevation of Point $C = 100.69 - 2.50 = 98.19$ m

The results of the computations performed above are summarized in Table 5.10:

TABLE 5.10
Elevations for a Differential Leveling Project Using the HI Method for Problem 5.10

Point	BS (m)	IS (m)	FS (m)	HI (m)	Elevation (m)
BM	2.45			102.45	100.00
A		−1.65			104.10
B			−0.86		103.31
	−2.62			100.69	
C			2.50		98.19
Total	−0.17	−1.65	1.64		

The following three verification checks are performed to double-check the computations in the HI method:

1. No. of BS Readings = No. of FS Readings

$\Rightarrow$

$$2 = 2$$

$\Rightarrow$ OK

2. RL of Last Point − RL of 1st Point = $\sum$ BS Readings − $\sum$ FS Readings

$\Rightarrow$

$$98.19 - 100.00 = -0.17 - 1.64$$

$$-1.81 = -1.81$$

$\Rightarrow$ OK

3. $\sum$ RL's except 1st Point = $\sum_{i=1}^{n} (HI_i)(\text{No. of IS and FS Readings}_i)$

$$- \sum \text{IS Readings} - \sum \text{FS Readings}$$

$\Rightarrow$

$$104.10 + 103.31 + 98.19 = \left[(102.45 \times 2) + (100.69 \times 1) \right] - (-1.65) - 1.64$$

⇒

$$305.60 = 305.59 + 1.65 - 1.64 = 305.60$$

⇒ OK

Figure 5.14 shows the MS Excel worksheet used to determine the elevations by the HI method in this problem.

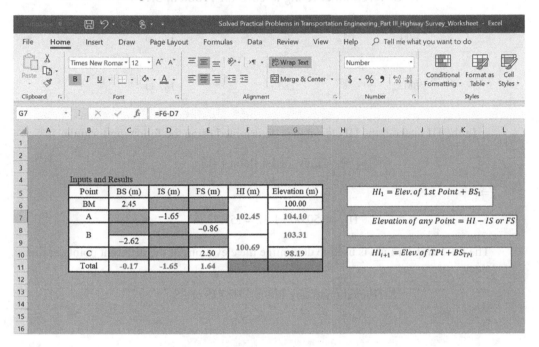

FIGURE 5.14 Image of MS Excel worksheet used for determining the elevations by the HI Problem 5.10.

5.11 For Figure 5.15, complete the differential leveling record in Table 5.11 using the *height of instrument (HI) method*.

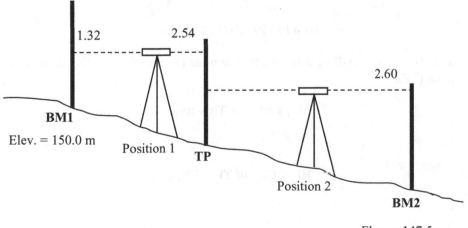

FIGURE 5.15 HI differential leveling project for Problem 5.11.

Solution:

The level readings at the different points along with the elevations of the benchmarks in the project are summarized in the differential leveling record shown in Table 5.11:

TABLE 5.11
HI Differential Leveling Record for Problem 5.11

Point	BS (m)	FS (m)	HI (m)	Elevation (m)
BM1	1.32			150.00
TP		2.54		
BM2		2.60		147.50

In the HI method, the HI elevation is determined using the following formula:

$$HI_1 = \text{Elev. of 1st Point} + BS_1$$

$\Rightarrow$

$$HI_1 = 150.00 + 1.32 = 151.32 \text{ m}$$

The following formula is used to determine the elevations of the different points in this stage:

$$\text{Elevation of any Point} = HI - IS \text{ or } FS$$

$\Rightarrow$

$$\text{Elevation of TP} = 151.32 - 2.54 = 148.78 \text{ m}$$

Similarly:

$$\text{Elevation of BM2} = HI_2 - FS \text{ at BM2}$$

$\Rightarrow$

$$HI_2 = 147.50 + 2.60 = 150.10 \text{ m}$$

The HI_2 elevation will be now used to determine the backsight reading at the turning point (TP):

$$HI_{i+1} = \text{Elev. of } TP_i + BS_{TPi}$$

$\Rightarrow$

$$HI_2 = \text{Elev. of } TP1 + BS_{TP1}$$

$\Rightarrow$

$$BS_{TP} = 150.10 - 148.78 = 1.32 \text{ m}$$

The results are summarized in Table 5.12:

TABLE 5.12
Elevations for a Differential Leveling Project Using the HI Method for Problem 5.11

Point	BS (m)	FS (m)	HI (m)	Elevation (m)
BM1	1.32		151.32	150.00
TP		2.54		148.78
	1.32		150.10	
BM2		2.60		147.50
Total	2.64	5.14		

Figure 5.16 shows the MS Excel worksheet used to determine the elevations by the HI method in this problem.

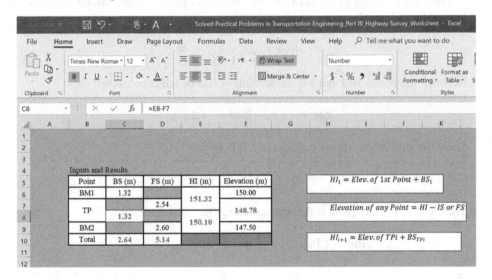

FIGURE 5.16 Image of MS Excel worksheet used for determining the elevations by the HI method for Problem 5.11.

5.12 Use the rise and fall (R&F) method to determine the elevations of the points in the following leveling job accomplished as shown in Figure 5.17. Perform the two verification checks to verify the computations in the R&F method.

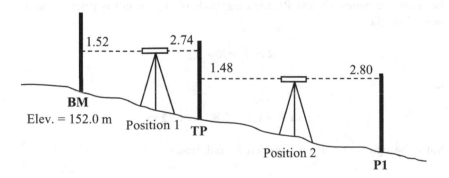

FIGURE 5.17 R&F differential leveling project for Problem 5.12.

Solution:

The level readings and the given elevation of the bench mark (BM) are tabulated in Table 5.13:

TABLE 5.13
R&F Differential Leveling Record for Problem 5.12

Point	BS (m)	FS (m)	Rise (m)	Fall (m)	Elevation (m)
BM	1.52				152.00
TP		2.74			
	1.48				
P1		2.80			

The amount of rise or fall between two successive points is determined using the following formula:

$$R \text{ or } F = R_i - R_{i+1}$$

Between the points BM and TP, the amount of rise or fall is computed as:

$$R \text{ or } F = 1.52 - 2.74 = -1.22 \text{ m}$$

Since the sign is −ve, this is considered a fall, hence:

$$F = 1.22 \text{ m}$$

The elevation of any point is determined using the formula below:

$$\text{Elev. of Point}_{i+1} = \text{Elev. of Point}_i + R \text{ or } F$$

$\Rightarrow$

$$\text{Elev. of TP} = \text{Elev. of BM} + R \text{ or } F$$

$\Rightarrow$

$$\text{Elcv. of TP} = 152.000 - 1.22 = 150.78 \text{ m}$$

Between the points TP and P1, the magnitude of rise or fall is determined using the same formula:

$$R \text{ or } F = R_i - R_{i+1}$$

$\Rightarrow$

$$R \text{ or } F = 1.48 - 2.80 = -1.32 \text{ m}$$

Since the sign is −ve, this is considered a fall, hence:

$$F = 1.32 \text{ m}$$

The elevation of point $P1$ is calculated as:

$$\text{Elev. of Point}_{i+1} = \text{Elev. of Point}_i + R \text{ or } F$$

$\Rightarrow$

$$\text{Elev. of } P1 = \text{Elev. of TP} + R \text{ or } F$$

$\Rightarrow$

$$\text{Elev. of } P1 = 150.78 - 1.32 = 149.46 \text{ m}$$

The results of all the points in the record of differential levels are summarized in Table 5.14:

TABLE 5.14

Elevations for a Differential Leveling Project Using the R&F Method for Problem 5.12

Point	BS (m)	FS (m)	Rise (m)	Fall (m)	Elevation (m)
BM	1.52				152.00
TP		2.74		1.22	150.78
	1.48				
P1		2.80		1.32	149.46
Total			0.00	2.54	

The following two verification checks are performed to double-check the computations in the R&F method:

1. The number of backsight (BS) readings should be equal to the number of foresight (FS) readings.

$$\text{No. of BS Readings} = \text{No. of FS Readings}$$

$\Rightarrow$

$$2 = 2$$

$\Rightarrow$ OK

2. The reduced level (RL) or elevation of the last point – the reduced level or elevation of the first point (BM) = summation of rise values – summation of fall values.

$$\text{RL of Last Point} - \text{RL of 1st Point} = \sum \text{Rises} - \sum \text{Falls}$$

$\Rightarrow$

$$149.46 - 152.00 = 0.00 - 2.54$$

$$-2.54 = -2.54$$

⇒ OK

Figure 5.18 shows the MS Excel worksheet used to determine the elevations by the R&F method in this problem.

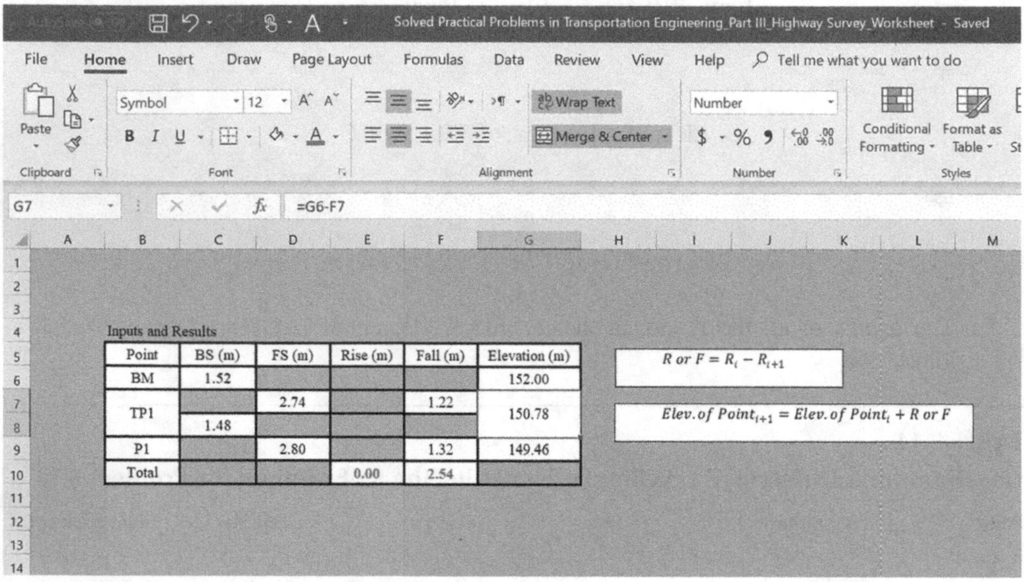

Point	BS (m)	FS (m)	Rise (m)	Fall (m)	Elevation (m)
BM	1.52				152.00
TP1		2.74		1.22	150.78
	1.48				
P1		2.80		1.32	149.46
Total			0.00	2.54	

$R \text{ or } F = R_i - R_{i+1}$

$Elev. \text{ of } Point_{i+1} = Elev. \text{ of } Point_i + R \text{ or } F$

FIGURE 5.18 Image of MS Excel worksheet used for determining the elevations by the R&F method for Problem 5.12.

5.13 The readings (m) at points *A*, *B*, and *C* on the ceiling of a building are taken using inverted level staff, as shown in Figure 5.19. If the elevation of point A is 102.00 m, determine the elevations of points *B* and *C* using the HI method.

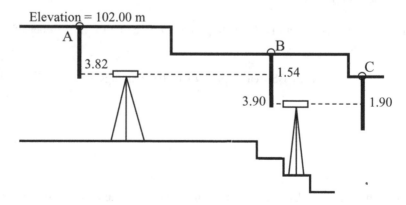

FIGURE 5.19 HI differential leveling project for Problem 5.13.

Solution:

The level readings and the elevation of point A are summarized in Table 5.15:

TABLE 5.15
HI Differential Leveling Record for Problem 5.13

Point	BS (m)	FS (m)	HI (m)	Elevation (m)
A	−3.82			102.00
B		−1.54		
	−3.90			
C		−1.90		

The readings at points A, B, and C are all recorded as −ve values in Table 5.15 because they are taken using an inverted level staff.

In the HI method, the HI elevation is determined using the following formula:

$$HI_1 = \text{Elev. of 1st Point} + BS_1$$

⇒

$$HI_1 = 102.00 + (-3.82) = 98.18 \text{ m}$$

The following formula is used to determine the elevations of the different points in this stage:

$$\text{Elevation of any Point} = HI - IS \text{ or } FS$$

⇒

$$\text{Elevation of } B = 98.18 - (-1.54) = 99.72 \text{ m}$$

Point B is a turning point; therefore, a new height of instrument (HI) elevation is determined:

$$HI_{i+1} = \text{Elev. of } TP_i + BS_{TPi}$$

⇒

$$HI_2 = \text{Elev. of } B + BS_B$$

⇒

$$HI_2 = 99.72 + (-3.90) = 95.82 \text{ m}$$

The elevation of point C is calculated using the formula below:

$$\text{Elevation of any Point} = HI - IS \text{ or } FS$$

⇒

$$\text{Elevation of } C = 95.82 - (-1.90) = 97.72 \text{ m}$$

Table 5.16 summarizes all the results computed above.

TABLE 5.16

Elevations for a Differential Leveling Project Using the HI Method for Problem 5.13

Point	BS (m)	FS (m)	HI (m)	Elevation (m)
A	−3.82		98.18	102.00
B		−1.54		99.72
	−3.90		95.82	
C		−1.90		97.72
Total	−7.72	−3.44		

Figure 5.20 shows the MS Excel worksheet used to determine the elevations by the HI method in this problem.

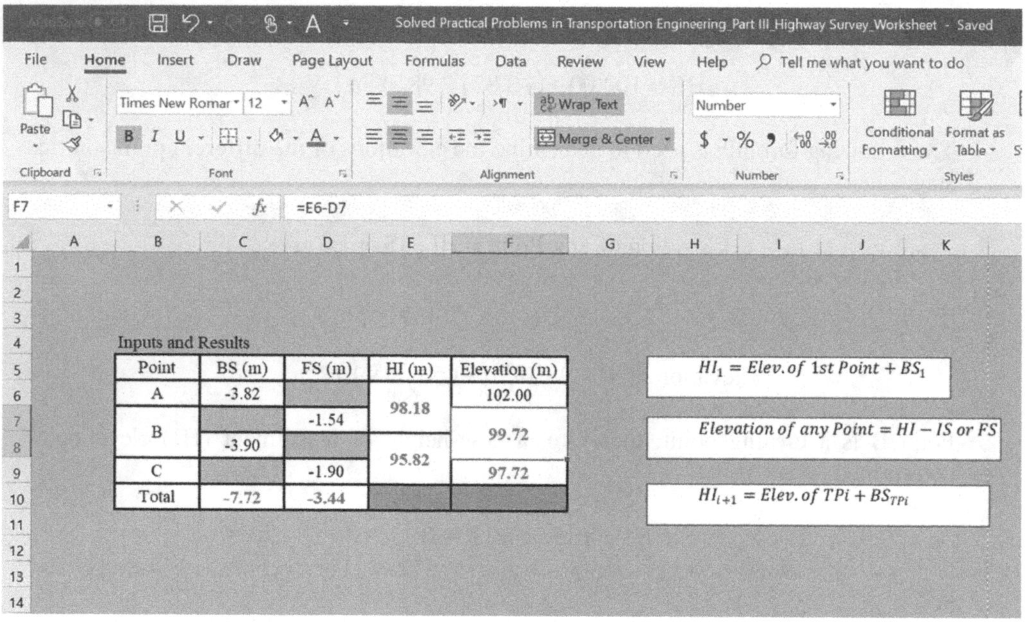

FIGURE 5.20 Image of MS Excel worksheet used for determining the elevations by the HI method for Problem 5.13.

5.14   The differential leveling readings at points A, B, and C are taken from two different positions of the level device, as shown in Figure 5.21. Determine the elevations of points A, B, and D using the HI method.

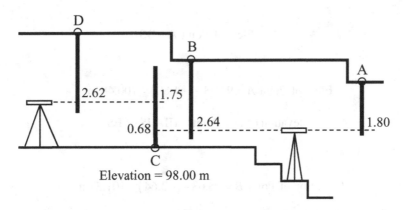

FIGURE 5.21   HI differential leveling project for Problem 5.14.

Solution:

The level readings and the elevation of point C are recorded in Table 5.17:

TABLE 5.17
HI Differential Leveling Record for Problem 5.14

Point	BS (m)	IS (m)	FS (m)	HI (m)	Elevation (m)
A	−1.80				
B		−2.64			
C			0.68		98.00
	1.75				
D			−2.62		

Using the III method, the elevation of any point is determined using the following formula:

$$\text{Elevation of any Point} = \text{HI} - \text{IS or FS}$$

Since the elevation of point C is provided, the HI elevation at the instrument location #1 can be determined by knowing the foresight (FS) reading at point C as described below:

$$\text{Elevation of Point } C = \text{HI}_1 - \text{FS}_C$$

$\Rightarrow$

$$98.00 = \text{HI}_1 - 0.68$$

$\Rightarrow$

$$\text{HI}_1 = 98.68 \text{ m}$$

The elevations of points A and B are now determined using the HI1 determined above:

$$HI_1 = \text{Elev. of 1st Point} + BS_1$$

$\Rightarrow$

$$HI_1 = \text{Elev. of Point } A + BS_A$$

$\Rightarrow$

$$\text{Elev. of Point } A = 98.68 - (-1.80) = 100.48 \text{ m}$$

$$\text{Elevation of any Point} = HI - IS \text{ or } FS$$

$\Rightarrow$

$$\text{Elevation of Point } B = 98.68 - (-2.64) = 101.32 \text{ m}$$

Since the elevation of point C is given, the HI elevation at instrument location #2 is determined using the following formula:

$$HI_{i+1} = \text{Elev. of } TP_i + BS_{TPi}$$

$\Rightarrow$

$$HI_2 = \text{Elev. of } C + BS_C$$

$\Rightarrow$

$$HI_2 = 98.00 + 1.75 = 99.75 \text{ m}$$

The elevation of point D is then determined using the formula shown below:

$$\text{Elevation of any Point} = HI - IS \text{ or } FS$$

$\Rightarrow$

$$\text{Elevation of } D = 99.75 - (-2.62) = 102.37 \text{ m}$$

The results determined in the procedure above are summarized in Table 5.18:

TABLE 5.18

Elevations for a Differential Leveling Project Using the HI Method for Problem 5.14

Point	BS (m)	IS (m)	FS (m)	HI (m)	Elevation (m)
A	−1.80			98.68	100.48
B		−2.64			101.32
C			0.68		98.00
	1.75			99.75	
D			−2.62		102.37
Total	−0.05	−2.64	−1.94		

The MS Excel worksheet used to determine the elevations by the HI method in this problem is shown in Figure 5.22.

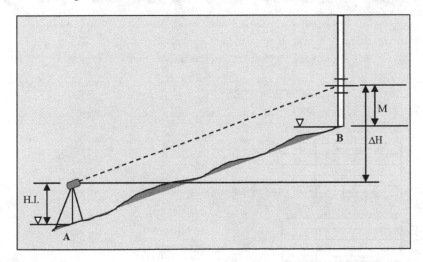

Point	BS (m)	IS (m)	FS (m)	HI (m)	Elevation (m)
A	−1.80				100.48
B		−2.64		98.68	101.32
C			0.68		98.00
	1.75			99.75	
D			−2.62		102.37
Total	−0.05	−2.64	−1.94		

$HI_1 = Elev. of\ 1st\ Point + BS_1$

$Elevation\ of\ any\ Point = HI − IS\ or\ FS$

$HI_{i+1} = Elev. of\ TPi + BS_{TPi}$

FIGURE 5.22 Image of MS Excel worksheet used for determining the elevations by the HI method for Problem 5.14.

5.15   In Figure 5.23, the rod readings of the three cross-hairs using the stadia method are 1.650 m, 1.450 m, and 1.350 m, for the upper, middle, and lower hairs, respectively. If the reduced level of point A is 310.244 m, the height of instrument (HI) is 1.625 m, and the vertical angle (α) is 30°, determine the reduced level (elevation) of point B.

FIGURE 5.23 Stadia method for determining elevations for Problem 5.15.

Solution:

In stadia leveling, the vertical distance is given by the following formula:

$$\Delta H = \frac{1}{2} kr \sin 2\alpha \qquad (5.26)$$

Where:
ΔH=the vertical distance
k=the stadia coefficient (a typical value of the stadia coefficient is equal to 100)
r=the rod interval or intercept (the difference between the upper and lower cross-hair reading)
α=the vertical angle measured by the instrument

$\Rightarrow$

$$\Delta H = \frac{1}{2}(100)(1.650-1.350)\sin(2\times30°) = 12.990 \text{ m}$$

The reduced level (elevation) of point B can be determined using the following formula:

$$\text{RL}(B) = \text{RL}(A) + \text{HI} + \Delta H - M \tag{5.27}$$

Where:
RL(B) and RL(A)=the reduced levels (elevations) of points B and A, respectively
HI=the height of instrument
ΔH=the vertical distance (shown in Figure 5.23)
M=the middle cross-hair reading

$\Rightarrow$

$$\text{RL}(B) = 310.244 + 1.625 + 12.990 - 1.450 = 323.409 \text{ m}$$

Figure 5.24 shows the MS Excel worksheet used to determine the elevations by the stadia method in this problem.

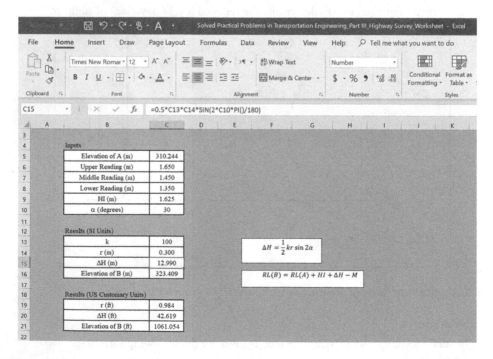

FIGURE 5.24 Image of MS Excel worksheet used for determining the elevations by the stadia method for Problem 5.15.

6 Angle Measurement and Traverse

This chapter deals with a major type of highway survey called a traverse survey. The traverse survey is used to determine the relative positioning of survey points from a known position. Traverses are classified into two main types: closed traverses and open traverses. This section presents problems on traverse computations, including azimuth determination and coordinates calculation. In addition, azimuth closure and position closure independent checks on the accuracy of field measurements are performed for closed traverse computations. Traverse computations include the following major steps: (1) the angle measurements are corrected for closure error in azimuth, (2) the preliminary coordinates of the traverse stations are computed, (3) the position closure error at the last point of the traverse is computed, (4) corrections to the preliminary coordinates are applied to compensate for the position closure error, and finally (5) the azimuth (or bearing) and length of each course of the traverse using the final coordinates are computed.

6.1   Given the lengths and the reduced true bearings for the sides of the traverse shown in Figure 6.1, compute the position closure errors in X and Y, the linear closure error (ε_d), and determine the relative precision (relative closure error) for the traverse.

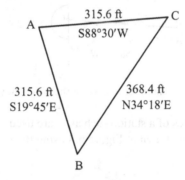

FIGURE 6.1   A closed traverse for Problem 6.1.

Solution:

The azimuth angle of each side of the three sides of the traverse is first determined. The azimuth angle is defined as the angle from the north at the beginning point of the side in the clockwise direction to the end point of the side. It is also called the whole circle bearing (WCB) as illustrated in Figure 6.2:

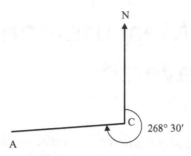

FIGURE 6.2   Azimuth for line *AC* of the closed traverse in Problem 6.1.

In this diagram, the azimuth angle (WCB) of the side (line) *CA* is equal to the angle from the north direction at point *C* of line *CA* rotated in the clockwise direction to point *A* of line *CA*. Since the reduced bearing (RB) of line *CA* is S 88° 30′ W, the azimuth is equal to the RB plus 180° (see Figure 6.3). In a similar manner, the azimuth is calculated for each side of the traverse.

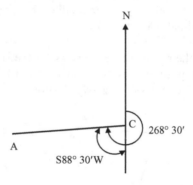

FIGURE 6.3   Determination of the azimuth for line *AC* of the closed traverse in Problem 6.1.

The known coordinates of a station such as A are used to compute the coordinates of another station such as B shown in Figure 6.4 using the following equations:

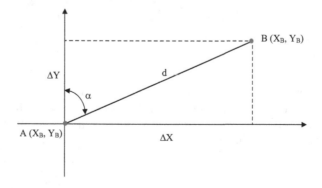

FIGURE 6.4   Illustration for ΔX and ΔY between points *A* and *B* for Problem 6.1.

$$X_B = X_A + \Delta X \qquad (6.1)$$

Where:

X_A and X_B = the x-coordinates of points A and B, respectively

ΔX = the difference in the x-coordinate between points A and B

$$Y_B = Y_A + \Delta Y \tag{6.2}$$

Where:

Y_A and Y_B = the Y-coordinates of points A and B, respectively

ΔY = the difference in the y-coordinate between points A and B

$$\Delta X = d \sin \alpha \tag{6.3}$$

$$\Delta Y = d \cos \alpha \tag{6.4}$$

Where:

d = the length of the side of the traverse

α = the azimuth angle of the line

Sample Calculation:

For line (side) AB:

The azimuth angle = α = 160° 15′ = 160.25°

$$\Delta X = d \sin \alpha \tag{6.5}$$

$\Rightarrow$

$$\Delta X = (315.6)\sin(160.25°) = 106.65 \ ft$$

$$\Delta Y = d \cos \alpha \tag{6.6}$$

$\Rightarrow$

$$\Delta Y = (315.6)\cos(160.25°) = -297.04 \ \text{ft}$$

In a similar manner, the azimuth angle, ΔX and ΔY are computed for the other sides of the traverse. The results are summarized in Table 6.1.

TABLE 6.1

Azimuth Angles, ΔX, and ΔY for the Sides of the Closed Traverse in Problem 6.1

Point	Side	Reduced Bearing	Azimuth (WCB)	Azimuth (°)	Length (ft)	ΔX (ft)	ΔY (ft)
A							
B	AB	S 19° 45′ E	160° 15′	160.30	315.6	106.65	−297.04
C	BC	N 34° 18′ E	34° 18′	34.30	368.4	207.60	304.33
A	CA	S 88° 30′ W	268° 30′	268.50	315.6	−315.49	−8.26
		Sum			999.6	−1.24	−0.96

The position closure errors in X and Y are computed using the following two formulas, respectively:

$$\varepsilon_X = \sum \Delta X \tag{6.7}$$

Where:
   ε_X = the position closure error in X
   ΔX = the difference in the x-coordinate between successive points of the traverse

$\Rightarrow$

$$\varepsilon_X = 106.65 + 207.60 + (-315.49) = -1.24 \text{ ft}$$

$$\varepsilon_Y = \sum \Delta Y \tag{6.8}$$

Where:
   ε_Y = the position closure error in Y
   ΔX = the difference in the x-coordinate between successive points of the traverse

$\Rightarrow$

$$\varepsilon_Y = -297.04 + 304.33 + (-8.26) = -0.96 \text{ ft}$$

The linear closure error (ε_d) is determined using the formula below:

$$\varepsilon_d = \sqrt{\left(\varepsilon_X\right)^2 + \left(\varepsilon_Y\right)^2} \tag{6.9}$$

Where:
   ε_d = the linear closure error
   ε_X = the position closure error in X
   ε_Y = the position closure error in Y

$\Rightarrow$

$$\varepsilon_d = \sqrt{\left(-1.50\right)^2 + \left(-1.05\right)^2} = 1.57 \text{ ft}$$

The relative precision (relative closure error) for the traverse is determined using the following formula:

$$\varepsilon_r = \frac{1}{\left(\dfrac{D}{\varepsilon_d}\right)} \tag{6.10}$$

Where:
   ε_r = the relative closure error
   D = the summation of all the lengths of the sides of the traverse
   ε_d = the linear closure error

$$\varepsilon_r = \frac{1}{\left(\dfrac{999.6}{1.57}\right)} \cong 1/636$$

Figure 6.5 shows the MS Excel worksheet used for the closed traverse computations of this problem.

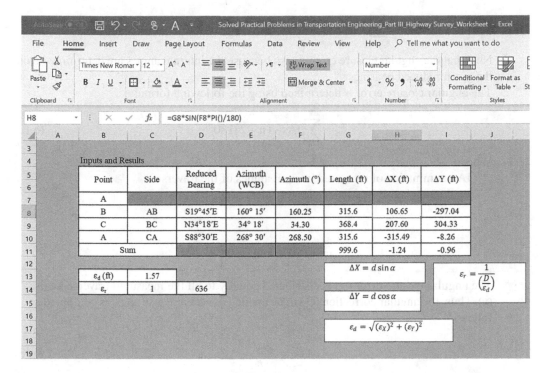

FIGURE 6.5 Image of MS Excel worksheet used for the closed traverse computations of Problem 6.1.

6.2   Using Problem 6.1 above, determine the internal angles of the traverse at *A*, *B*, and *C*, compute the angular closure error (ε_a), determine the corrected azimuth for each side, correct for the position closure errors in *X* and *Y* using the compass rule, and compute the corrected coordinates of the traverse points *A*, *B*, and *C* if the coordinates (ft) of point *A* are (200, 300).

Solution:

The summation of internal angles of any traverse is given by the following formula:

$$IA = (n-2)180° \qquad (6.11)$$

Where:
   IA = summation of internal angles of the traverse
   n = the number of internal angles (or sides) of the traverse

⇒

$$IA = (3-2)180° = 180°$$

The summation of the measured traverse internal angles is calculated, and the angular closure error is computed accordingly, as shown in the procedure below:

$$\text{The Internal Angle at } A = 180° - (88°30' + 19°45') = 71°45'$$

$$\text{The Internal Angle at } B = 19°45' + 34°18' = 54°03'$$

$$\text{The Internal Angle at } C = 88°30' - 34°18' = 54°12'$$

$$\sum \text{IAs} = 71°45' + 54°03' + 54°12' = 180°$$

Note that the angular closure error is zero because the internal angles are determined from the given reduced bearings.

The angular closure error is determined using the following formula:

$$\varepsilon_a = \sum \text{IAs} - (n-2)180° \tag{6.12}$$

Where:
   IA = summation of measured internal angles of the traverse
   n = the number of internal angles (or sides) of the traverse

$\Rightarrow$

$$\varepsilon_a = 180° - 180° = 0°$$

The angular closure error (ε_a) is distributed equally to all the measured traverse angles (n). Thus, the angular correction (C_a) to be applied to each traverse angle is computed as follows:

$$C_a = -\frac{\varepsilon_a}{n} \tag{6.13}$$

Where:
   C_a = the angular correction
   ε_a = the angular closure error
   n = the number of internal angles (or sides) of the traverse

$\Rightarrow$

$$C_a = -\frac{0}{3} = 0$$

Therefore, there is no angular correction since the angular closure error is zero. The results are summarized in Table 6.2.

TABLE 6.2

Corrected Internal Angles for the Closed Traverse in Problem 6.2

Point	Side	Reduced Bearing	Internal Angle	Corrected Internal Angle
A			71° 45′	71° 45′
B	AB	S 19° 45′ E	54° 03′	54° 03′
C	BC	N 34° 18′ E	54° 12′	54° 12′
A	CA	S 88° 30′ E		
	Sum		180°	180°

The position closure error in X (ε_X) and the position closure error in Y (ε_Y) are determined in Problem 6.1 earlier. Based on the position closure errors, a correction will be applied to both errors, as shown below:

$$\text{Correction in } \Delta X = -\varepsilon_X \frac{d_i}{\sum d_i} \quad (6.14)$$

Where:

ε_X = the position closure error in X

d_i = the length of the side i of the traverse

$\sum d_i$ = the summation of the lengths of all sides of the traverse

$$\text{Correction in } \Delta Y = -\varepsilon_Y \frac{d_i}{\sum d_i} \quad (6.15)$$

Where:

ε_Y = the position closure error in Y

d_i = the length of the side i of the traverse

$\sum d_i$ = the summation of the lengths of all sides of the traverse

Sample Calculation:

For side AB:

$$\varepsilon_X = -1.24 \text{ ft}$$

$$\text{Correction in } \Delta X = -\varepsilon_X \frac{d_i}{\sum d_i}$$

$\Rightarrow$

$$\text{Correction in } \Delta X = -(-1.24)\frac{315.6}{999.6} = 0.39 \text{ ft}$$

$$\text{Correction in } \Delta Y = -\varepsilon_Y \frac{d_i}{\sum d_i}$$

$\Rightarrow$

$$\text{Correction in } \Delta Y = -(-0.96)\frac{315.6}{999.6} = 0.30 \text{ ft}$$

The X and Y coordinates are determined using the following formulas:

$$X_B = X_A + \Delta X + \text{Correction in } \Delta X \quad (6.16)$$

$$Y_B = Y_A + \Delta Y + \text{Correction in } \Delta Y \quad (6.17)$$

$\Rightarrow$

$$X_B = 200.00 + 106.65 + 0.39 = 307.04 \text{ ft}$$

$$Y_B = 300.00 + (-297.04) + 0.30 = 3.27 \text{ ft}$$

Similarly, the corrections for the other sides of the traverse and the coordinates of the other points are computed. The results are summarized in Table 6.3.

TABLE 6.3
Corrected Coordinates of the Sides of the Closed Traverse in Problem 6.2

Point	Side	ΔX (ft)	ΔY (ft)	Correction in ΔX (ft)	Correction in ΔY (ft)	Corrected Coordinates	
						X (ft)	Y (ft)
A						200.00	300.00
B	AB	106.65	−297.04	0.39	0.30	307.04	3.27
C	BC	207.60	304.33	0.46	0.35	515.10	307.96
A	CA	−315.49	−8.26	0.39	0.30	200.00	300.00
Sum		−1.24	−0.96	1.24	0.96		

Figure 6.6 shows the MS Excel worksheet used for determining the corrected coordinates of the closed traverse in this problem.

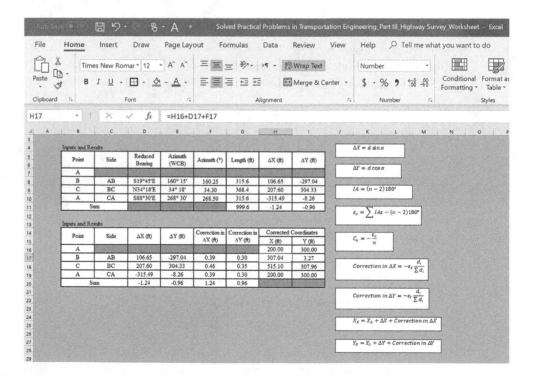

FIGURE 6.6   Image of MS Excel worksheet used for determining the corrected coordinates of the closed traverse in Problem 6.2.

6.3   For the closed loop traverse shown in Figure 6.7, if the lengths and the reduced true bearings (corrected) for the sides of the traverse are as shown in the figure, compute the position closure errors in X and Y, correct for the position closure error using the compass rule, and compute the coordinates of the traverse points A, B, C, and D if the coordinates (ft) of point A are (220, 410).

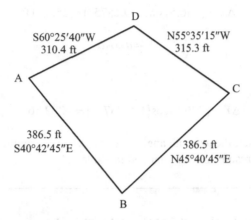

FIGURE 6.7   A four-sided closed traverse for Problem 6.3.

Solution:

The azimuth angle of each side of the four sides of the traverse is determined. In Figure 6.8, the azimuth angle (WCB) of the side (line) DA is equal to the angle from the north direction at point D of line DA rotated in the clockwise direction to point A of line DA. Since the reduced bearing (RB) of line DA is S 60° 25′ 40″ W, the azimuth is equal to the RB plus 180°. In other words, the azimuth angle of line DA is 240° 25′ 40″. In a similar manner, the azimuth is calculated for each side of the traverse.

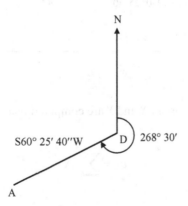

FIGURE 6.8   The azimuth of line AD of the closed traverse in Problem 6.3.

Sample Calculation:
For line (side) AB:

$$\text{The azimuth Angle of } AB = 180° - (40° 42′ 45″) = 139° 17′ 15″$$

$$\text{The azimuth Angle of } AB = \alpha = 139.2875°$$

$$\Delta X = d \sin \alpha$$

$\Rightarrow$

$$\Delta X = (386.5) \sin (139.2875°) = 252.10 \text{ ft}$$

$$\Delta Y = d \cos \alpha$$

$\Rightarrow$

$$\Delta Y = (386.5) \cos (139.2875°) = -292.96 \text{ ft}$$

In a similar manner, the azimuth angle, ΔX and ΔY are computed for the other sides of the traverse. The results are summarized in Table 6.4.

TABLE 6.4

Azimuth Angles, ΔX, and ΔY for the Sides of the Closed Traverse in Problem 6.3

Point	Side	Reduced Bearing	Azimuth (WCB)	Azimuth (°)	Length (ft)	ΔX (ft)	ΔY (ft)
A							
B	AB	S 40° 42′ 45″ E	139° 17′ 15″	139.29	386.5	252.10	−292.96
C	BC	N 45° 40′ 45″ E	45° 40′ 45″	45.68	386.5	276.52	270.04
D	CD	N 55° 35′ 15″ W	304° 24′ 45″	304.41	315.3	−260.12	178.19
A	DA	S 60° 25′ 40″ W	240° 25′ 40″	240.43	310.4	−269.97	−153.19
		Sum			1398.7	−1.47	2.08

The position closure errors in X and Y are computed using the following two formulas, respectively:

$$\varepsilon_X = \sum \Delta X$$

$\Rightarrow$

$$\varepsilon_X = 252.10 + 276.52 + (-260.12) + (-269.97) = -1.47 \text{ ft}$$

$$\varepsilon_Y = \sum \Delta Y$$

$\Rightarrow$

$$\varepsilon_Y = -292.96 + 270.04 + 178.19 + (-153.19) = 2.08 \text{ ft}$$

The position closure error in X (ε_X) and the position closure error in Y (ε_Y) are determined using the following formulas. Based on the position closure errors, a correction will be applied to both errors, as shown below:

$$\text{Correction in } \Delta X = -\varepsilon_X \frac{d_i}{\sum d_i}$$

$$\text{Correction in } \Delta Y = -\varepsilon_Y \frac{d_i}{\sum d_i}$$

Sample Calculation:
For side AB:

$$\varepsilon_X = -1.47 \text{ ft}$$

$$\text{Correction in } \Delta X = -\varepsilon_X \frac{d_i}{\sum d_i}$$

$\Rightarrow$

$$\text{Correction in } \Delta X = -(-1.47)\frac{386.5}{1398.7} = 0.41 \text{ ft}$$

$$\text{Correction in } \Delta Y = -\varepsilon_Y \frac{d_i}{\sum d_i}$$

$\Rightarrow$

$$\text{Correction in } \Delta Y = -2.08\frac{386.5}{1398.7} = -0.57 \text{ ft}$$

The X and Y coordinates are determined using the following formulas:

$$X_B = X_A + \Delta X + \text{Correction in } \Delta X$$

$$Y_B = Y_A + \Delta Y + \text{Correction in } \Delta Y$$

$\Rightarrow$

$$X_B = 220.00 + 252.10 + 0.41 = 472.51 \text{ ft}$$

$$Y_B = 410.00 + (-292.96) + (-0.57) = 116.47 \text{ ft}$$

Similarly, the corrections for the other sides of the traverse and the coordinates of the other points are computed. The results are summarized in Table 6.5.

TABLE 6.5
Corrected Internal Angles for the Closed Traverse in Problem 6.3

Point	Side	ΔX (ft)	ΔY (ft)	Correction in ΔX (ft)	Correction in ΔY (ft)	Corrected Coordinates X (ft)	Y (ft)
A						220.00	410.00
B	AB	252.1	−293.0	0.41	−0.57	472.50	116.46
C	BC	276.5	270.0	0.41	−0.57	749.43	385.92
D	CD	−260.1	178.2	0.33	−0.47	489.64	563.64
A	DA	−270.0	−153.2	0.33	−0.46	220.00	410.00
Sum		−1.47	2.08	1.47	−2.08		

Figure 6.9 shows the MS Excel worksheet used for the closed traverse computations of this problem.

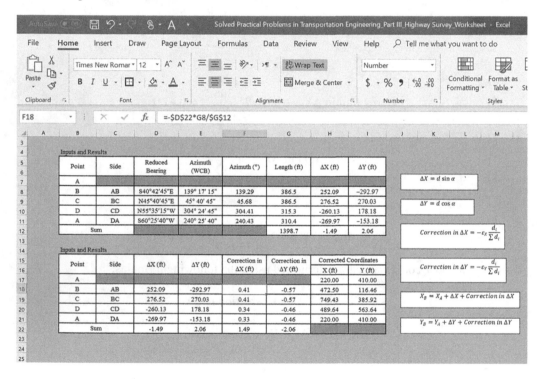

FIGURE 6.9   Image of MS Excel worksheet used for the closed traverse computations of Problem 6.3.

6.4　Compute the azimuth of all the lines (legs) of the hexagon (six-sided) traverse shown in Figure 6.10.

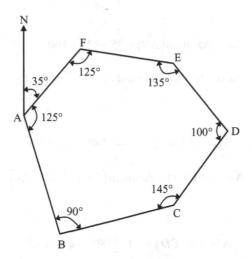

FIGURE 6.10　A six-sided closed traverse for Problem 6.4.

Solution:

The azimuth angle of each side of the six sides of the traverse is determined. In Figure 6.11, the azimuth angle (WCB) of the side (line) *FA* is equal to the angle from the north direction at point *F* of line *FA* rotated in the clockwise direction to point *A* of line *FA*. Therefore:

$$\text{Azimuth}(FA) = 180 + 35 = 215°$$

Similarly, the azimuth of the other legs (sides) of the traverse is determined as described in the following procedure:

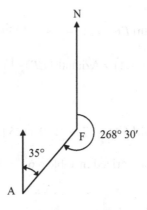

FIGURE 6.11　The azimuth of line *AF* of the closed traverse in Problem 6.4.

$$\text{Azimuth}\left(AB\right) = \text{Azimuth}\left(AF\right) + I_A$$

$\Rightarrow$

$$\text{Azimuth}\left(AB\right) = 35 + 125 = 160°$$

$$\text{Azimuth}\left(BC\right) = \text{Azimuth}\left(AB\right) - \left[180 - I_B\right]$$

$\Rightarrow$

$$\text{Azimuth}\left(BC\right) = 160 - \left[180 - 90\right] = 70°$$

$$\text{Azimuth}\left(CD\right) = \text{Azimuth}\left(BC\right) - \left[180 - I_C\right]$$

$\Rightarrow$

$$\text{Azimuth}\left(CD\right) = 70 - \left[180 - 145\right] = 35°$$

$$\text{Azimuth}\left(DE\right) = \text{Azimuth}\left(CD\right) + 180 + I_D$$

Or:

$$\text{Azimuth}\left(DE\right) = 360 - \left[180 - \left(\text{Azimuth}\left(CD\right) + I_D\right)\right]$$

$\Rightarrow$

$$\text{Azimuth}\left(DE\right) = 35 + 180 + 100 = 315°$$

$$\text{Azimuth}\left(EF\right) = \text{Azimuth}\left(DE\right) - \left[180 - I_E\right]$$

$\Rightarrow$

$$\text{Azimuth}\left(EF\right) = 315 - \left[180 - 135\right] = 270°$$

$$\text{Azimuth}\left(FA\right) = \text{Azimuth}\left(EF\right) - \left[180 - I_F\right]$$

$\Rightarrow$

$$\text{Azimuth}\left(FA\right) = 270 - \left[180 - 125\right] = 215°$$

The azimuth results are summarized in Table 6.6.

TABLE 6.6
The Azimuth Angles of the Sides of the Closed Traverse in Problem 6.4

Point	Side	Internal Angle (°)	Azimuth (°)
A		125	
B	AB	90	160
C	BC	145	70
D	CD	100	35
E	DE	135	315
F	EF	125	270
A	FA		215

Figure 6.12 shows the MS Excel worksheet used to determine the azimuth of the sides of the closed traverse in this problem.

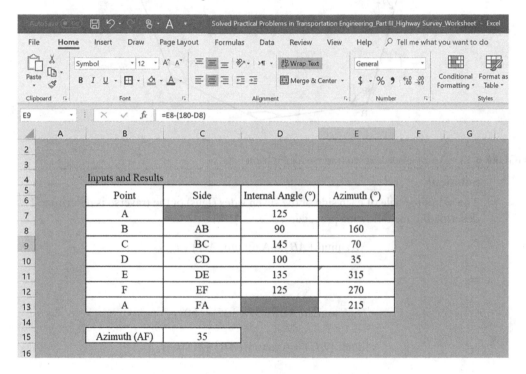

FIGURE 6.12   Image of MS Excel worksheet used for determining the azimuth of the sides of the closed traverse in Problem 6.4.

6.5   Compute the azimuth of all the lines (legs) of the seven-sided closed traverse shown in Figure 6.13.

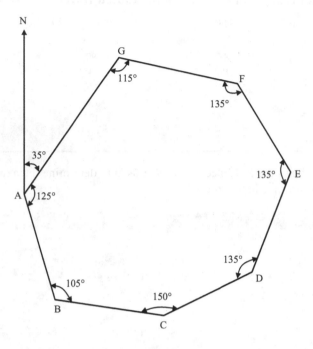

FIGURE 6.13   A seven-sided closed traverse for Problem 6.5.

Solution:

The azimuth angle of each side of the seven sides (legs) of the traverse is determined as described in the following procedure:

$$\text{Azimuth}(AB) = \text{Azimuth}(AG) + I_A$$

$\Rightarrow$

$$\text{Azimuth}(AB) = 35 + 125 = 160°$$

$$\text{Azimuth}(BC) = \text{Azimuth}(AB) - \left[180 - I_B\right]$$

$\Rightarrow$

$$\text{Azimuth}(BC) = 160 - \left[180 - 105\right] = 85°$$

$$\text{Azimuth}(CD) = \text{Azimuth}(BC) - \left[180 - I_C\right]$$

$\Rightarrow$

$$\text{Azimuth}(CD) = 85 - \left[180 - 150\right] = 55°$$

$$\text{Azimuth}(DE) = \text{Azimuth}(CD) - \left[180 - I_D\right]$$

⇒

$$\text{Azimuth}(DE) = 55 - \left[180 - 135\right] = 10°$$

$$\text{Azimuth}(EF) = \text{Azimuth}(DE) + 180 + I_E$$

⇒

$$\text{Azimuth}(EF) = 10 + 180 + 135 = 325°$$

$$\text{Azimuth}(FG) = \text{Azimuth}(EF) - \left[180 - I_F\right]$$

⇒

$$\text{Azimuth}(FG) = 325 - \left[180 - 135\right] = 280°$$

$$\text{Azimuth}(GA) = \text{Azimuth}(FG) - \left[180 - I_G\right]$$

⇒

$$\text{Azimuth}(GA) = 280 - \left[180 - 115\right] = 215°$$

The azimuth results are summarized in Table 6.7.

TABLE 6.7

The Azimuth Angles of the Sides of the Closed Traverse in Problem 6.5

Point	Side	Internal Angle (°)	Azimuth (°)
A		125	
B	AB	105	160
C	BC	150	85
D	CD	135	55
E	DE	135	10
F	EF	135	325
G	FG	115	280
A	GA		215

Figure 6.14 shows the MS Excel worksheet used to determine the azimuth of the sides of the closed traverse in this problem.

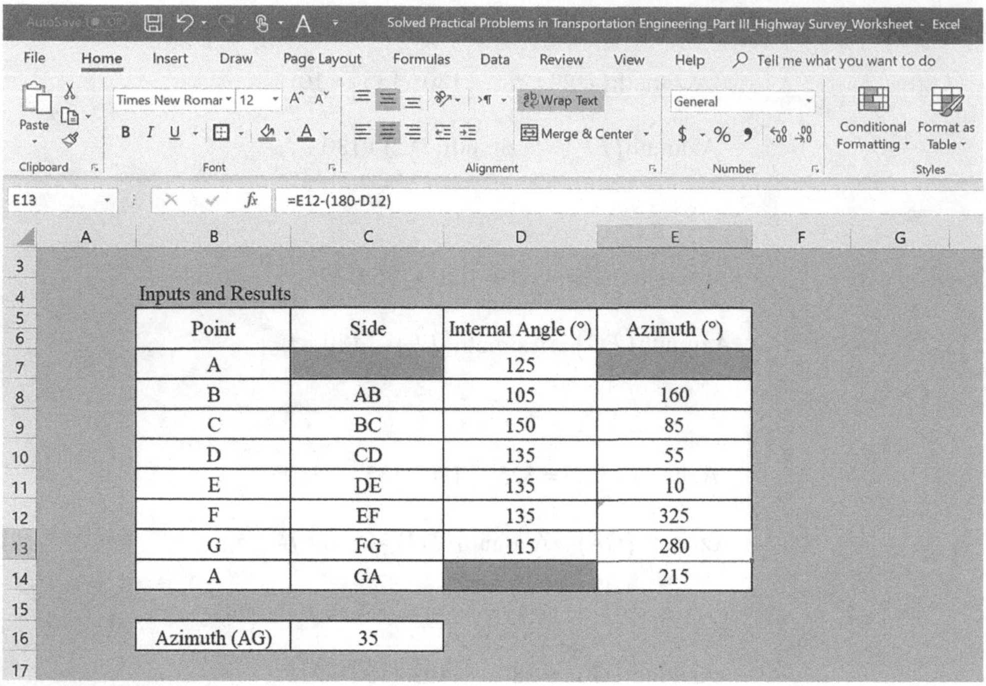

FIGURE 6.14   Image of MS Excel worksheet used for determining the azimuth of the sides of the closed traverse in Problem 6.5.

6.6   If the true bearings (reduced bearings) are given for the lines *AB* and *DA* of the parallelogram shown in Figure 6.15, determine the reduced bearings of the lines *BC* and *CD* and the back bearings, azimuths, and back azimuths of all lines.

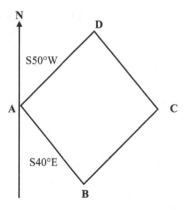

FIGURE 6.15   A parallelogram for Problem 6.6.

Solution:
The reduced bearings of the lines *BC* and *CD* are calculated below:
Since the line *BC* is parallel to the line *AD*, the angle between the north direction and the line *BC* is equal to the angle between the north direction and the line *AD*, which is equal to 50°. Hence, the reduced bearing of the line *BC* is N 50° E. Similarly, the line *DC* is also parallel to the line *AB*, the angle between the south direction and the line *DC* is equal to the angle between the south direction and the line *AB*, that is 40°. This is the

same angle between the north direction and the line *CD*. In other words, the reduced bearing of the line *CD* is N 40° W. The results are summarized in Table 6.8.

TABLE 6.8

Calculated Reduced Bearings for the Sides of the Parallelogram in Problem 6.6

Line	Reduced Bearing
AB	S 40° E
BC	N 50° E
CD	N 40° W
DA	S 50° W

The back bearing of any line starting from point *A* and ending at point *B* is the reduced bearing of the same line but starting from *B* and ending at *A*. Therefore, the back bearing of line *BC* is the bearing of line *CB*, and so on. The azimuth of any line is the angle from the north direction at the beginning point rotating clockwise to the end point of the line. In other words, the azimuth of line *AB* is the angle from the north direction at point *A* rotating clockwise to point *B*. On the other hand, the back azimuth of any line starting at point *A* and ending at point *B* is the azimuth of the same line but starting at point *B* and ending at point *A*. Based on these definitions, the back bearing, the azimuth, and the back azimuth of the required lines (sides) of the parallelogram (four-sided traverse) shown in the diagram above (Figure 6.15) are shown in Table 6.9.

TABLE 6.9

Calculated Back Bearings, Azimuth Angles, and Back Azimuth Angles for the Sides of the Parallelogram in Problem 6.6

Line	Back Bearing	Azimuth	Back Azimuth
AB	N 40° W	140°	320°
BC	S 50° W	50°	230°
CD	S 40° E	320°	140°
DA	N 50° E	230°	50°

6.7 For the closed loop traverse (*A–B–C–D–E–A*) shown in Figure 6.16:

(1) Determine the angular closure error and correct the measured angle.
(2) Compute the corrected azimuths of the sides of the traverse.
(3) Compute the position closure error in X (ε_X) and the position closure error in Y (ε_Y).
(4) Balance the traverse using the compass rule and compute the corrected coordinates of the traverse stations.

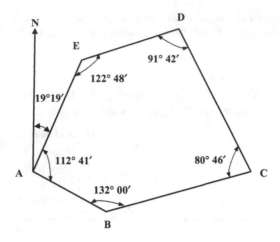

FIGURE 6.16   A five-sided closed traverse for problem 6.7.

The lengths of the sides of the traverse are given in Table 6.10.

TABLE 6.10
Given Lengths for the Sides of the
Closed Traverse in Problem 6.7

Side	Length (ft)
AB	219.78
BC	417.26
CD	551.40
DE	318.25
EA	374.63

Point	X_A (ft)	Y_A (ft)
A	1200.5	1100.4

Solution:

(1) The angular closure error and the corrected measured angles:

The angular closure error is determined using the following formula:

$$\varepsilon_a = \sum \text{IAs} - (n-2)180°$$

$\Rightarrow$

$$\varepsilon_a = \left(112°\,41' + 132°\,00' + 80°\,46' + 91°\,42' + 122°\,48'\right)$$
$$- \left[(5-2)180°\right]$$

$\Rightarrow$

$$\varepsilon_a = \left(539°\,57'\right) - 540° = -03'$$

The angular closure error (ε_a) is distributed equally to all the measured traverse angles (*n*). Thus, the angular correction (C_a) to be applied to each traverse angle is computed as follows:

$$C_a = -\frac{\varepsilon_a}{n}$$

$\Rightarrow$

$$C_a = -\frac{(-3)}{5} = 0.6' = 36''$$

Therefore, 36″ will be added to each internal angle to correct for the angular closure error determined above. The results are shown in Tables 6.11 and 6.12.

TABLE 6.11

Angular Error and Correction for the Closed Traverse in Problem 6.7

Angular Error	−3 minutes = −180 seconds
Angular Correction	36 seconds

TABLE 6.12

Corrected Internal Angles for the Closed Traverse in Problem 6.7

Angle	Corrected Value (° ′ ″)
A	= 112° 41′ 36″
B	= 132° 00′ 36″
C	80.78 = 80° 46′ 36″
D	91.71 = 91° 42′ 36″
E	122.81 = 122° 48′ 36″

(2) The corrected azimuths of the sides of the traverse:

The corrected azimuth of each line (leg) of the traverse is computed based on the corrected internal angle determined above. The results are summarized in Table 6.13:

TABLE 6.13

Corrected Azimuth Angles for the Closed Traverse in Problem 6.7

Side	Corrected Azimuth (° ′ ″)
AB	132.01 = 132° 00′ 36″
BC	84.02 = 84° 01′ 12″
CD	344.80 = 344° 47′ 48″
DE	256.51 = 256° 30′ 24″
EA	199.32 = 199° 19′ 00″

(3) The position closure error in X (ε_X) and the position closure error in Y (ε_Y):

The position closure error in X (ε_X) and the position closure error in Y (ε_Y) are computed as described in the procedure below:

$$\Delta X = d \sin \alpha$$

Sample Calculation:
For line AB:

⇒

$$\Delta X = (219.78)\sin(132.01°) = 163.30 \text{ ft}$$

$$\Delta Y = d \cos \alpha$$

⇒

$$\Delta Y = (219.78)\cos(132.01°) = -147.09 \text{ ft}$$

In a similar manner, the azimuth angle, ΔX and ΔY are computed for the other sides of the traverse. The results are summarized in Table 6.14.

TABLE 6.14

X and Y Coordinates of the Sides of the Closed Traverse in Problem 6.7

Station	Side	Corrected Azimuth (WCB)	Length (ft)	ΔX (ft)	ΔY (ft)	Coordinates X (ft)	Coordinates Y (ft)
A						1200.50	1100.40
B	AB	132° 00′ 36″	219.78	163.30	−147.09	1363.80	953.31
C	BC	84° 01′ 12″	417.26	414.99	43.47	1778.79	996.78
D	CD	344° 47′ 48″	551.40	−144.60	532.10	1634.19	1528.88
E	DE	256° 30′ 24″	318.25	−309.47	−74.26	1324.72	1454.62
A	EA	199° 19′ 00″	374.63	−123.92	−353.54	1200.80	1101.08
	Sum		1881.32	0.30	0.68		

$$\varepsilon_X = \sum \Delta X$$

⇒

$$\varepsilon_X = 163.30 + 414.99 + (-144.60) + (-309.47) + (-123.92)$$

$$= 0.30 \text{ ft}$$

$$\varepsilon_Y = \sum \Delta Y$$

⇒

$$\varepsilon_Y = -147.09 + 43.47 + 532.10 + (-74.26) + (-353.54)$$

$$= 0.68 \text{ ft}$$

(4) Balancing the traverse using the *compass rule* and computing the corrected coordinates of the traverse stations:

Based on the position closure errors (ε_X and ε_Y), a correction will be applied to both errors, as shown below:

$$\text{Correction in } \Delta X = -\varepsilon_X \frac{d_i}{\sum d_i}$$

$$\text{Correction in } \Delta Y = -\varepsilon_Y \frac{d_i}{\sum d_i}$$

Sample Calculation:
For line AB:

$$\varepsilon_X = 0.30 \text{ ft}$$

$$\text{Correction in } \Delta X = -\varepsilon_X \frac{d_i}{\sum d_i}$$

$\Rightarrow$

$$\text{Correction in } \Delta X = -0.30 \frac{219.78}{1881.32} = -0.04 \text{ ft}$$

$$\text{Correction in } \Delta Y = -\varepsilon_Y \frac{d_i}{\sum d_i}$$

$\Rightarrow$

$$\text{Correction in } \Delta Y = -0.68 \frac{219.78}{1881.32} = -0.08 \text{ ft}$$

The X and Y coordinates are determined using the following formulas:

$$X_B = X_A + \Delta X + \text{Correction in } \Delta X$$

$$Y_B = Y_A + \Delta Y + \text{Correction in } \Delta Y$$

$\Rightarrow$

$$X_B = 1200.50 + 163.30 + (-0.04) = 1363.76 \text{ ft}$$

$$Y_B = 1100.40 + (-147.09) + (-0.08) = 953.23 \text{ ft}$$

Similarly, the corrections for the other sides of the traverse and the coordinates of the other points are computed. The results are summarized in Table 6.15.

TABLE 6.15
Corrected Coordinates of the Sides of the Closed Traverse in Problem 6.7

Station	Side	ΔX (ft)	ΔY (ft)	Correction in ΔX (ft)	Correction in ΔY (ft)	Corrected Coordinates X (ft)	Corrected Coordinates Y (ft)
A						1200.50	1100.40
B	AB	163.30	−147.09	−0.04	−0.08	1363.77	953.23
C	BC	414.99	43.47	−0.07	−0.15	1778.69	996.55
D	CD	−144.60	532.10	−0.09	−0.20	1634.00	1528.45
E	DE	−309.47	−74.26	−0.05	−0.12	1324.48	1454.08
A	EA	−123.92	−353.54	−0.06	−0.14	1200.50	1100.40
Sum		0.30	0.68	−0.30	−0.68		

Figure 6.17 shows the MS Excel worksheet used for the closed traverse computations of this problem.

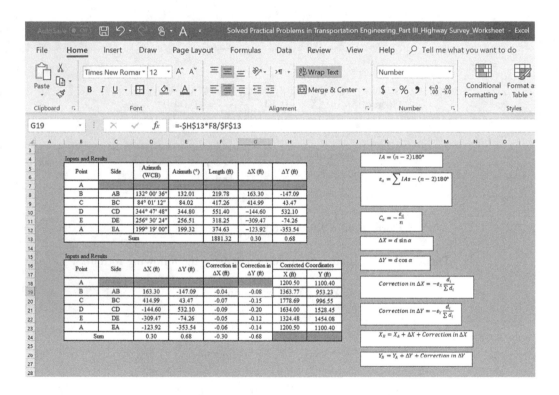

FIGURE 6.17 Image of MS Excel worksheet used for the closed traverse computations of Problem 6.7.

6.8   A highway segment taking the form of a connecting open traverse (A–B–C–D–E) is shown in Figure 6.18. The coordinates of points A and E, and the lengths of all the sides (legs) of the traverse are given in Tables 6.16 and 6.17, respectively. The known azimuth of the segment $A'A$ is $109° 45'$ and the true (reduced) bearing of side DE is also given as N $59° 08'$ W. Based on this data:

(1) Determine the azimuth of the sides of the traverse.
(2) Determine the azimuth error and correction.
(3) Determine the corrected azimuth of all sides.
(4) Determine the coordinates of all points of the traverse.
(5) Compute the position closure error in X (ε_X) and the position closure error in Y (ε_Y).
(6) Determine the corrected coordinates of all points of the traverse based on the corrections in ΔX and ΔY.

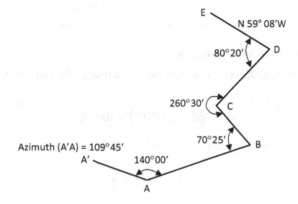

FIGURE 6.18   A connecting open traverse for Problem 6.8.

TABLE 6.16
Given Coordinates for Points A and E
of the Open Traverse in Problem 6.8

Point	Coordinates (m)
A	(820, 460)
E	(1105, 1252)

TABLE 6.17
Given Lengths of the Sides of the
Open Traverse in Problem 6.8

Side	Length (m)
AB	500
BC	240
CD	360
DE	310

Solution:

(1) Determine the azimuth of the sides of the traverse:

The azimuth of the sides (legs) of the traverse are computed based on the measured (given) angles and the azimuth of line $A'A$. The results are summarized in Table 6.18.

TABLE 6.18
Calculated Azimuth Angles of the Sides
of the Open Traverse in Problem 6.8

Side	Azimuth (° ′ ″)
AB	69° 45′ = 69.750°
BC	320° 10′ = 320.167°
CD	40° 40′ = 40.667°
DE	301° 00′ = 301.000°

(2) The azimuth error and correction:

The azimuth of line DE is determined using the true (reduced) bearing that is given, as shown below:

$$\alpha = 360° - (59°08') = 300°52'$$

The azimuth error is computed using the following formula:

$$\varepsilon_\alpha = \alpha' - \alpha \qquad (6.18)$$

Where:
ε_α = the closure error in azimuth
α = the known azimuth of the reference line at the end of the traverse
α' = the azimuth of the same reference line computed from the traverse angles

$\Rightarrow$

$$\varepsilon_\alpha = 301° - (300°52') = 08'$$

The angular correction is determined based on the azimuth closure error, as shown in the formula below (see Table 6.19).

$$C_\alpha = -\frac{\varepsilon_\alpha}{n} \qquad (6.19)$$

Where:
C_α = the angular (azimuth) correction
ε_α = the angular closure error (the azimuth closure error)
n = the number of internal angles (or sides) of the traverse

TABLE 6.19
Azimuth Error and Correction for the
Open Traverse in Problem 6.8

Azimuth Error	08′
Azimuth Correction	−08/4 = −02′

⇒

$$C_\alpha = -\frac{08}{4} = -02'$$

This correction can be applied to each angle of the traverse before computing the corrected azimuth angles of the different lines (legs) of the traverse. Alternately, the angular (azimuth) correction (C_α) can be applied directly to the computed azimuth of the traverse lines (legs) instead of applying the correction to the traverse angles. In this case, the corrected azimuth of line i can be computed as:

$$\alpha_i'' = \alpha_i' - i\left(\frac{\varepsilon_\alpha}{n}\right) \qquad (6.20)$$

Where:
α_i'' = the corrected azimuth of line i
α_i' = the computed azimuth of line i (before correction)
ε_α = the angular (azimuth) closure error
n = the number of traverse angles

The corrected azimuths of the sides of the traverse:
The corrected azimuth of each line (leg) of the traverse is computed based on the corrected internal angle determined above or using directly the formula shown above. The results are summarized in Table 6.20.

TABLE 6.20

Corrected Azimuth of the Sides of the Open Traverse in Problem 6.8

Side	Corrected Azimuth (° ′ ″)
AB	69° 43′ = 69.72°
BC	320° 06′ = 320.10°
CD	40° 34′ = 40.57°
DE	300° 52′ = 300.87°

(4) The coordinates of the points of the traverse:

The coordinates of the points of the traverse are computed using the corrected azimuth and the given length for each side (leg) of the traverse. A sample calculation is provided below:

Sample Calculation:
For line AB:

$$\Delta X = d\sin\alpha$$

⇒

$$\Delta X = (500)\sin(69.72°) = 469.00 \text{ m}$$

$$\Delta Y = d\cos\alpha$$

⇒

$$\Delta Y = (500)\cos(69.72°) = 173.30 \text{ m}$$

⇒

$$X_B = X_A + \Delta X$$

⇒

$$X_B = 820.00 + 469.00 = 1289.00 \text{ m}$$

$$Y_B = Y_A + \Delta Y$$

⇒

$$Y_B = 460.00 + 173.30 = 633.30 \text{ m}$$

Similarly, the coordinates of the other points of the traverse are computed. The results of all the points of the traverse are summarized in Table 6.21.

TABLE 6.21

X and Y Coordinates of the Sides of the Open Traverse in Problem 6.8

Station	Side	Corrected Azimuth (WCB)	Length (m)	ΔX (m)	ΔY (m)	Coordinates X (m)	Y (m)
A						820.00	460.00
B	AB	69° 43′	500	468.99	173.33	1288.99	633.33
C	BC	320° 06′	240	−153.95	184.12	1135.05	817.45
D	CD	40° 34′	360	234.12	273.47	1369.17	1090.92
E	DE	300° 52′	310	−266.09	159.04	1103.07	1249.97
	Sum		1410				

(5) The position closure error in X (ε_X) and the position closure error in Y (ε_Y):

The position closure error in X (ε_X) and the position closure error in Y (ε_Y) are computed based on the difference between the given coordinates and the computed coordinates for the last point of the traverse (point E). Hence, the following formulas are used to determine these two errors, respectively:

The azimuth error is computed using the following formula.

$$\varepsilon_X = X' - X \tag{6.21}$$

$$\varepsilon_Y = Y' - Y \tag{6.22}$$

Where:

    ε_X and ε_Y = the closure errors in X and Y, respectively

    X' and X = the computed and given x-coordinates of the last point of the traverse, respectively

    Y' and Y = the computed and given Y-coordinates of the last point of the traverse, respectively

$\Rightarrow$

$$\varepsilon_X = 1103.07 - 1105 = -1.93 \text{ m}$$

$$\varepsilon_Y = 1249.97 - 1252 = -2.03 \text{ m}$$

The azimuth closure errors in X and Y are shown in Table 6.22.

TABLE 6.22

X and Y Azimuth Errors of the Open Traverse in Problem 6.8.

ε_X (m)	−1.93
ε_Y (m)	−2.03

(6) Balancing the traverse using the compass rule and the corrected coordinates of the traverse stations:

Based on the position closure errors (ε_X and ε_Y), the corrections in ΔX and ΔY are computed using the following two formulas, respectively:

$$\text{Correction in } \Delta X = -\varepsilon_X \frac{d_i}{\sum d_i}$$

$$\text{Correction in } \Delta Y = -\varepsilon_Y \frac{d_i}{\sum d_i}$$

Sample Calculation:

For line AB:

$$\varepsilon_X = -1.93 \text{ m}$$

$$\varepsilon_Y = -2.03 \text{ m}$$

$$\text{Correction in } \Delta X = -\varepsilon_X \frac{d_i}{\sum d_i}$$

$\Rightarrow$

$$\text{Correction in } \Delta X = -(-1.93)\frac{500}{1410} = 0.68 \text{ m}$$

$$\text{Correction in } \Delta Y = -\varepsilon_Y \frac{d_i}{\sum d_i}$$

$\Rightarrow$

$$\text{Correction in } \Delta Y = -(-2.03)\frac{500}{1410} = 0.72 \text{ m}$$

The X and Y coordinates are determined using the following formulas:

$$X_B = X_A + \Delta X + \text{Correction in } \Delta X$$

$$Y_B = Y_A + \Delta Y + \text{Correction in } \Delta Y$$

$\Rightarrow$

$$X_B = 820.00 + 468.99 + 0.68 = 1289.67 \text{ m}$$

$$Y_B = 460.00 + 173.33 + 0.72 = 634.05 \text{ m}$$

Similarly, the corrections for the other side (legs) of the traverse and the coordinates of the other points are computed. The results are summarized in Table 6.23.

TABLE 6.23
Corrected Coordinates of the Sides of the Open Traverse in Problem 6.8

Station	Side	ΔX (m)	ΔY (m)	Correction in ΔX (m)	Correction in ΔY (m)	Corrected Coordinates X (m)	Corrected Coordinates Y (m)
A						820.00	460.00
B	AB	468.99	173.33	0.68	0.72	1289.68	634.05
C	BC	−153.95	184.12	0.33	0.35	1136.06	818.52
D	CD	234.12	273.47	0.49	0.52	1370.67	1092.51
E	DE	−266.09	159.04	0.42	0.45	1105.00	1252.00
Sum				1.93	2.03		

Figure 6.19 shows the MS Excel worksheet used for the connecting open traverse computations of this problem.

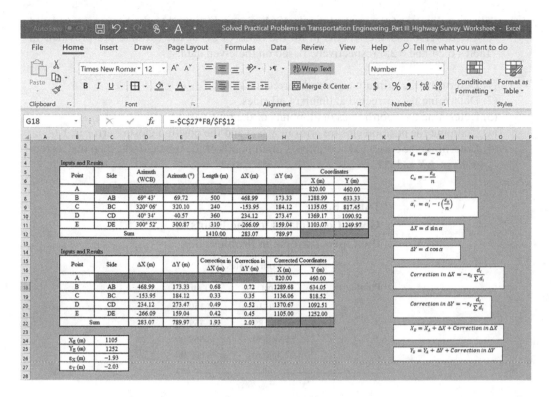

FIGURE 6.19 Image of MS Excel worksheet used for the connecting open traverse computations of Problem 6.8.

6.9   The highway segment shown in Figure 6.20 takes the form of a connecting open traverse *A–B–C–D–E–F.*

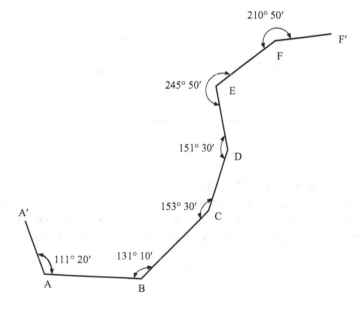

FIGURE 6.20 A connecting open traverse for Problem 6.9.

The coordinates of the control points A', A, F, and F' linked to the beginning and end of the traverse and the lengths of all the sides (legs) of the traverse are provided in Tables 6.24 and 6.25, respectively. Based on the given data:

(1) Determine the azimuth error and correction.
(2) Determine the corrected azimuth of all sides.
(3) Determine the coordinates of all points of the traverse.
(4) Compute the position closure error in X (ε_X) and the position closure error in Y (ε_Y).
(5) Determine the corrected coordinates of all points of the traverse based on the corrections in ΔX and ΔY.

TABLE 6.24
Given Coordinates for the Points of the Open Traverse in Problem 6.9

| Point | Coordinates (m) | |
	X (m)	Y (m)
A'	50,240.10	32,150.25
A	50,392.89	31,759.03
F	51,948.50	33,559.00
F'	52,358.50	33,663.00

TABLE 6.25
Given Lengths for the Sides of the Open Traverse in Problem 6.9

Side	Length (m)
$A'A$	420.00
AB	720.00
BC	720.00
CD	480.00
DE	480.00
EF	570.00
FF'	420.00

Solution:

(1) Determine the azimuth of the sides of the traverse:

The azimuth of the sides (legs) of the traverse are computed based on the measured (given) angles and the azimuth of line $A'A$. The azimuth of side AB is computed from the given coordinates as described in the procedure below (see Figure 6.21).

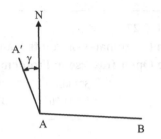

FIGURE 6.21 The azimuth of line AB of the open traverse in Problem 6.9.

The angle γ in the diagram above is determined as:

$$\tan\gamma = \frac{\Delta X_{A'-A}}{\Delta Y_{A'-A}} \tag{6.23}$$

$\Rightarrow$

$$\tan\gamma = \frac{(50390.89 - 50240.10)}{(32150.25 - 31759.03)} = 0.3854$$

$\Rightarrow$

$$\gamma = 21.078° = 21°05'$$

Therefore, the azimuth of line AB is computed as:

$$\text{Azimuth}(AB) = (111°20') - (21°05') = 90°15'$$

Based on this and following the same procedure described in previous problems, the azimuth of the other sides (legs) of the traverse are computed accordingly. The results are summarized in Table 6.26.

TABLE 6.26
Calculated Azimuth Angles of the Sides of the Open Traverse in Problem 6.9

Side	Azimuth (° ′ ″)
AB	90° 15′ = 90.25°
BC	41° 25′ = 41.417°
CD	14° 55′ = 14.917°
DE	336° 25′ = 336.417°
EF	42° 15′ = 42.25°
FF′	72° 50′ = 72.833°

The known azimuth of FF' can be determined from the given coordinates of the control points F and F' linked to the end of the traverse (see Table 6.27). Consequently, the following procedure describes this (see Figure 6.22):

TABLE 6.27

Given Coordinates of Points *F* and *F'*
of the Open Traverse in Problem 6.9

| *F* | 52180.10 | 33500.30 |
| *F'* | 52411.70 | 33559.00 |

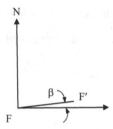

FIGURE 6.22   The Angle between line *FF'* of the open traverse and the horizontal line for Problem 6.9.

$$\tan \beta = \frac{\Delta Y_{F-F'}}{\Delta X_{F-F'}} \tag{6.24}$$

⇒

$$\tan \beta = \frac{(33663.00 - 33559.00)}{(52358.50 - 51948.50)} = 0.25366$$

⇒

$$\beta = 14°14'$$

Therefore, the known azimuth of the line *FF'* based on the given *X*- and *Y*-coordinates of the points *F* and *F'* is equal to:

$$\text{Azimuth}(FF') = 90° - (14°14') = 75°46'$$

(2) The azimuth error and correction:

The azimuth error is computed using the following formula:

$$\varepsilon_\alpha = \alpha' - \alpha$$

⇒

$$\varepsilon_\alpha = (72°50') - (75°46') = -02°56' = -176'$$

The angular correction is determined based on the azimuth closure error, as shown in the formula below:

$$C_\alpha = -\frac{\varepsilon_\alpha}{n}$$

This correction can be applied to each angle of the traverse before computing the corrected azimuth angles of the different lines (legs) of the traverse. Alternately, the angular (azimuth) correction (C_a) shown in Table 6.28 can be applied directly to the computed azimuth of the traverse lines (legs) instead of applying the correction to the traverse angles. In this case, the corrected azimuth of line i can be computed as:

$$\alpha_i'' = \alpha_i' - i\left(\frac{\varepsilon_a}{n}\right)$$

Sample Calculation:
For line BC:
$\Rightarrow$

$$\alpha_i'' = 41°25' - 2\left(\frac{-176'}{6}\right) = 42°24'$$

TABLE 6.28
Azimuth Error and Correction for the Open Traverse in Problem 6.9

Azimuth Error	−02° 56′
Azimuth Correction	−176/6 = −29.3′

---

(3) The corrected azimuths of the sides of the traverse:

The corrected azimuth of each line (leg) of the traverse is computed based on the corrected internal angle determined above or using directly the formula shown above. The results are summarized in Table 6.29.

TABLE 6.29
Corrected Azimuth Angles of the Sides of the Open Traverse in Problem 6.9

Side	Corrected Azimuth (° ′ ″)
AB	90° 44′ = 90.73°
BC	42° 24′ = 42.40°
CD	16° 23′ = 16.38°
DE	338° 22′ = 338.37°
EF	44° 42′ = 44.70°
FF′	75° 46′ = 75.77°

---

(4) The coordinates of the points of the traverse:

The coordinates of the points of the traverse are computed using the corrected azimuth and the given length for each side (leg) of the traverse. A sample calculation is provided below:

Sample Calculation:
For line AB:

$$\Delta X = d \sin \alpha$$

$\Rightarrow$

$$\Delta X = (720)\sin(90°\,44') = 719.94 \text{ m}$$

$$\Delta Y = d \cos \alpha$$

$\Rightarrow$

$$\Delta Y = (720)\cos(90°\,44') = -9.22 \text{ m}$$

$\Rightarrow$

$$X_C = X_B + \Delta X \qquad\qquad\qquad (6.25)$$

$\Rightarrow$

$$X_C = 50392.89 + 719.94 = 51111.89 \text{ m}$$

$$Y_C = Y_B + \Delta Y \qquad\qquad\qquad (6.26)$$

$\Rightarrow$

$$Y_B = 31759.03 + (-9.22) = 31749.81 \text{ m}$$

Similarly, the coordinates of the other points of the traverse are computed. The results of all the points of the traverse are summarized in Table 6.30.

TABLE 6.30
X and Y Coordinates of the Sides of the Open Traverse in Problem 6.9

Station	Side	Corrected Azimuth (WCB)	Length (m)	ΔX (m)	ΔY (m)	Coordinates X (m)	Coordinates Y (m)
A						50,392.89	31,759.03
B	AB	90° 44′	720	719.94	−9.22	51,112.83	31,749.81
C	BC	42° 24′	720	485.50	531.69	51,598.33	32,281.50
D	CD	16° 23′	480	135.39	460.51	51,733.72	32,742.01
E	DE	338° 22′	480	−176.96	446.19	51,556.76	33,188.20
F	EF	44° 42′	570	400.93	405.16	51,957.69	33,593.36
	Sum		2970.00	1564.80	1834.33		

(5) The position closure error in X (ε_X) and the position closure error in Y (ε_Y):

The position closure error in X (ε_X) and the position closure error in Y (ε_Y) are computed based on the difference between the given coordinates and the computed coordinates for the last point of the traverse (point F). Hence, the following formulas are used to determine these two errors, respectively:

The azimuth error is computed using the following formula (see Table 6.31):

$$\varepsilon_X = X' - X$$

$$\varepsilon_Y = Y' - Y$$

$\Rightarrow$

$$\varepsilon_X = 51957.69 - 51948.50 = 9.19 \text{ m}$$

$$\varepsilon_Y = 33593.36 - 33559.00 = 34.36 \text{ m}$$

TABLE 6.31

X and Y Azimuth Errors of the Open Traverse in Problem 6.9

ε_X (m)	9.19
ε_Y (m)	34.36

(6) Balancing the traverse using the compass rule and the corrected coordinates of the traverse stations:

Based on the position closure errors (ε_X and ε_Y), the corrections in ΔX and ΔY are computed using the following two formulas, respectively:

$$\text{Correction in } \Delta X = -\varepsilon_X \frac{d_i}{\sum d_i}$$

$$\text{Correction in } \Delta Y = -\varepsilon_Y \frac{d_i}{\sum d_i}$$

Sample Calculation:
For line AB:

$$\varepsilon_X = 9.19 \text{ m}$$

$$\varepsilon_Y = 34.26 \text{ m}$$

$$\text{Correction in } \Delta X = -\varepsilon_X \frac{d_i}{\sum d_i}$$

$\Rightarrow$

$$\text{Correction in } \Delta X = -9.19 \frac{720}{2970} = -2.23 \text{ m}$$

$$\text{Correction in } \Delta Y = -\varepsilon_Y \frac{d_i}{\sum d_i}$$

$\Rightarrow$

$$\text{Correction in } \Delta Y = -34.36 \frac{720}{2970} = -8.33 \text{ m}$$

The X and Y coordinates are determined using the following formulas:

$$X_B = X_A + \Delta X + \text{Correction in } \Delta X$$

$$Y_B = Y_A + \Delta Y + \text{Correction in } \Delta Y$$

⇒

$$X_B = 50392.89 + 719.94 + (-2.23) = 51110.60 \text{ m}$$

$$Y_B = 31759.03 + (-9.22) + (-8.33) = 31741.48 \text{ m}$$

Similarly, the corrections for the other side (legs) of the traverse and the coordinates of the other points are computed. The results are summarized in Table 6.32.

Figure 6.23 shows the MS Excel worksheet used for the connecting open traverse computations of this problem.

TABLE 6.32

Corrected Coordinates of the Sides of the Open Traverse in Problem 6.9

Station	Side	ΔX (m)	ΔY (m)	Correction in ΔX (m)	Correction in ΔY (m)	Corrected Coordinates X (m)	Corrected Coordinates Y (m)
A						50,392.89	31,759.03
B	AB	719.94	−9.22	−2.23	−8.33	51,110.60	31,741.49
C	BC	485.50	531.69	−2.23	−8.33	51,593.87	32,264.84
D	CD	135.39	460.51	−1.49	−5.55	51,727.77	32,719.80
E	DE	−176.96	446.19	−1.49	−5.55	51,549.33	33,160.44
F	EF	400.93	405.16	−1.76	−6.59	51,948.50	33,559.00
	Sum	1564.80	1834.33	−9.19	−34.36		

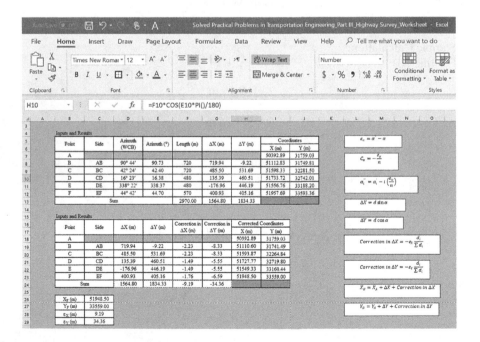

FIGURE 6.23 Image of MS Excel worksheet used for the connecting open traverse computations of Problem 6.9.

7 Areas, Earthwork Volumes, and Mass Haul Diagrams

Chapter 7 deals with the key elements of highway surveys and designs. First of these is the computation of areas in general, particularly the irregular shapes of different areas, as an introduction to the methods used to compute the areas of cross-sections at stations along a highway or roadway. Second, the computation of the earthwork volumes between consecutive stations along highway segments. This section introduces the different methods used to calculate these areas and the volumes needed for highway projects. In addition, the methodology for computing cumulative earthwork volumes at successive stations over an entire highway project is introduced. Using earthwork volumes at all the stations of the highway project starting from the 0+00 station to the end point of the project, a mass haul diagram (MHD) is plotted. The MHD is a figure that shows the relationship between the horizontal distance (chainage) along the highway centerline versus the cumulative earthwork volumes (cut and fill). The +ve side of the MHD indicates an excess in the cut volumes, and the −ve side indicates an excess in the fill volumes or shortage in the cut volumes. The +ve slope of the MHD shows an increase in cut volumes and a −ve slope indicates an increase in the fill volumes. And finally, the costs of excavation and hauling are presented and illustrated for loops on the MHD. The free haul distance (FHD) and the economic haul distance (EHD) and the balance lines in loops of the MHD are also introduced and used in the computation of the cost of hauling.

7.1　Determine how many circles are needed to cover the entire area of the square shown in Figure 7.1:

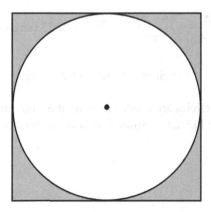

FIGURE 7.1　A circle inscribed in a square for Problem 7.1.

Solution:

If the radius of the circle $= r$, then the side of the square $= 2r$. The area of the square is given by:

$$A_s = a^2 \tag{7.1}$$

Where:
A_s = the area of the square
a = the length of the side of the square

$\Rightarrow$

$$A_s = (2r)^2 = 4r^2$$

On the other hand, the area of the circle is given by:

$$A_c = \pi r^2 \tag{7.2}$$

Where:
A_c = the area of the circle
r = the radius of the circle

$\Rightarrow$

$$A_c = \pi r^2$$

The area of the extra parts, beyond the area of the circle inside the square that are not covered (A_{s-c}), can be computed as shown below:

$$A_{s-c} = 4r^2 - \pi r^2 = (4-\pi)r^2$$

This area divided by the area of the circle plus one whole circle provides the number of circles that are needed to cover the entire area of the square. Therefore:

$$\frac{A_{s-c}}{A_c} = \frac{(4-\pi)r^2}{\pi r^2} = \frac{4-\pi}{\pi} = 0.273 \text{ circle}$$

$$0.273 \text{ circle} + 1 \text{ circle} = 1.273 \text{ circle}$$

And therefore, 1.273 circles are needed to cover the whole area of the square.
Figure 7.2 shows the MS Excel worksheet used to perform the computations of this problem.

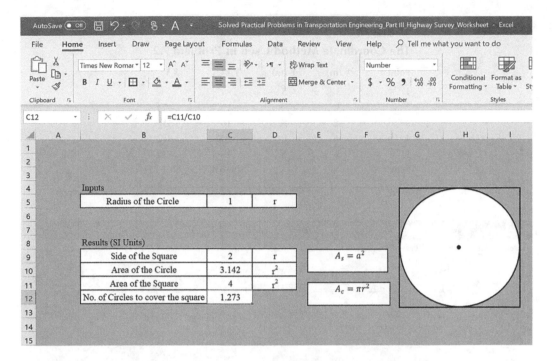

FIGURE 7.2   Image of MS Excel worksheet used for the computations of Problem 7.1.

7.2   If the area enclosed in a loop traverse of 1–2–3–4–1 that is composed of the points: point 1 (3, 5), point 2 (3, –8), point 3 (–2, Y_3), and point 4 (–6, 6) is 71 m², then determine the y-coordinate (m) of point 3.

Solution:

The coordinate method is used as described in the following procedure:

$$A = \frac{1}{2}\left(\text{sum } 1 - \text{sum } 2\right) \tag{7.3}$$

Where:
   A = the area of enclosed shape
   Sum 1 = the summation of the product of Y_i times X_{i+1} (in order)
   Sum 2 = the summation of the product of X_i times Y_{i+1} (in order)

By moving in the clockwise direction from point 1 to point 2 to point 3 to point 4 and going back to point 1 to close the traverse, the above formula applies with the two sums (sum 1 and sum 2) computed as defined above (see Table 7.1). Figure 7.3 shows the MS Excel worksheet used to perform the computations of this problem.
⇒

$$A = \frac{1}{2}\left[49 - 6Y_3 - \left(-66 + 3Y_3\right)\right] = \frac{1}{2}\left[115 - 9Y_3\right] = 71$$

⇒

$$Y_3 = -3$$

TABLE 7.1

The Coordinate Method Used in Problem 7.2

Point	X (m)	Y (m)	Y_iX_{i+1}	X_iY_{i+1}
1	3	5		
2	3	−8	15	−24
3	−2	Y_3	16	$3Y_3$
4	−6	6	$-6Y_3$	−12
1	3	5	18	−30
Sum			$49-6Y_3$	$-66+3Y_3$

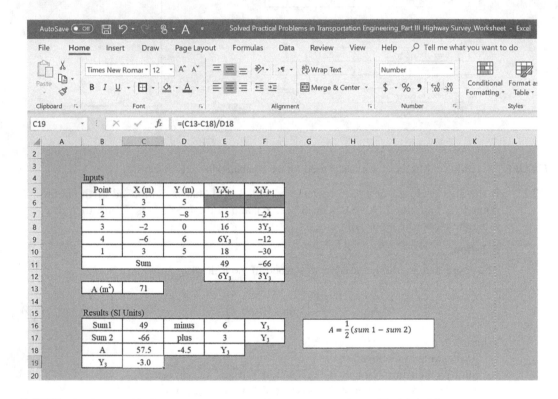

FIGURE 7.3 Image of MS Excel worksheet used for the computations of Problem 7.2.

7.3   Determine the area of the circular sector shown in Figure 7.4.

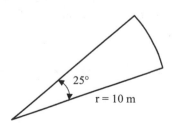

FIGURE 7.4 A circular sector with a central angle of 25° for Problem 7.3.

Solution:

This sector is part of a circle. The central angle of the circle is 360° (2π). The central angle of this sector given in this problem is 25°. Therefore, the area of the sector can be calculated by the following formula:

$$A_{\text{sector}} = \frac{\Delta_{\text{sector}}}{\Delta_{\text{circle}}} \pi r^2 \qquad (7.4)$$

Where:

A_{sector} = the area of the sector
r = the radius of the circle/sector
Δ_{sector} = the central angle of the sector
Δ_{circle} = the central angle of the circle (360°)

$\Rightarrow$

$$A_{\text{sector}} = \frac{25}{360} \pi (10)^2 = 21.82 \text{ m}^2$$

Figure 7.5 shows the MS Excel worksheet used to do the computations of this problem.

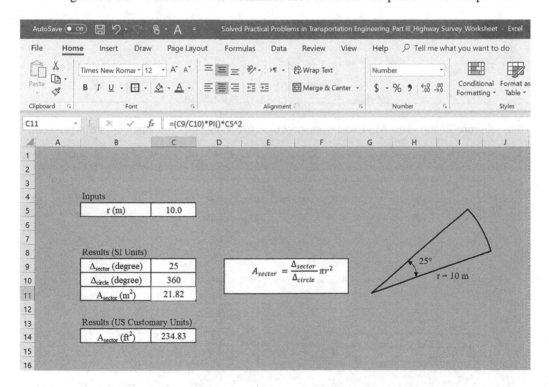

FIGURE 7.5   Image of MS Excel worksheet used for the computations of Problem 7.3.

7.4   Compute the area of circular sector A (in gray) shown in Figure 7.6.

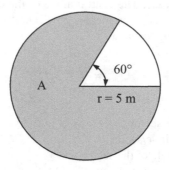

FIGURE 7.6   A circular sector with a central angle of 60° for Problem 7.4.

Solution:

This sector is part of a circle. The central angle of the circle is 360° (2π). The central angle of this sector according to the data given in this problem is 360° − 60°=300°. The area of the sector can be calculated using the same formula in Equation 7.4 used in Problem 7.3:

$$A_{sector} = \frac{\Delta_{sector}}{\Delta_{circle}} \pi r^2$$

⇒

$$A_{sector} = \frac{(360-60)}{360} \pi (5)^2 = 65.45 \text{ m}^2$$

Figure 7.7 shows the MS Excel worksheet used to conduct the computations of this problem.

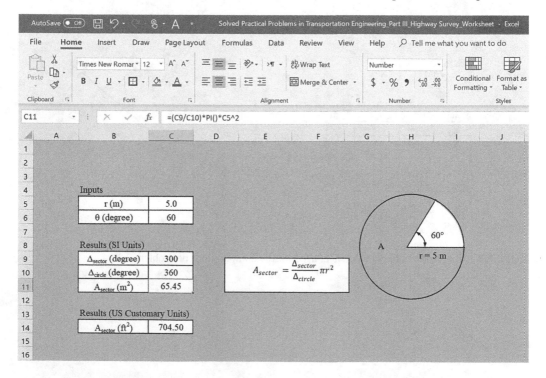

FIGURE 7.7   Image of MS Excel worksheet used for the computations of Problem 7.4.

7.5 If the length of the side of the equilateral triangle shown in Figure 7.8 is 4 m, determine the area of the shaded portion.

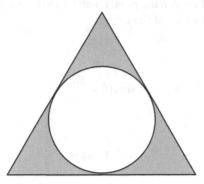

FIGURE 7.8 A circle inscribed in an equilateral triangle for Problem 7.5.

Solution:

The shaded area is equal to the area of the triangle minus the area of the circle inscribed inside the triangle. The area of an equilateral triangle with side a is given by the following formula:

$$A_{triangle} = \frac{\sqrt{3}}{4} a^2 \tag{7.5}$$

Where:

$A_{triangle}$ = the area of the equilateral triangle
a = the length of the side of the triangle

$\Rightarrow$

$$A_{triangle} = \frac{\sqrt{3}}{4}(4)^2 = 6.928 \text{ m}^2$$

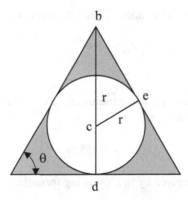

FIGURE 7.9 Geometry used to determine the radius (r) of the circle for Problem 7.5.

Based on the geometry of Figure 7.9, the following relationships apply:

$$\sin\theta = \frac{\overline{bd}}{a} \tag{7.6}$$

Where:

θ=the angle of the equilateral triangle (60°)

$\overline{bd}$=the distance between the points b and d in the diagram

a=the length of the side of the triangle

⇒

$$\sin 60° = \frac{\overline{bd}}{4}$$

⇒

$$\overline{bd} = 3.464 \text{ m}$$

Also, based on the geometry and the Pythagorean theorem, the following formula can be written:

$$r^2 + \left(\frac{a}{2}\right)^2 = \overline{bc}^2 \tag{7.7}$$

Where:

r=the radius of the circle inscribed inside the triangle

a=the length of the side of the equilateral triangle

bc=the distance from the point b (the head of the triangle) to the center of the circle c

$a/2$=the distance between the points b and e in the diagram

But the distance bc is also given by:

$$\overline{bc} = \overline{bd} - r$$

⇒

$$\overline{bc} = 3.464 - r$$

Substituting this value into Equation 7.7 and the value of a=4 provides:

$$r^2 + \left(\frac{4}{2}\right)^2 = \left(3.464 - r\right)^2$$

Solving the above equation for r (the radius of the circle) using the MS Excel solver tool provides the following solution:

$$r = 1.155 \text{ m}$$

The area of the circle is given by the following formula:

$$A_c = \pi r^2$$

⇒

$$A_c = \pi \left(1.155\right)^2 = 4.189 \text{ m}^2$$

The shaded area is equal to the area of the triangle minus the area of the circle . Hence:

$$A = 6.928 - 4.189 = 2.739 \text{ m}^2$$

Figure 7.10 shows the MS Excel worksheet used to perform the computations of this problem.

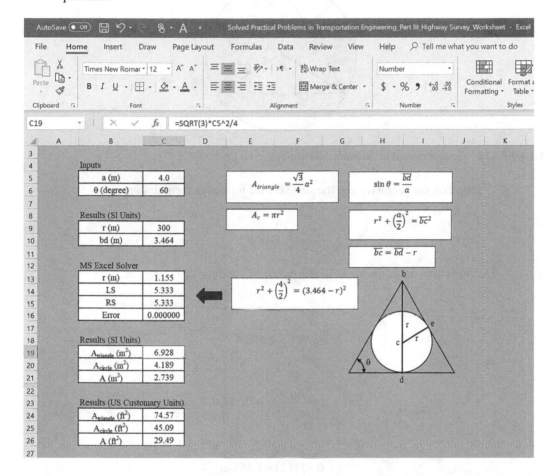

FIGURE 7.10   Image of MS Excel worksheet used for the computations of Problem 7.5.

7.6   An equilateral triangle is inscribed inside a circle, as shown in Figure 7.11. If the length of the side for the equilateral triangle in this diagram is 2 m, determine the shaded (gray) area of the diagram.

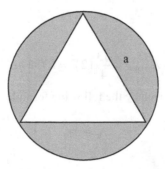

FIGURE 7.11   An equilateral triangle inscribed in a circle for Problem 7.6.

Solution:

The center point of the circle and the triangle and the radius of the circle are plotted, as shown in Figure 7.12.

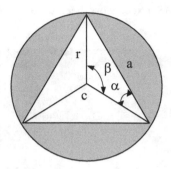

FIGURE 7.12 Geometry used to determine the distance, r for Problem 7.6.

Based on the geometry and the sine law, the following formulas apply:

$$\frac{r}{\sin \alpha} = \frac{a}{\sin \beta} \tag{7.8}$$

Due to symmetry in the above diagram, $\alpha = 30°$, and $\beta = 120°$.

$\Rightarrow$

$$\frac{r}{\sin 30°} = \frac{a}{\sin 120°}$$

$\Rightarrow$

$$r = a \times \frac{\sin 30°}{\sin 120°} = 0.577a$$

$\Rightarrow$

$$r = 0.577(2) = 1.155 \text{ m}$$

The area of an equilateral triangle with side a is given by the following formula:

$$A_{\text{triangle}} = \frac{\sqrt{3}}{4} a^2$$

$\Rightarrow$

$$A_{\text{triangle}} = \frac{\sqrt{3}}{4}(2)^2 = 1.732 \text{ m}^2$$

The area of the circle is given by the following formula:

$$A_c = \pi r^2$$

$\Rightarrow$

$$A_c = \pi (1.155)^2 = 4.189 \text{ m}^2$$

Therefore, the shaded (gray) area is computed as the area of the circle minus the area of the equilateral triangle:

$$A = 4.189 - 1.732 = 2.457 \text{ m}^2$$

Figure 7.13 shows the MS Excel worksheet used to conduct the computations of this problem.

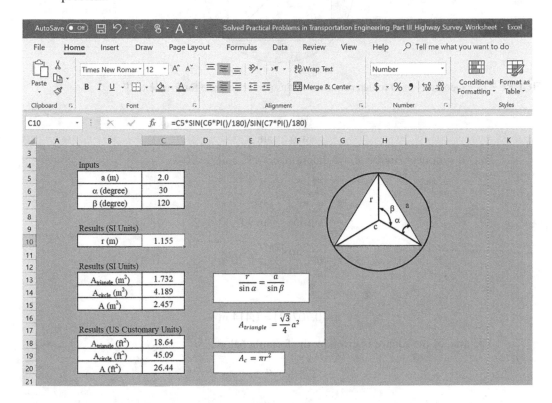

FIGURE 7.13 Image of MS Excel worksheet used for the computations of Problem 7.6

7.7   If the length of the side (a) of the square inscribed inside the circle in Figure 7.14 is 2 m, determine the shaded area of the figure.

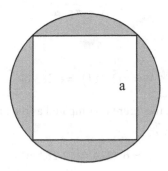

FIGURE 7.14 A square inscribed in a circle for Problem 7.7.

Solution:

Based on the geometry of Figure 7.15, plotted in detail as shown below, the following formula using the sine law applies:

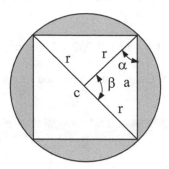

FIGURE 7.15 Geometry used to determine the radius (r) of the circle for Problem 7.7.

$$\frac{r}{\sin \alpha} = \frac{a}{\sin \beta}$$

Due to symmetry in the above diagram, $\alpha=45°$, and $\beta=90°$.

⇒

$$\frac{r}{\sin 45°} = \frac{a}{\sin 90°}$$

$$r = a \times \frac{\sin 45°}{\sin 90°} = 0.707a$$

⇒

$$r = 0.707(2) = 1.414 \text{ m}$$

The area of the square is given by the formula below:

$$A_s = a^2$$

⇒

$$A_s = (2)^2 = 4 \text{ m}^2$$

And the area of the circle is given by the following formula:

$$A_c = \pi r^2$$

⇒

$$A_c = \pi (1.414)^2 = 6.283 \text{ m}^2$$

The shaded (gray) area in the figure is computed as the area of the circle minus the area of the square. Therefore:

$$A = 6.283 - 4 = 2.283 \text{ m}^2$$

Figure 7.16 shows the MS Excel worksheet used to perform the computations of this problem.

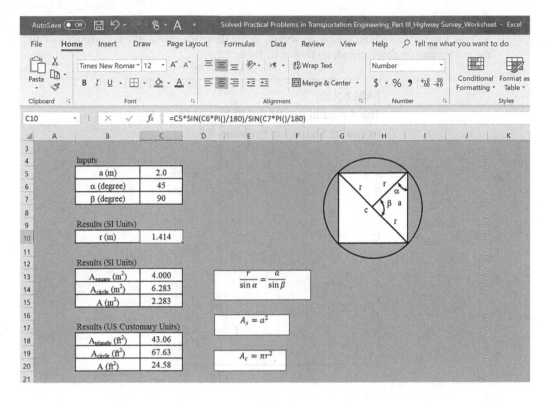

FIGURE 7.16 Image of MS Excel worksheet used for the computations of Problem 7.7.

7.8 Compute the approximate area of the irregular shape shown in Figure 7.17 using the *coordinate method*.

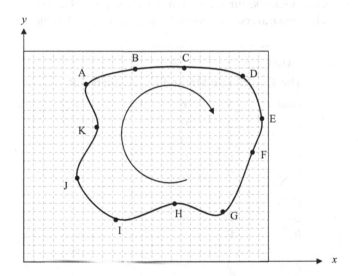

FIGURE 7.17 An irregular shape for Problem 7.8.

Solution:

Setting the x-axis and the y-axis, as shown in the diagram above, the coordinates of all the points can be determined, as shown in Table 7.2.

TABLE 7.2
The Coordinates of the Points of the Irregular Shape for Problem 7.8

Point	X	Y
A	3	21
B	8	23
C	13	23
D	19	22
E	21	17
F	20	13
G	17	6
H	12	7
I	6	5
J	2	10
K	4	16
A	3	21

The coordinate method is used as described in the following procedure:

$$A = \frac{1}{2}\left(\text{sum } 1 - \text{sum } 2\right)$$

By moving in the clockwise direction from point A to point B and so on and going back to point A to close the traverse, the above formula applies (see Table 7.3).

TABLE 7.3
The Coordinate Method for Problem 7.8

Point	X	Y	Y_iX_{i+1}	X_iY_{i+1}
A	3	21		
B	8	23	168	69
C	13	23	299	184
D	19	22	437	286
E	21	17	462	323
F	20	13	340	273
G	17	6	221	120
H	12	7	72	119
I	6	5	42	60
J	2	10	10	60
K	4	16	40	32
A	3	21	48	84
Sum			2139	1610

$$\text{sum } 1 = 2139$$

$$\text{sum } 2 = 1610$$

$\Rightarrow$

$$A = \frac{1}{2}(2139 - 1610) = 264.5$$

Figure 7.18 shows the MS Excel worksheet used to perform the computations of this problem.

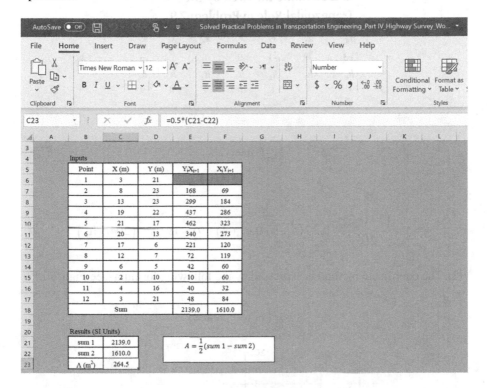

FIGURE 7.18 Image of MS Excel worksheet used for determining the area in Problem 7.8.

7.9 Compute the area enclosed between the x-axis, $x = -3$, $x = 3$, and the curve given by the function $f(x) = x^2$ using the three methods: (1) the trapezoidal rule using $\Delta x = 1$, (2) the coordinate method, and (3) an exact method.

Solution:

(1) The trapezoidal rule is expressed by the following formula:

$$A = \frac{(x_n - x_0)}{2n}\left(h_0 + 2\sum_{i=1}^{n-1} h_i + h_n\right) \qquad (7.9)$$

Where:

x_i = the x-value of the points starting from $i = 0$ to n

$h_i = f(x_i)$ = the height (the y-value) at the points starting from $i = 0$ to n

n = number of intervals (segments)

$n+1$ = number of x-values or points (heights)

$\dfrac{(x_n - x_0)}{n}$ = the value of the interval

The values of x and h ($f(x)$) are shown in Table 7.4 based on $\Delta x = 1$ and using the given function $f(x) = x^2$:

TABLE 7.4

x and h Values for the Points Needed to Determine the Area by the Trapezoidal Rule in Problem 7.9

x_0	x_1	x_2	x_3	x_4	x_5	x_6
−3	−2	−1	0	1	2	3
h_0	h_1	h_2	h_3	h_4	h_5	h_6
9	4	1	0	1	4	9

$n = 6$ intervals

$\Rightarrow$

$$A = \frac{(3-(-3))}{2(6)}\left(9 + 2(4+1+0+1+4) + 9\right) = 19.0 \text{ m}^2 \left(204.5 \text{ ft}^2\right)$$

(2) The coordinate method is expressed by the following formula (see Figure 7.19):

$$A = \frac{1}{2}\left(\text{sum } 1 - \text{sum } 2\right)$$

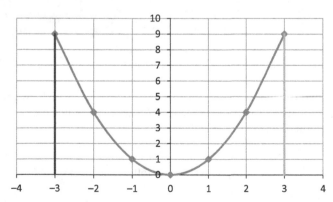

FIGURE 7.19   The area enclosed between the x-axis, $x=-3$, $x=3$, and the function $f(x)=x^2$ for Problem 7.9.

Starting from point (−3, 9) and moving in a clockwise direction to point (−2, 4), then to the next point until the point (−3, 0) is reached, and going back to the first point (−3, 9), the traverse is closed (Figure 7.19), and the coordinate method can be used properly as below (Table 7.5):

TABLE 7.5

The Coordinate Method for Problem 7.9

Point	X (m)	Y (m)	$Y_i X_{i+1}$	$X_i Y_{i+1}$
1	−3	9		
2	−2	4	−18	−12
3	−1	1	−4	−2
4	0	0	0	0
5	1	1	0	0
6	2	4	2	4
7	3	9	12	18
8	3	0	27	0
9	2	0	0	0
10	1	0	0	0
11	0	0	0	0
12	−1	0	0	0
13	−2	0	0	0
14	−3	0	0	0
1	−3	9	0	−27
Sum			19	−19

$$\text{sum } 1 = 19.0$$

$$\text{sum } 2 = -19.0$$

$$\Rightarrow$$

$$A = \frac{1}{2}\left(19.0 - (-19.0)\right) = 19.0 \text{ m}^2 \left(204.5 \text{ ft}^2\right)$$

(3) An exact method (by integrating the function using calculus) is used. The following result is obtained:

$$A = \int_{-3}^{3} x^2 dx = \left[\frac{x^3}{3}\right]_{-3}^{3} = 9.0 - (-9.0) = 18.0 \text{ m}^2 \left(193.8 \text{ ft}^2\right)$$

Figures 7.20 and 7.21 show the MS Excel worksheets used to estimate the area in this problem by the trapezoidal rule and by the coordinate method, respectively.

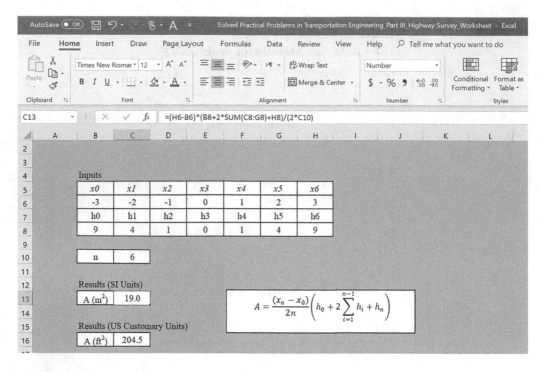

FIGURE 7.20 Image of MS Excel worksheet used for determining the area by the trapezoidal rule for Problem 7.9.

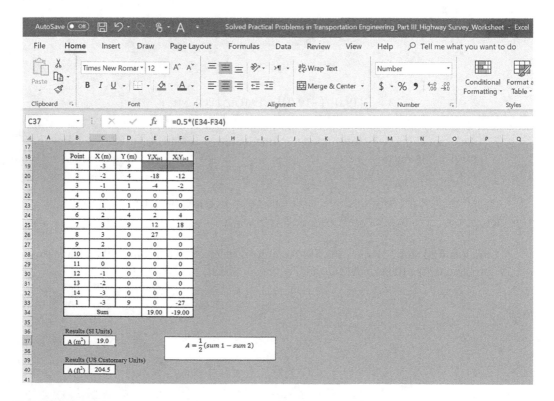

FIGURE 7.21 Image of MS Excel worksheet used for determining the area by the coordinate method for Problem 7.9.

7.10 Compute the area enclosed between the x-axis, $x=3$, and the curve given by the function $f(x)=x^3$ using the trapezoidal rule ($\Delta x=1$) and the coordinate method. Compare the results of the two methods with the exact solution.

Solution:

(1) The trapezoidal rule is expressed by the following formula:

$$A = \frac{(x_n - x_0)}{2n}\left(h_0 + 2\sum_{i=1}^{n-1}h_i + h_n\right)$$

The values of x and h ($f(x)$) are shown in Table 7.6 based on $\Delta x=1$ and using the given function $f(x)=x^3$:

TABLE 7.6
x and h Values for the Points Needed to Determine the Area by the Trapezoidal Rule in Problem 7.10

x_0	x_1	x_2	x_3
0	1	2	3
h_0	h_1	h_2	h_3
0	1	8	27

$n=3$ intervals

$\Rightarrow$

$$A = \frac{(3-0)}{2(3)}\left(0+2(1+8)+27\right) = 22.5$$

(2) The coordinate method is expressed by the following formula (see Figure 7.22):

$$A = \frac{1}{2}\left(\text{sum}\,1 - \text{sum}\,2\right)$$

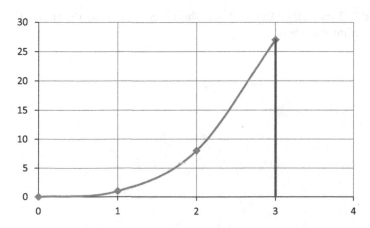

FIGURE 7.22 The area enclosed between the x-axis, $x=3$, and the function $f(x)=x^3$ for Problem 7.10.

Starting from point (0, 0) and moving in a clockwise direction to point (1, 1), then to the next point until the point (1, 0) is reached, and going back to the first point (0, 0), the traverse is closed (Figure 7.22), and the coordinate method can be used properly as below (Table 7.7):

TABLE 7.7
The Coordinate Method for Problem 7.10

Point	X	Y	Y_iX_{i+1}	X_iY_{i+1}
1	0	0		
2	1	1	0	0
3	2	8	2	8
4	3	27	24	54
5	3	0	81	0
6	2	0	0	0
7	1	0	0	0
1	0	0	0	0
Sum			107	62

$$\text{sum}\,1 = 107$$

$$\text{sum}\,2 = 62$$

$\Rightarrow$

$$A = \frac{1}{2}\left(107 - 62\right) = 22.5$$

The exact solution can be determined by integrating the function using calculus, as shown below:

$$A = \int_0^3 x^3 dx = \left[\frac{x^4}{4}\right]_0^3 = 20.25 - 0 = 20.25$$

Figure 7.23 shows the MS Excel worksheet used to estimate the area in this problem by the coordinate method.

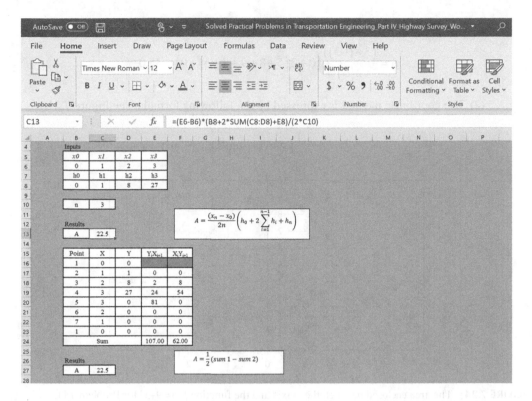

FIGURE 7.23 Image of MS Excel worksheet used for determining the area by the coordinate method for Problem 7.10.

7.11 Compute the area enclosed between the x-axis and the curve given by the function $f(x) = 9 - x^2$ using the trapezoidal rule ($\Delta x = 1$) and the coordinate method. Compare the results with the exact solution.

(1) The trapezoidal rule is expressed by the following formula:

$$A = \frac{(x_n - x_0)}{2n}\left(h_0 + 2\sum_{i=1}^{n-1} h_i + h_n\right)$$

The values of x and h ($f(x)$) are shown in Table 7.8 based on $\Delta x = 1$ and using the given function $f(x) = 9 - x^2$:

TABLE 7.8

x and h Values for the Points Needed to Determine the Area by the Trapezoidal Rule in Problem 7.11

x_0	x_1	x_2	x_3	x_4	x_5	x_6
-3	-2	-1	0	1	2	3
h_0	h_1	h_2	h_3	h_4	h_5	h_6
0	5	8	9	8	5	0

$n = 6$ intervals

$\Rightarrow$

$$A = \frac{(3-(-3))}{2(6)}\left(0+2(5+8+9+8+5)+0\right) = 35.0$$

(2) The coordinate method is expressed by the following formula (see Figure 7.24):

$$A = \frac{1}{2}\left(\text{sum}\,1 - \text{sum}\,2\right)$$

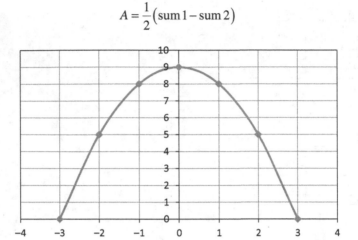

FIGURE 7.24 The area enclosed between the x-axis and the function $f(x) = 9 - x^2$ for Problem 7.11.

Starting from point (−3, 0) and moving in a clockwise direction to point (−2, 5), then to the next point until point (−2, 0) is reached, and going back to the first point (−3, 0), the traverse is closed (Figure 7.24), and the coordinate method can be used properly as below (Table 7.9):

TABLE 7.9
The Coordinate Method for Problem 7.11

Point	X	Y	$Y_i X_{i+1}$	$X_i Y_{i+1}$
1	−3	0		
2	−2	5	0	−15
3	−1	8	−5	−16
4	0	9	0	−9
5	1	8	9	0
6	2	5	16	5
7	3	0	15	0
8	2	0	0	0
9	1	0	0	0
10	0	0	0	0
11	−1	0	0	0
12	−2	0	0	0
1	−3	0	0	0
Sum			35	−35

$$\text{sum}\,1 = 35$$

$$\text{sum}\,2 = -35$$

⇒

$$A = \frac{1}{2}\big(35-(-35)\big) = 35.0$$

The exact solution can be determined by integrating the function using calculus, as shown below:

$$A = \int_{-3}^{3}(9-x^2)dx = \left[9x - \frac{x^3}{3}\right]_{-3}^{3} = 18.0 - (-18.0) = 36.0$$

Figures 7.25 and 7.26 show the MS Excel worksheets used to estimate the area in this problem by the trapezoidal rule and by the coordinate method, respectively.

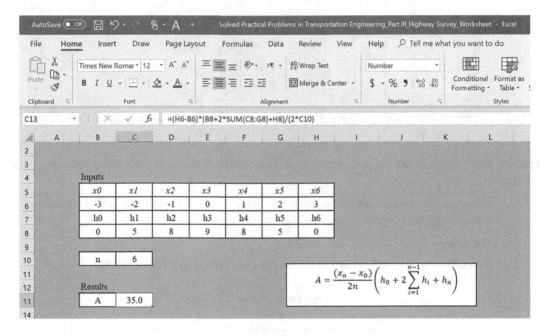

FIGURE 7.25 Image of MS Excel worksheet used for determining the area by the trapezoidal rule for Problem 7.11.

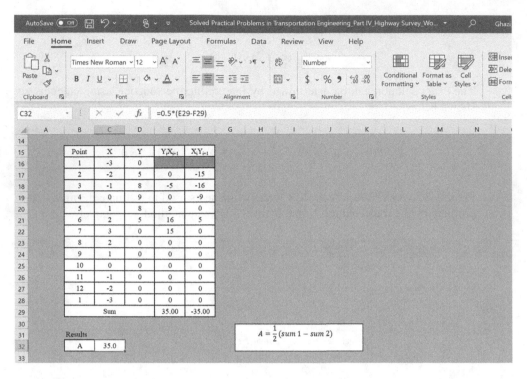

FIGURE 7.26   Image of MS Excel worksheet used for determining the area by the coordinate method for Problem 7.11.

7.12   Compute the area enclosed between the lines $y = 6 - 1.5x$ and $y = 6 + 1.5x$ and the x-axis using the coordinate method if $\Delta x = 1.0$.

Solution:

The following formula is used in the coordinate method (see Figure 7.27):

$$A = \frac{1}{2}\left(\text{sum}\,1 - \text{sum}\,2\right)$$

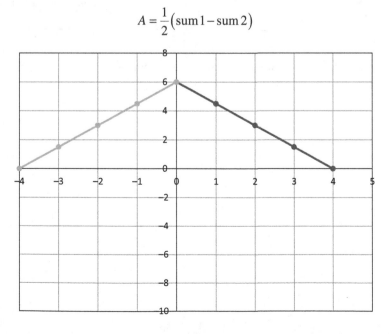

FIGURE 7.27   The area enclosed between the lines $y = 6 - 1.5x$ and $y = 6 + 1.5x$ and the x-axis for Problem 7.12.

Starting from point (−4, 0) and moving in a clockwise direction to point (−3, 1.5), then to the next point until point (−3, 0) is reached, and going back to the first point (−4, 0), the traverse is closed (Figure 7.27), and the coordinate method can be used properly as below (Table 7.10):

TABLE 7.10

The Coordinate Method for Problem 7.12

Point	X (m)	Y (m)	Y_iX_{i+1}	X_iY_{i+1}
1	−4	0		
2	0	−6	0	−6
3	−3	−9	−3	−9
4	−3	−9	−3	−9
5	0	−6	0	−6
6	6	0	6	0
7	9	3	9	3
8	9	3	9	3
9	6	0	6	0
10	0	0	0	0
11	0	0	0	0
12	0	0	0	0
13	0	0	0	0
14	0	0	0	0
15	0	0	0	0
16	0	0	0	0
1	0	0	0	0
Sum			24.00	−24.00

$$\text{sum} 1 = 24$$

$$\text{sum} 2 = -24$$

$$\Rightarrow$$

$$A = \frac{1}{2}\left(24 - (-24)\right) = 24.0$$

Figure 7.28 shows the MS Excel worksheet used to compute the area in this problem by the coordinate method.

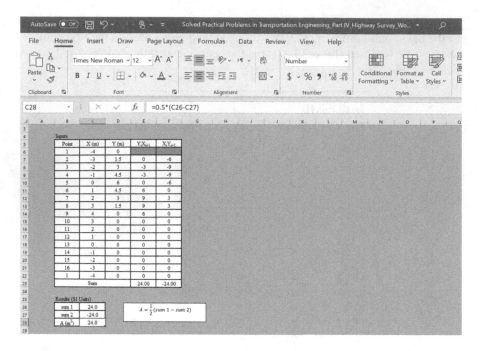

FIGURE 7.28 Image of MS Excel worksheet used for determining the area by the coordinate method for Problem 7.12.

7.13 Use the *coordinate method* to determine the cross-sectional area of the following cut cross-section at a specific station of a highway segment (Figure 7.29).

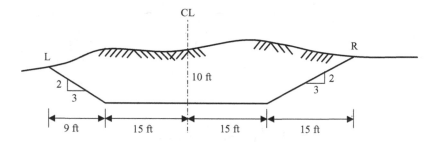

FIGURE 7.29 A cross-section of a highway segment for Problem 7.13.

Solution:

The same formula is used in the coordinate method:

$$A = \frac{1}{2}\left(\text{sum}1 - \text{sum}2\right)$$

Starting from point (−24, 6) and moving in the clockwise direction to point (0, 10), then to the next point until point (−15, 0) is reached, and going back to the first point (−24, 6), the traverse is closed, and the coordinate method can be used properly as below (Table 7.11):

TABLE 7.11

The Coordinate Method for Problem 7.13

Point	X (ft)	Y (ft)	$Y_i X_{i+1}$	$X_i Y_{i+1}$
1	−24	6		
2	0	10	0	−240
3	30	10	300	0
4	15	0	150	0
5	0	0	0	0
6	−15	0	0	0
1	−24	6	0	−90
Sum			450.0	−330.0

$$\text{sum}\,1 = 450.0$$

$$\text{sum}\,2 = -330.0$$

$$\Rightarrow$$

$$A = \frac{1}{2}\left(450.0 - \left(-330.0\right)\right) = 390.0 \text{ ft}^2 \left(36.2 \text{ m}^2\right)$$

Figure 7.30 shows the MS Excel worksheet used to compute the cross-sectional area in this problem by the coordinate method.

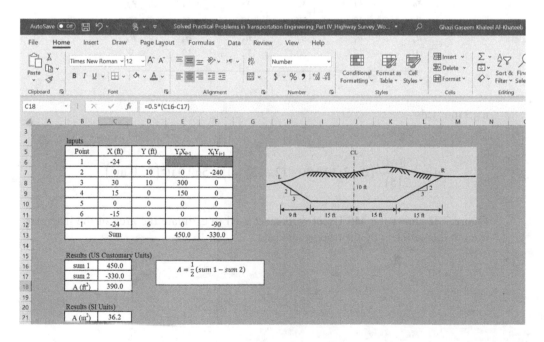

FIGURE 7.30 Image of MS Excel worksheet used for determining the area by the coordinate method for Problem 7.13.

7.14   Using the *trapezoidal rule*, determine the area of the irregular tract shown in Figure 7.31.

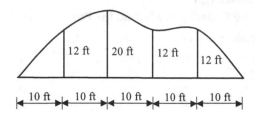

FIGURE 7.31   An irregular tract for Problem 7.14.

Solution:

The trapezoidal rule is expressed by the following formula:

$$A = \frac{(x_n - x_0)}{2n}\left(h_0 + 2\sum_{i=1}^{n-1}h_i + h_n\right)$$

The values of x and h ($f(x)$) of the given tract are shown in Table 7.12 based on $\Delta x = 10$:

TABLE 7.12

x and h Values for the Points Needed to Determine the Area by the Trapezoidal Rule in Problem 7.14

x_0	x_1	x_2	x_3	x_4	x_5
0	10	20	30	40	50
h_0	h_1	h_2	h_3	h_4	h_5
0	12	20	12	12	0

$n = 5$ intervals

⇒

$$A = \frac{(50-0)}{2(5)}\left(0 + 2(12 + 20 + 12 + 12) + 0\right) = 560.0\ \text{ft}^2\left(52.0\ \text{m}^2\right)$$

Figure 7.32 shows the MS Excel worksheet used to estimate the area of the irregular tract in this problem by the trapezoidal rule.

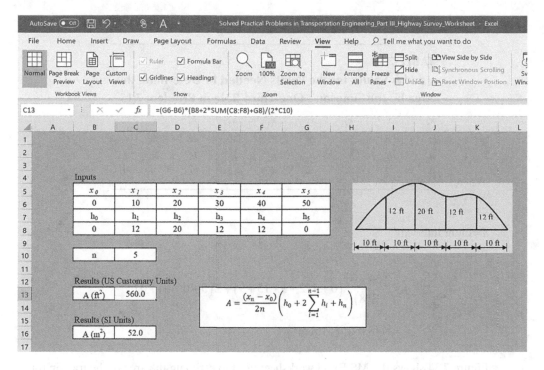

FIGURE 7.32 Image of MS Excel worksheet used for determining the area by the trapezoidal rule for Problem 7.14.

7.15 Using *Simpson's one-third rule*, determine the area enclosed in the tract shown in Figure 7.33.

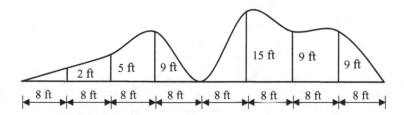

FIGURE 7.33 An irregular tract for Problem 7.15.

Solution:

Simpson's one-third rule is expressed by the following formula:

$$A = \frac{(x_n - x_0)}{3n}\left(h_0 + 4\sum_{i=1,3,5,\dots}^{n-1} h_i + 2\sum_{j=2,4,6,\dots}^{n-2} h_j + h_n\right) \qquad (7.10)$$

Where:

x_i = the x-value of the points starting from $i = 0$ to n

$h_i = f(x_i)$ = the height (the y-value) at the points starting from $i = 0$ to n

n = number of intervals (segments)

$n + 1$ = number of x-values or points (heights)

$\dfrac{(x_n - x_0)}{n}$ = the value of the interval

The values of x and h ($f(x)$) of the given tract are shown in Table 7.13 based on $\Delta x = 8$:

TABLE 7.13
x and h Values for the Points Needed to Determine the
Area by the Simpson's One-Third Rule in Problem 7.15

x_0	x_1	x_2	x_3	x_4	x_5	x_6	x_7	x_8
0	8	16	24	32	40	48	56	64
h_0	h_1	h_2	h_3	h_4	h_5	h_6	h_7	h_8
0	2	5	9	0	15	9	9	0

n = eight intervals
$\Rightarrow$

$$A = \frac{(64-0)}{3(8)}\left(0+4(2+9+15+9)+2(5+0+9)+0\right)$$

$$= 448.0 \text{ ft}^2\left(41.6 \text{ m}^2\right)$$

Figure 7.34 shows the MS Excel worksheet used to estimate the area of the tract in this problem by Simposon's one-third rule.

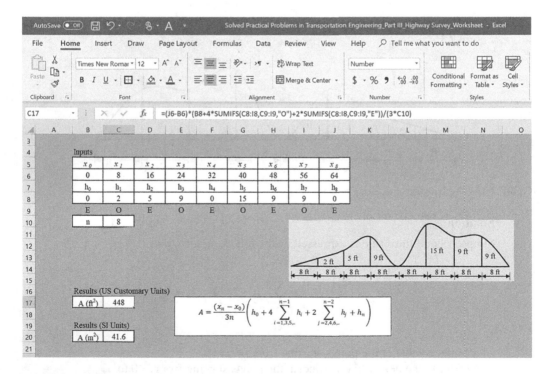

FIGURE 7.34 Image of MS Excel worksheet used for determining the area by Simpson's one-third rule for Problem 7.15.

7.16   Compute the area of the irregular tract of land shown in Figure 7.35 using the trapezoi-
        dal rule and Simpson's one-third rule.

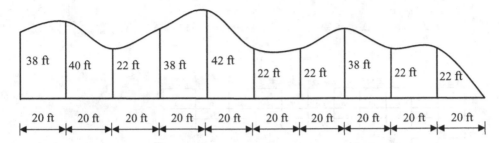

FIGURE 7.35   An irregular tract for Problem 7.16.

The trapezoidal rule is expressed by the following formula:

$$A = \frac{(x_n - x_0)}{2n}\left(h_0 + 2\sum_{i=1}^{n-1} h_i + h_n \right)$$

The values of x and h ($f(x)$) of the given tract are shown in Table 7.14 based on $\Delta x = 20$:

TABLE 7.14
x and h Values for the Points Needed to Determine the Area in Problem 7.16

x_0	x_1	x_2	x_3	x_4	x_5	x_6	x_7	x_8	x_9	x_{10}
0	20	40	60	80	100	120	140	160	180	200
h_0	h_1	h_2	h_3	h_4	h_5	h_6	h_7	h_8	h_9	h_{10}
38	40	22	38	42	22	22	38	22	22	0

$n = 10$ intervals

$\Rightarrow$

$$A = \frac{(200-0)}{2(10)}\left(38 + 2(40 + 22 + 38 + 42 + 22 + 22 + 38 + 22 + 22) + 0\right)$$

$$= 5740.0 \text{ ft}^2 \left(533.3 \text{ m}^2\right)$$

Simpson's one-third rule is expressed by the following formula:

$$A = \frac{(x_n - x_0)}{3n}\left(h_0 + 4\sum_{i=1,3,5,\dots}^{n-1} h_i + 2\sum_{j=2,4,6,\dots}^{n-2} h_j + h_n \right)$$

$\Rightarrow$

$$A = \frac{(200-0)}{3(10)}\left(38 + 4(40 + 38 + 22 + 38 + 22) + 2(22 + 42 + 22 + 22) + 0\right)$$

$$= 5960.0 \text{ ft}^2 \left(553.7 \text{ m}^2\right)$$

Figure 7.36 shows the MS Excel worksheet used to estimate the area of the irregular
tract in this problem by Simposon's one-third rule.

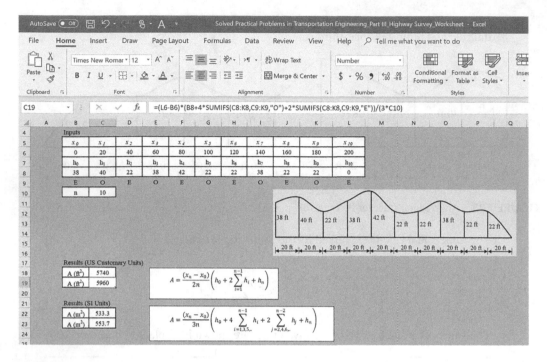

FIGURE 7.36 Image of MS Excel worksheet used for determining the area by the trapezoidal rule and Simpson's one-third rule for Problem 7.16.

7.17 Compute the area enclosed between the lines $y = 3 - 2x$ and $y = 3 + 2x$ and the x-axis using the coordinate method if $\Delta x = 0.5$.

Solution:

The following formula is used in the coordinate method (see Figure 7.37):

$$A = \frac{1}{2}\left(\text{sum}1 - \text{sum}2\right)$$

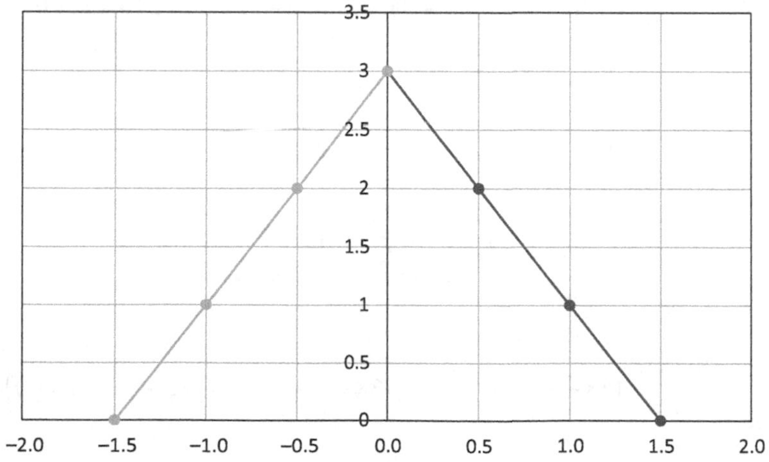

FIGURE 7.37 The area enclosed between the lines $y = 3 - 2x$ and $y = 3 + 2x$ and the x-axis for Problem 7.17.

Starting from point (−1.5, 0) and moving in a clockwise direction to point (−1, 1), then to the next point until point (−1, 0) is reached, and going back to the first point (−1.5, 0), the traverse is closed (Figure 7.37), and the coordinate method can be used properly as below (Table 7.15):

TABLE 7.15
The Coordinate Method for Problem 7.17

Point	X (m)	Y (m)	Y_iX_{i+1}	X_iY_{i+1}
1	−1.5	0		
2	−1	1	0	−1.5
3	−0.5	2	−0.5	−2
4	0	3	0	−1.5
5	0.5	2	1.5	0
6	1	1	2	0.5
7	1.5	0	1.5	0
8	1	0	0	0
9	0.5	0	0	0
10	0	0	0	0
11	−0.5	0	0	0
12	−1	0	0	0
1	−1.5	0	0	0
Sum			4.5	−4.5

$$\text{sum}\,1 = 4.5$$

$$\text{sum}\,2 = -4.5$$

$\Rightarrow$

$$A = \frac{1}{2}\left(4.5 - \left(-4.5\right)\right) = 4.5$$

The MS Excel worksheet used to estimate the area in this problem by the coordinate method is shown in Figure 7.38.

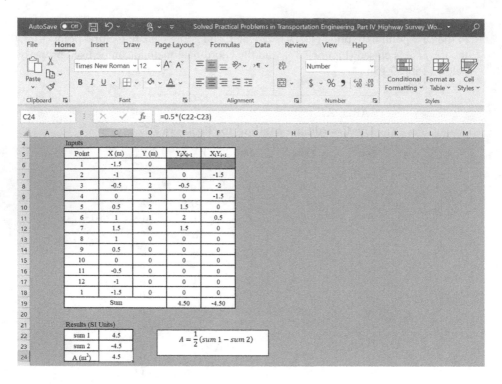

Cell C24 formula bar: `=0.5*(C22-C23)`

Inputs				
Point	X (m)	Y (m)	Y_iX_{i+1}	X_iY_{i+1}
1	-1.5	0		
2	-1	1	0	-1.5
3	-0.5	2	-0.5	-2
4	0	3	0	-1.5
5	0.5	2	1.5	0
6	1	1	2	0.5
7	1.5	0	1.5	0
8	1	0	0	0
9	0.5	0	0	0
10	0	0	0	0
11	-0.5	0	0	0
12	-1	0	0	0
1	-1.5	0	0	0
	Sum		4.50	-4.50

Results (SI Units)

sum 1	4.5
sum 2	-4.5
A (m²)	4.5

$$A = \frac{1}{2}(sum\ 1 - sum\ 2)$$

FIGURE 7.38 Image of MS Excel worksheet used for determining the area by the coordinate method for Problem 7.17.

7.18 Compute the earthwork volume between stations 50 and 51 for a road with a width of 30 ft (9.14 m) using the *average-end-area method* if the depths of cut at the center line (CL), at a point right to the CL, and at a point left to the CL are as shown in Table 7.16.

TABLE 7.16
Depth of Cut Across Road Cross-Sections at Stations 50 + 00 and 51 + 00 for Problem 7.18

Station	Left	Center	Right
50 + 00	C 5.0	C 9.0	C 8.0
	25.0	0.0	31.0
51 + 00	C 6.0	C 8.0	C 4.0
	27.0	0.0	23.0

Solution:

A sketch of the road cross-section at station 50 + 00 is shown in Figure 7.39.

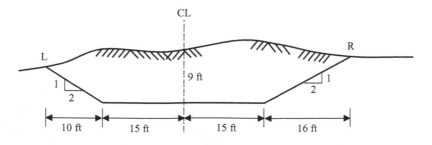

FIGURE 7.39 Sketch for the road cross-section at station 50 + 00 for Problem 7.18.

Starting from the first point to the left of CL (0, 5) and moving in a clockwise direction to the point at CL (25, 9), then to the next point until point (10, 0) is reached, and going back to the first point (0, 5), the traverse is closed, and the coordinate method can be used properly as below (Table 7.17).

TABLE 7.17

The Coordinate Method for the Road Cross-Sectional Area at Station 50 + 00 for Problem 7.18

Point	X	Y	Y_iX_{i+1}	X_iY_{i+1}
1: L	0	5		
2: CL	25	9	125	0
3: R	56	8	504	200
4	40	0	320	0
5	25	0	0	0
6	10	0	0	0
1: L	0	5	0	50
Sum			949.0	250.0

The following formula is used in the coordinate method:

$$A = \frac{1}{2}\left(\text{sum}1 - \text{sum}2\right)$$

Therefore, the area of the cross-section at station 50 + 00:

$$A = \frac{1}{2}\left(949.0 - 250.0\right) = 349.5 \text{ ft}^2 \left(32.5 \text{ m}^2\right)$$

In a similar manner, the closed loop for station 51 + 00 is established, as shown in Table 7.18 that includes the coordinates of the points:

TABLE 7.18

The Coordinate Method for the Road Cross-Sectional Area at Station 51 + 00 for Problem 7.18

Point	X	Y	Y_iX_{i+1}	X_iY_{i+1}
1: L	0	6		
2: CL	27	8	162	0
3: R	50	4	400	108
4	42	0	168	0
5	27	0	0	0
6	12	0	0	0
1: L	0	6	0	72
Sum			730	180

The following formula is used in the coordinate method:

$$A = \frac{1}{2}\left(\text{sum}1 - \text{sum}2\right)$$

Therefore, the area of the cross-section at station 51 + 00:

$$A = \frac{1}{2}(730.0 - 180.0) = 275.0 \text{ ft}^2 \left(25.5 \text{ m}^2\right)$$

The earthwork volume is calculated using the average-end-area method that is given by the following expression:

$$V_C = \left(\frac{C_1 + C_2}{2}\right) L \tag{7.11}$$

Where:

V_C=the volume of earthwork between two successive stations along the highway
C_1 and C_2=the cut areas of the two cross-sections at the two successive stations, respectively
L=the distance between the two stations

$\Rightarrow$

$$V = \left(\frac{349.5 + 275}{2}\right)100 = 31225 \text{ ft}^3 = 884.2 \text{ m}^3$$

The MS Excel worksheet used to compute the cross-sectional areas and the earthwork volume in this problem is shown in Figure 7.40.

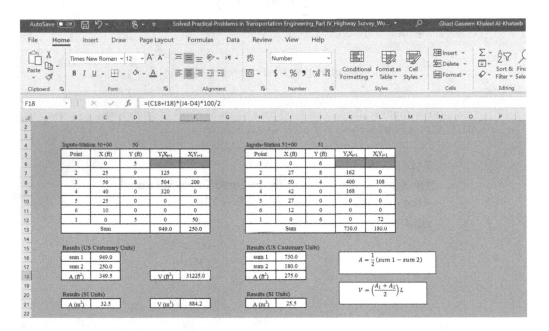

FIGURE 7.40 Image of MS Excel worksheet used for computing the cross-sectional areas and the earthwork volume between stations 50 + 00 and 51 + 00 for Problem 7.18.

7.19   The cross-sectional areas of three successive stations along a highway segment (stations $40+00$, $41+00$, and $42+00$) are given in Table 7.19. Based on this, determine the accumulative volume of cut or fill for the three stations. *Assume that the accumulative volume at station $40+00$ is zero.*

TABLE 7.19

Highway Cross-Sectional Areas at Stations 40 + 00, 41 + 00, and 42 + 00 for Problem 7.19

Station	Cross-Sectional Area	
	Cut (ft²)	Fill (ft²)
40+00	32	0
41+00	0	20
42+00	12	16

Solution:

The average-end-area method is used to calculate the volume between the successive stations using the following formulas:

For cut–cut cross-sections, Equation 7.11 used earlier in Problem 7.18 can be used:

$$V_C = \left(\frac{C_1 + C_2}{2} \right) L \tag{7.11}$$

For fill–fill cross-sections:

$$V_F = \left(\frac{F_1 + F_2}{2} \right) L \tag{7.12}$$

For cut–fill cross-sections:

$$V_C = \frac{1}{2} \left(\frac{C^2}{C + F} \right) L \tag{7.13}$$

$$V_F = \frac{1}{2} \left(\frac{F^2}{C + F} \right) L \tag{7.14}$$

For mixed–mixed cross-sections:

$$V_C = \left(\frac{C_1 + C_2}{2} \right) L \tag{7.15}$$

$$V_F = \left(\frac{F_1 + F_2}{2} \right) L \tag{7.16}$$

For mixed–cut cross-sections:

$$V_C = \left(\frac{C_1 + C_2}{2} \right) L \tag{7.17}$$

$$V_F = \frac{1}{3}\left(F_i\right)L \tag{7.18}$$

For mixed–fill cross-sections:

$$V_C = \frac{1}{3}\left(C_i\right)L \tag{7.19}$$

$$V_F = \left(\frac{F_1 + F_2}{2}\right)L \tag{7.20}$$

Where:

V_C=the volume of cut between the two cross-sections
V_F=the volume of fill between the two cross-sections
C_1 and C_2=the areas of the cut in the cross-sections 1 and 2 at the two stations, respectively
F_1 and F_2=the areas of the fill in the cross-sections 1 and 2 at the two stations, respectively
C and F=the areas of cut and fill for pure cut cross-sections and pure fill cross-sections, respectively
C_i and F_i=the area of cut and the area of fill for mixed cross-sections for the cases mixed-fill cross-sections and mixed-cut cross-sections, respectively
L=the distance between the two stations

In this case, station 40 has a cut cross-section, station 41 has a fill cross-section, and station 42 has a mixed cross-section. Therefore, the volume between stations 40 and 41 is determined using the formulas for cut-fill cross-sections, as shown below:
For cut-fill cross-sections:

$$V_C = \frac{1}{2}\left(\frac{C^2}{C+F}\right)L$$

$$V_F = \frac{1}{2}\left(\frac{F^2}{C+F}\right)L$$

$\Rightarrow$

$$V_C = \frac{1}{2}\left(\frac{(32)^2}{32+20}\right)100 = 984.6 \text{ ft}^3\left(27.9 \text{ m}^3\right)$$

$$V_F = \frac{1}{2}\left(\frac{(20)^2}{32+20}\right)100 = 384.6 \text{ ft}^3\left(10.9 \text{ m}^3\right)$$

In a similar manner, the volume between stations 41 and 42 is determined using the formulas for fill-mixed cross-sections, as shown below:
For mixed-fill cross-sections:

$$V_C = \frac{1}{3}\left(C_i\right)L$$

$$V_F = \left(\frac{F_1 + F_2}{2}\right)L$$

$$\Rightarrow \quad V_C = \frac{1}{3}(12)100 = 400.0 \text{ ft}^3 \left(11.3 \text{ m}^3\right)$$

$$V_F = \left(\frac{20+16}{2}\right)100 = 1800.0 \text{ ft}^3 \left(51.0 \text{ m}^3\right)$$

Therefore, the accumulative volumes from station 40 + 00 to station 42 + 00 are computed and summarized in Table 7.20 (+ve is used for cut volumes and −ve is used for fill volumes).

TABLE 7.20

Earthwork Volumes and Accumulative Volume between Highway Stations 40 + 00 and 42 + 00 for Problem 7.19

	Earthwork Volume (ft³)			Accumulative
	Cut	Fill	Station	Volume (ft³)
From–To			40 + 00	0
40 + 00 to 41 + 00	984.6	384.6	41 + 00	600.0
41 + 00 to 42 + 00	400.0	1800.0	42 + 00	−800.0

The MS Excel worksheet used to compute the earthwork volumes and the accumulative volume is shown in Figure 7.41.

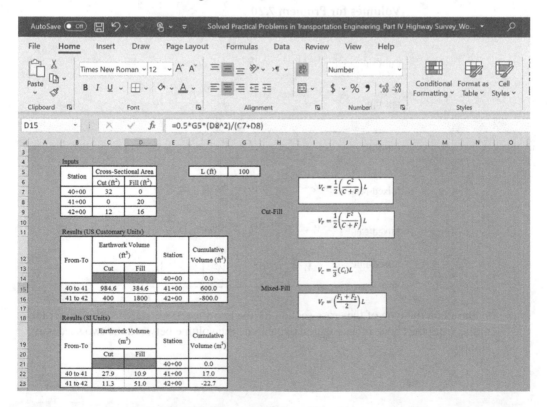

FIGURE 7.41 Image of MS Excel worksheet used for computing the earthwork volumes and accumulative volume between stations 40 + 00 and 42 + 00 for Problem 7.19.

7.20   The cut and fill areas for the cross-sections at five stations along the centerline of a highway segment are given in Table 7.21. Compute the accumulative earthwork volumes and plot the mass haul diagram (MHD) for this highway segment.

TABLE 7.21

Cut and Fill Areas for Five Cross-Sections along the Centerline of Highway Segment for Problem 7.20

Station	Area (m²)	
	Cut	Fill
0 + 00	24	0
1 + 00	0	26
2 + 00	0	18
3 + 00	12	18
4 + 00	22	0

Solution:

The average-end-area method is used to calculate the volume between two successive stations using the following formulas summarized in Table 7.22.

TABLE 7.22

Formulas for Computing Cut and Fill Volumes for Problem 7.20

Case	Volume of Cut	Volume of Fill
Cut-Cut:	$V_C = \left(\dfrac{C_1 + C_2}{2} \right) L$	
Fill-Fill:		$V_F = \left(\dfrac{F_1 + F_2}{2} \right) L$
Cut-Fill:	$V_C = \dfrac{1}{2} \left(\dfrac{C^2}{C + F} \right) L$	$V_F = \dfrac{1}{2} \left(\dfrac{F^2}{C + F} \right) L$
Mixed-Mixed:	$V_C = \left(\dfrac{C_1 + C_2}{2} \right) L$	$V_F = \left(\dfrac{F_1 + F_2}{2} \right) L$
Mixed-Cut:	$V_C = \left(\dfrac{C_1 + C_2}{2} \right) L$	$V_F = \dfrac{1}{3} \left(F_i \right) L$
Mixed-Fill:	$V_C = \dfrac{1}{3} \left(C_i \right) L$	$V_F = \left(\dfrac{F_1 + F_2}{2} \right) L$

Based on the type of the cross-section for the two successive stations, the proper formulas from the above table are used to compute the volume between the two successive stations.

Sample Calculation:

Between stations $1 + 00$ and $2 + 00$:

Since both cross-sections at station $1 + 00$ and station $2 + 00$ are fill sections, the case is "fill–fill" cross-sections. Hence, the following formula is used:

$$V_F = \left(\frac{F_1 + F_2}{2} \right) L$$

$\Rightarrow$

$$V_F = \left(\frac{26 + 18}{2} \right) 100 = 2200 \text{ m}^3 \left(2877.5 \text{ yd}^3 \right)$$

The computed volumes along with the cumulative volumes are summarized in Table 7.23.

TABLE 7.23

Calculated Earthwork Volumes and Cumulative Volume for Highway Segment for Problem 7.20

	Earthwork Volume (m³)		Station	Cumulative Volume (m³)
	Cut	Fill		
From–To			0 + 00	0
0 + 00 to 1 + 00	576	676	1 + 00	−100
1 + 00 to 2 + 00	0	2200	2 + 00	−2300
2 + 00 to 3 + 00	400	1800	3 + 00	−3700
3 + 00 to 4 + 00	1700	600	4 + 00	−2600

The mass haul diagram is simply the diagram that plots the cumulative volumes versus the station (horizontal distance) from the beginning point to the end point of the highway project. According to the results in the above table, the following MHD is obtained (Figure 7.42).

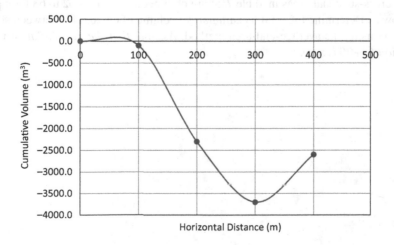

FIGURE 7.42 Mass haul diagram (MHD) for the highway segment of Problem 7.20.

Figure 7.43 shows the MS Excel worksheet used to perform the computations and plot the MHD in this problem.

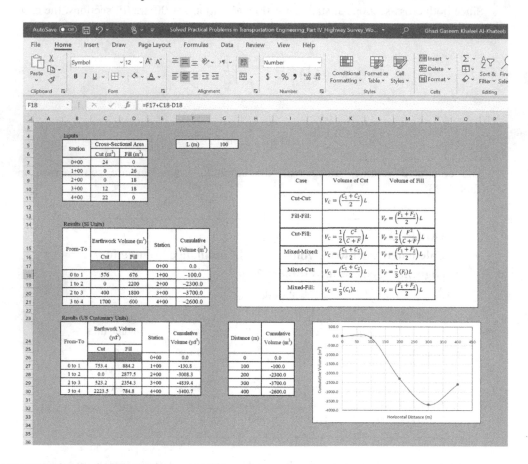

FIGURE 7.43 Image of MS Excel worksheet used for computing the earthwork volumes and accumulative volume and drawing the mass haul diagram (MHD) for Problem 7.20.

7.21   The cross-sectional areas in Table 7.24 are obtained for stations 62 to 68 for a proposed highway. Determine the total (accumulative) volume of cut and fill between stations 62 and 68 using the average-end-area method. *Assume that the accumulative volume at station 62+00 is zero.*

TABLE 7.24

Cut and Fill Areas for Seven Cross-Sections along the Centerline of Proposed Highway for Problem 7.21

Station	Area (ft²)	
	Cut	Fill
62 + 00		44
63 + 00		52
64 + 00		60
65 + 00	0	0
66 + 00	28	
67 + 00	50	
68 + 00	32	

Solution:

The following formulas summarized in Table 7.25 are used in the average-end-area method to calculate the volume between two successive stations:

TABLE 7.25

Formulas for Computing Cut and Fill Volumes for Problem 7.21

Case	Volume of Cut	Volume of Fill
Cut-Cut:	$V_C = \left(\dfrac{C_1 + C_2}{2}\right)L$	
Fill-Fill:		$V_F = \left(\dfrac{F_1 + F_2}{2}\right)L$
Cut-Fill:	$V_C = \dfrac{1}{2}\left(\dfrac{C^2}{C+F}\right)L$	$V_F = \dfrac{1}{2}\left(\dfrac{F^2}{C+F}\right)L$
Mixed-Mixed:	$V_C = \left(\dfrac{C_1 + C_2}{2}\right)L$	$V_F = \left(\dfrac{F_1 + F_2}{2}\right)L$
Mixed-Cut:	$V_C = \left(\dfrac{C_1 + C_2}{2}\right)L$	$V_F = \dfrac{1}{3}\left(F_i\right)L$
Mixed-Fill:	$V_C = \dfrac{1}{3}\left(C_i\right)L$	$V_F = \left(\dfrac{F_1 + F_2}{2}\right)L$

The type of the cross-sections of the two successive stations controls which formulas to use. The following sample calculation is given for stations 66 + 00 and 67 + 00.

Sample Calculation:

Between stations 66 + 00 and 67 + 00:

Both cross-sections at station 66 + 00 and station 67 + 00 are cut sections; therefore, the case is "cut–cut" cross-sections. Hence, the following formula is used:

$$V_C = \left(\frac{C_1 + C_2}{2}\right) L$$

⇒

$$V_C = \left(\frac{28 + 50}{2}\right) 100 = 3900 \text{ ft}^3 \left(110.4 \text{ m}^3\right)$$

The computed volumes along with the cumulative volumes for all the stations of the highway are summarized in Table 7.26.

TABLE 7.26
Calculated Earthwork Volumes and Cumulative
Volume for Highway Segment for Problem 7.21

	Earthwork Volume (ft³)			Cumulative
	Cut	Fill	Station	Volume (ft³)
From–To			62 + 00	0.0
62 + 00 to 63 + 00	0	4800.0	63 + 00	−4800.0
63 + 00 to 64 + 00	0	5600.0	64 + 00	−10,400.0
64 + 00 to 65 + 00	0	3000.0	65 + 00	−13,400.0
65 + 00 to 66 + 00	1400	0	66 + 00	−12,000.0
66 + 00 to 67 + 00	3900	0	67 + 00	−8100.0
67 + 00 to 68 + 00	4100	0	68 + 00	−4000.0

The MS Excel worksheet used to conduct the computations in this problem is shown in Figure 7.44.

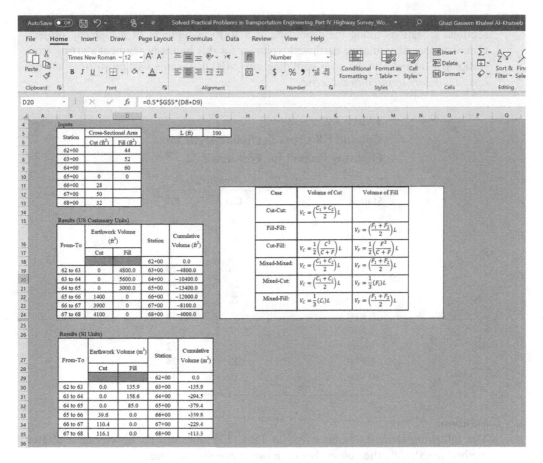

FIGURE 7.44 Image of MS Excel worksheet used for computing the earthwork volumes and accumulative volume for Problem 7.21.

7.22 The cut and fill cross-sectional areas shown in Table 7.27 are computed for the cross-sections at stations along the centerline of a 1000-m segment of a highway project. Calculate the accumulative earthwork volumes for this highway segment starting from station 0 + 00 and plot the mass haul diagram (MHD). *Assume that no swell or shrinkage factors are applied in this case.*

TABLE 7.27

Cut and Fill Cross-Sectional Areas for Highway Project for Problem 7.22

Station	Area (m²)	
	Cut	Fill
0 + 00	36	0
1 + 00	42	0
2 + 00	0	30
3 + 00	45	24
4 + 00	28	36
5 + 00	22	0
6 + 00	0	18
7 + 00	0	26
8 + 00	0	32
9 + 00	20	24
10 + 00	36	48

Solution:

The following formulas summarized in Table 7.28 are used in the average-end-area method to calculate the volume between two successive stations:

TABLE 7.28

Formulas for Computing Cut and Fill Volumes for Problem 7.22

Case	Volume of Cut	Volume of Fill
Cut-Cut:	$V_C = \left(\dfrac{C_1 + C_2}{2} \right) L$	
Fill-Fill:		$V_F = \left(\dfrac{F_1 + F_2}{2} \right) L$
Cut-Fill:	$V_C = \dfrac{1}{2} \left(\dfrac{C^2}{C + F} \right) L$	$V_F = \dfrac{1}{2} \left(\dfrac{F^2}{C + F} \right) L$
Mixed-Mixed:	$V_C = \left(\dfrac{C_1 + C_2}{2} \right) L$	$V_F = \left(\dfrac{F_1 + F_2}{2} \right) L$
Mixed-Cut:	$V_C = \left(\dfrac{C_1 + C_2}{2} \right) L$	$V_F = \dfrac{1}{3} \left(F_i \right) L$
Mixed-Fill:	$V_C = \dfrac{1}{3} \left(C_i \right) L$	$V_F = \left(\dfrac{F_1 + F_2}{2} \right) L$

The sample calculation shown below is performed for stations 3 + 00 and 4 + 00.

Sample Calculation:
Between stations 3 + 00 and 4 + 00:
Since both cross-sections at station 3 + 00 and station 4 + 00 are mixed sections; therefore, the case is "mixed–mixed" cross-sections. Hence, the following formulas are used:

$$V_C = \left(\frac{C_1 + C_2}{2} \right) L$$

$\Rightarrow$

$$V_C = \left(\frac{45 + 28}{2} \right) 100 = 3650 \text{ m}^3 \left(4774.0 \text{ yd}^3 \right)$$

$$V_F = \left(\frac{F_1 + F_2}{2} \right) L$$

$\Rightarrow$

$$V_F = \left(\frac{24 + 36}{2} \right) 100 = 3000 \text{ m}^3 \left(3923.9 \text{ yd}^3 \right)$$

The computed volumes along with the cumulative volumes for all the stations of the highway are summarized in Table 7.29.

TABLE 7.29
Calculated Earthwork Volumes and Cumulative Volume for Highway Project for Problem 7.22

	Earthwork Volume (m³)			Cumulative
	Cut	Fill	Station	Volume (m³)
From–To			0 + 00	0.0
0 + 00 to 1 + 00	3900.0	0.0	1 + 00	3900.0
1 + 00 to 2 + 00	1225.0	625.0	2 + 00	4500.0
2 + 00 to 3 + 00	1500.0	2700.0	3 + 00	3300.0
3 + 00 to 4 + 00	3650.0	3000.0	4 + 00	3950.0
4 + 00 to 5 + 00	2500.0	1200.0	5 + 00	5250.0
5 + 00 to 6 + 00	605.0	405.0	6 + 00	5450.0
6 + 00 to 7 + 00	0.0	2200.0	7 + 00	3250.0
7 + 00 to 8 + 00	0.0	2900.0	8 + 00	350.0
8 + 00 to 9 + 00	666.7	2800.0	9 + 00	−1783.3
9 + 00 to 10 + 00	2800.0	3600.0	10 + 00	−2583.3

The mass haul diagram (MHD) for the highway segment is plotted as shown in Figure 7.45.

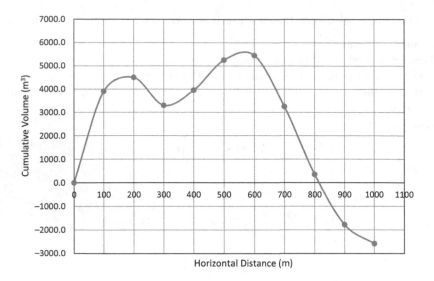

FIGURE 7.45 The mass haul diagram (MHD) for the highway project for Problem 7.22.

The MS Excel worksheet used to perform the computations and plot the MHD in this problem is shown in Figure 7.46.

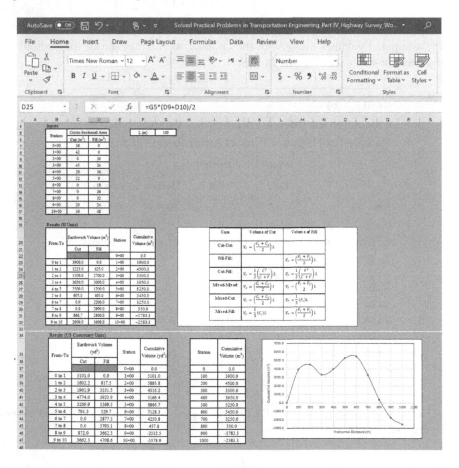

FIGURE 7.46 Image of MS Excel worksheet used for computing the earthwork volumes and accumulative volume and drawing the mass haul diagram (MHD) for Problem 7.22.

7.23 For the following mass haul diagram (MHD) loop (Figure 7.47), calculate the cost of earthwork, including excavation and hauling given that the cost of excavation = $2.0/m³ and the cost of over haul = $0.150/m³.station.
Note: Station = 100 m. Numbers shown on the loop are accumulative volumes (m³).

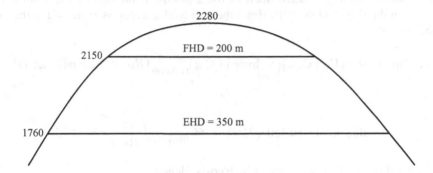

FIGURE 7.47 A mass haul diagram (MHD) loop for Problem 7.23.

Solution:

The average total haul distance is calculated, as shown below:

$$THD_{avg} = \frac{FHD + EHD}{2} \qquad (7.21)$$

Where:

THD_{avg} = the average total haul distance
FHD = the free haul distance
EHD = the economic haul distance

⇒

$$THD_{avg} = \frac{200 + 350}{2} = 275 \text{ m} \left(902.2 \text{ ft}\right)$$

The over haul distance is simply the average total haul distance minus the free haul distance. The following formula is used:

$$OHD = THD_{avg} - FHD \qquad (7.22)$$

Where:

OHD = the over haul distance
THD_{avg} = the average total haul distance
FHD = the free haul distance

⇒

$$OHD = 275 - 200 = 75 \text{ m} \left(246.1 \text{ ft}\right)$$

The cost of excavation is given per unit volume. Therefore, it is computed, as shown below:

$$\text{Excavation Cost} = \left(\text{Unit Cost}\right)\left(\text{Net Volume of Cut}\right) \qquad (7.23)$$

$\Rightarrow$

$$\text{Excavation Cost} = (2)(2280 - 1760) = \$1040$$

The cost of hauling is determined as the unit cost multiplied by the volume of cut between the FHD and the EHD times the over haul distance as expressed in the following formula:

$$\text{Hauling Cost} = (\text{Unit Cost})(\text{Volume of Cut})_{\text{FHD-EHD}}(\text{No. of Stations in OHD}) \quad (7.24)$$

$\Rightarrow$

$$\text{Hauling Cost} = (0.150)(2150 - 1760)_{\text{FHD-EHD}}\left(\frac{75}{100}\right) = \$43.875$$

The total cost is the summation of both costs. Hence,

$$\text{Total Cost} = \text{Excavation Cost} + \text{Hauling Cost} \quad (7.25)$$

$\Rightarrow$

$$\text{Total Cost} = 1040 + 43.875 = \$1083.875$$

Figure 7.48 shows the MS Excel worksheet used to estimate the cost of earthwork in this problem.

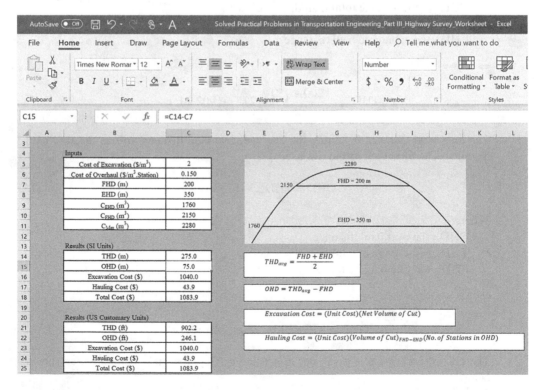

FIGURE 7.48 Image of MS Excel worksheet used for computing the cost of earthwork for a MHD loop for Problem 7.23.

7.24 Given the cross-sectional areas along a short segment of a proposed highway in Table 7.30:

(1) Compute the cumulative volumes along the highway segment.
(2) Draw the mass haul diagram (MHD) for this segment, assuming no shrinkage or swell in the soil.
(3) Quantify the amount of excess or shortage.
(4) Calculate the net earthwork volume between stations 1 + 00 and 4 + 00.
(5) If the free haul distance (FHD) is 160 m, and the economic haul distance (EHD) is 350 m, determine the cost of hauling excavations from station 0 + 00 to station 3 + 50 given that the cost of over haul = $0.30/m^3.station.

TABLE 7.30

Cross-sectional Areas along a Short Highway Segment for Problem 7.24

	Area (m²)	
Station (m)	Cut	Fill
0 + 00	10	0
1 + 00	6	0
2 + 00	10	12
3 + 00	6	20
4 + 00	0	8
5 + 00	0	8
6 + 00	24	0

Solution:

(1) The average-end-area method is used again to compute the volumes between successive stations. The formulas summarized in Table 7.31 are used in the average-end-area method:

TABLE 7.31

Formulas for Computing Cut and Fill Volumes for Problem 7.24

Case	Volume of Cut	Volume of Fill
Cut-Cut:	$V_C = \left(\dfrac{C_1 + C_2}{2}\right)L$	
Fill-Fill:		$V_F = \left(\dfrac{F_1 + F_2}{2}\right)L$
Cut-Fill:	$V_C = \dfrac{1}{2}\left(\dfrac{C^2}{C+F}\right)L$	$V_F = \dfrac{1}{2}\left(\dfrac{F^2}{C+F}\right)L$
Mixed-Mixed:	$V_C = \left(\dfrac{C_1 + C_2}{2}\right)L$	$V_F = \left(\dfrac{F_1 + F_2}{2}\right)L$
Mixed-Cut:	$V_C = \left(\dfrac{C_1 + C_2}{2}\right)L$	$V_F = \dfrac{1}{3}\left(F_i\right)L$
Mixed-Fill:	$V_C = \dfrac{1}{3}\left(C_i\right)L$	$V_F = \left(\dfrac{F_1 + F_2}{2}\right)L$

A sample calculation is shown below for the earthwork volume between stations 2 + 00 and 3 + 00.

Sample Calculation:

Between stations 2 + 00 and 3 + 00:

Both cross-sections at station 2 + 00 and station 3 + 00 are mixed sections; therefore, the case is "mixed-mixed" cross-sections. Hence, the following formulas are used:

$$V_C = \left(\frac{C_1 + C_2}{2} \right) L$$

$$\Rightarrow$$

$$V_C = \left(\frac{10 + 6}{2} \right) 100 = 800 \text{ m}^3 \left(4774.0 \text{ yd}^3 \right)$$

$$V_F = \left(\frac{F_1 + F_2}{2} \right) L$$

$$\Rightarrow$$

$$V_F = \left(\frac{12 + 20}{2} \right) 100 = 1600 \text{ m}^3 \left(3923.9 \text{ yd}^3 \right)$$

The computed volumes along with the cumulative volumes for all the stations of the highway segment are summarized in Table 7.32.

TABLE 7.32

Calculated Earthwork Volumes and Cumulative Volume for Highway Segment for Problem 7.24

	Earthwork Volume (m³)			Cumulative
	Cut	Fill	Station	Volume (m³)
From–To			0 + 00	0.0
0 + 00 to 1 + 00	800.0	0.0	1 + 00	800.0
1 + 00 to 2 + 00	800.0	400.0	2 + 00	1200.0
2 + 00 to 3 + 00	800.0	1600.0	3 + 00	400.0
3 + 00 to 4 + 00	200.0	1400.0	4 + 00	−800.0
4 + 00 to 5 + 00	0.0	800.0	5 + 00	−1600.0
5 + 00 to 6 + 00	900.0	100.0	6 + 00	−800.0

(2) The mass haul diagram (MHD) is plotted using the cumulative earthwork volumes computed above at the different stations of the highway segment. The MHD is shown in Figure 7.49.

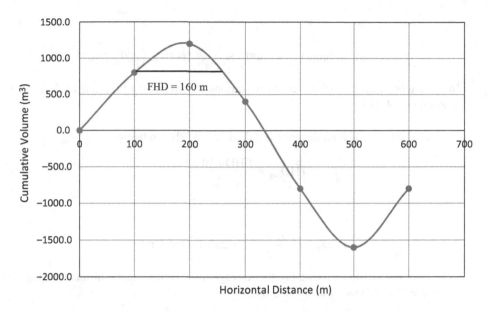

FIGURE 7.49 The mass haul diagram (MHD) for highway segment for Problem 7.24.

(3) The amount of excess or shortage is the cumulative volume at the last station of the project. In this case, it is equal to the cumulative earthwork volume at station 6 + 00.

$$\text{Excess or Shortage} = \text{Cumulative Volume in MHD at Last Station} \qquad (7.26)$$

⇒

$$\text{Excess or Shortage} = 800 \text{ m}^3 \left(1046.4 \text{ yd}^3\right)$$

Since the value has −ve sign, this is considered a shortage; therefore, the amount of shortage = 800 m³.

(4) The net earthwork volume between station 1 + 00 and station 4 + 00 is equal to the cumulative volume at station 4 + 00 minus the cumulative volume at station 1 + 00. Therefore, it is given by the following formula:

$$\left(\text{Earthwork Volume}\right)_{i-j} = \left(\text{Cumulative Volume}\right)_{j}$$
$$-\left(\text{Cumulative Volume}\right)_{i} \qquad (7.27)$$

Where:
Earthwork Volume$_{i-j}$ = the earthwork volume between station i and station j
Cumulative Volume$_i$ = the cumulative volume from the MHD at station i
Cumulative Volume$_j$ = the cumulative volume from the MHD at station j

⇒

$$\left(\text{Earthwork Volume}\right)_{1-4} = \left(\text{Cumulative Volume}\right)_{4}$$
$$-\left(\text{Cumulative Volume}\right)_{1}$$

$\Rightarrow$

$$\left(\text{Earthwork Volume}\right)_{1-4} = -800 - 800 = -1600 \text{ m}^3 \left(2092.7 \text{ yd}^3\right)$$

In other words, the net volume between stations 1 + 00 and 4 + 00 is a fill volume of the amount of 1600 m³.

(5) The average total haul distance is calculated, as shown below:

$$\text{THD}_{\text{avg}} = \frac{\text{FHD} + \text{EHD}}{2}$$

$\Rightarrow$

$$\text{THD}_{\text{avg}} = \frac{160 + 350}{2} = 255 \text{ m} \left(836.6 \text{ ft}\right)$$

The over haul distance is simply the average total haul distance minus the free haul distance. The following formula is used:

$$\text{OHD} = \text{THD}_{\text{avg}} - \text{FHD}$$

$\Rightarrow$

$$\text{OHD} = 255 - 160 = 95 \text{ m} \left(311.7 \text{ ft}\right)$$

The cost of hauling is determined as the unit cost multiplied by the volume of cut between the FHD and the EHD times the over haul distance as expressed in the following formula:

The free haul distance having a value of 160 m is plotted in the MHD, as shown above. The cumulative volume at the balance line (FHD) is equal to 800 m³ and the cumulative volume at the balance line that connects the stations 0 + 00 and 3 + 50 is 0. Therefore, the net volume is equal to 800 m³.

$$\text{Hauling Cost} = \left(\text{Unit Cost}\right)\left(\text{Volume of Cut}\right)_{\text{FHD-EHD}}$$

$$\times \left(\text{No. of Stations in OHD}\right)$$

$\Rightarrow$

$$\text{Hauling Cost} = \left(0.3\right)\left(800 - 0\right)_{\text{FHD-EHD}} \left(\frac{95}{100}\right) = \$228$$

Figure 7.50 shows the MS Excel worksheet used to perform the computations and plot the MHD in this problem.

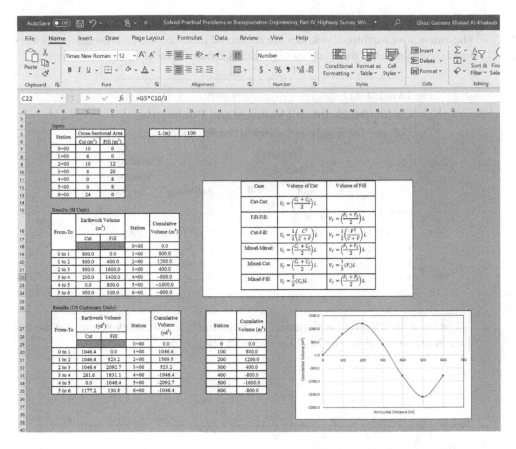

FIGURE 7.50 Image of MS Excel worksheet used for computing the earthwork volumes and accumulative volume and drawing the mass haul diagram (MHD) for Problem 7.24.

7.25   The earthwork volumes between successive stations for a roadway segment are given in Table 7.33. Based on that, plot the mass haul diagram (MHD) for this segment assuming no shrinkage or swell in the soil and determine the cost to transport the cut volume from station $0 + 00$ to station $1 + 80$ given that the cost of over haul = \$0.50/m^3.station and the free haul distance (FHD) is 100 m.

TABLE 7.33

The Earthwork Volumes between Successive Stations For Roadway Segment for Problem 7.25

From–To	Earthwork Volume (m^3)	
	Cut	Fill
$0 + 00$ to $1 + 00$	900	0
$1 + 00$ to $2 + 00$	900	400
$2 + 00$ to $3 + 00$	900	1500
$3 + 00$ to $4 + 00$	200	1200

Solution:

The cumulative earthwork volumes are computed based on the given volumes between successive stations of the roadway segment, as shown in Table 7.34.

TABLE 7.34

Calculated Earthwork Volumes and Cumulative Volume for Roadway Segment for Problem 7.25

From–To	Earthwork Volume (m³)		Station	Cumulative Volume (m³)
	Cut	Fill		
From–To			0 + 00	0
0 + 00 to 1 + 00	800.0	0.0	1 + 00	900
1 + 00 to 2 + 00	800.0	400.0	2 + 00	1400
2 + 00 to 3 + 00	800.0	1600.0	3 + 00	800
3 + 00 to 4 + 00	200.0	1400.0	4 + 00	−200

The mass haul diagram is plotted using the cumulative earthwork volumes at the stations in the above table. The MHD is shown in Figure 7.51.

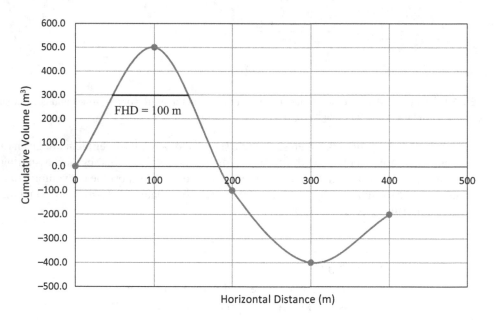

FIGURE 7.51   The mass haul diagram (MHD) for highway segment for Problem 7.25.

The average total haul distance is calculated, as shown below:

$$\text{THD}_{\text{avg}} = \frac{\text{FHD} + \text{EHD}}{2}$$

$\Rightarrow$

$$\text{THD}_{\text{avg}} = \frac{100 + 180}{2} = 140 \text{ m (459.3 ft)}$$

The over haul distance is simply the average total haul distance minus the free haul distance. The following formula is used:

$$\text{OHD} = \text{THD}_{\text{avg}} - \text{FHD}$$

$\Rightarrow$

$$\text{OHD} = 140 - 100 = 40 \text{ m (131.2 ft)}$$

The cost of hauling is determined as the unit cost multiplied by the volume of cut between the FHD and the EHD times the over haul distance as expressed in the following formula:

The free haul distance having a value of 100 m is plotted in the MHD, as shown above. The cumulative volume at the balance line (FHD) is equal to 800 m^3 and the cumulative volume at the balance line that connects stations $0 + 00$ and $1 + 80$ is 0. Therefore, the net volume is equal to 800 m^3.

$$\text{Hauling Cost} = (\text{Unit Cost})(\text{Volume of Cut})_{\text{FHD-EHD}}$$

$$\times (\text{No. of Stations in OHD})$$

$\Rightarrow$

$$\text{Hauling Cost} = (0.5)(300 - 0)_{\text{FHD-EHD}} \left(\frac{40}{100} \right) = \$60$$

Figure 7.52 shows the MS Excel worksheet used to perform the computations in this problem.

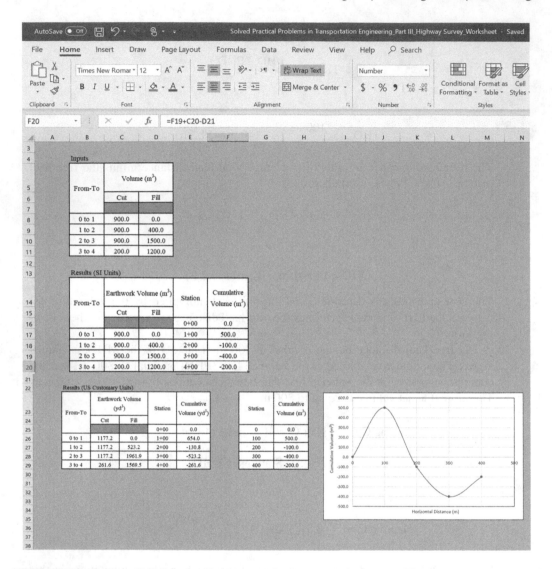

FIGURE 7.52 Image of MS Excel worksheet used for computing the accumulative earthwork volume and drawing the mass haul diagram (MHD) for Problem 7.25.

Multiple-Choice Questions and Answers for Chapter 3

1. The true value of a survey measurement:
   a. Is never known
   b. Can be determined exactly
   c. Can never be estimated.
   d. Can never be determined
   e. **a and d**
2. A blunder error is known when the estimated error is:
   a. **Higher than three times the standard error**
   b. Lower than three times the standard error
   c. Higher than the standard error
   d. Lower than two times the standard error
   e. Higher than 0.67 times the standard error
3. Systematic errors:
   a. Can be detected
   b. Can be of any sign (+ or −)
   c. May be mathematically modeled
   d. Are not predictable
   e. **a, b, and c**
4. Random errors can be detected when the estimated error is:
   a. Higher than three times the standard error
   b. **Lower than three times the standard error**
   c. Higher than the standard error
   d. Lower than two times the standard error
   e. Higher than 0.67 times the standard error
5. Since random errors reflect the limitations of the measurement device and the human operator, they can be minimized by:
   a. Advanced equipment
   b. Appropriate procedures
   c. Repeated measurements
   d. Calibrated instruments
   e. **All of the above**
6. One of the following is not among the characteristics of random errors:
   a. Positive and negative errors of the same magnitude occur with equal frequency
   b. Small errors occur more frequently than large errors
   c. The occurrence of such errors follows approximately the normal distribution
   d. Random errors can never be minimized
   e. **a, b, and c**
7. A distance was measured n times. One of the following parameters can be considered as an estimate of the true value of the distance measurement:
   a. **The mean**
   b. The standard deviation
   c. The variance
   d. The standard error of the mean
   e. None of the above

8. The probable error is equal to:
   a. Three times the standard error
   b. Two times the standard error
   c. The standard error
   d. **0.6745 times the standard error**
   e. 1.6449 times the standard error

9. The maximum error is equal to:
   a. **Three times the standard error**
   b. Two times the standard error
   c. The standard error
   d. 0.6745 times the standard error
   e. 1.6449 times the standard error

10. The standard deviation or how close individual survey measurements are to each other is related to which parameter:
   a. Probable error
   b. Maximum error
   c. **Precision**
   d. Accuracy
   e. None of the above

11. The parameter that indicates how close an individual survey measurement is to the estimated true value is:
   a. Probable error
   b. Maximum error
   c. Precision
   d. **Accuracy**
   e. None of the above

12. Relative precision in a highway survey is used to describe one of the following measurements:
   a. Horizontal angle
   b. Vertical angle
   c. Cross-sectional area
   d. **Distance**
   e. Earthwork volume

13. The effect of the increase in the number of repetitions of a survey measurement is reflected in:
   a. A decrease in the estimated standard error of the mean
   b. An increase in precision
   c. A decrease in accuracy
   d. **a and b**
   e. a, b, and c

14. The most probable value of a survey quantity of n independent measurements is given by:
   a. The standard error of the mean
   b. Standard deviation
   c. **The weighted mean of the n measurements**
   d. The arithmetic mean
   e. None of the above

15. Corrections for systematic errors in distance measurement include:
   a. Length correction
   b. Temperature correction
   c. Sag correction

    d. Tension correction
    e. **All of the above**

16. The elevation of a point can alternatively be replaced by the term:
    a. Longitude
    b. Leveling
    c. Latitude
    d. **Altitude**
    e. Mean sea level

17. A level surface is defined as:
    a. A non-curved surface
    b. A horizontal surface
    c. **A surface parallel to the curvature of the earth**
    d. a and b
    e. b and c

18. A level line has the following characteristics:
    a. A line of constant elevation
    b. A curved line
    c. A line that is normal to the plumb line
    d. A line that is parallel to the curvature of the earth
    e. **All of the above**

19. A horizontal line has the following characteristics:
    a. A line perpendicular to the plumb line
    b. A curved line
    c. **A line that is normal to the plumb line only at the point of observation**
    d. A line that is parallel to the curvature of the earth
    e. All of the above

20. Which two errors are typically involved in a leveling measurement:
    a. Refraction error
    b. Curvature error
    c. Tension error
    d. **a and b**
    e. b and c

21. A permanent object of known elevation in leveling is called:
    a. **Bench mark**
    b. Backsight (BS)
    c. Foresight (FS)
    d. Intermediate sight (IS)
    e. Turning point (TP)

22. The first reading taken on the leveling staff in a leveling project is called:
    a. Bench mark
    b. **Backsight (BS)**
    c. Foresight (FS)
    d. Intermediate sight (IS)
    e. Turning point (TP)

23. The last reading taken on the leveling staff in a leveling job is called:
    a. Bench mark
    b. Backsight (BS)
    c. **Foresight (FS)**
    d. Intermediate sight (IS)
    e. Turning point (TP)

24. The point at which a backsight reading and a foresight reading are taken from two different positions of the level is called:
    a. Bench mark
    b. Backsight (BS)
    c. Foresight (FS)
    d. Intermediate sight (IS)
    e. **Turning point (TP)**

25. The elevation of the line of sight of the leveling telescope is called:
    a. Backsight (BS) elevation
    b. **Height of instrument (HI)**
    c. Foresight (FS) elevation
    d. Intermediate sight (IS) elevation
    e. None of the above

26. To minimize errors and improve the accuracy involved in measuring the difference between elevations between two points, the best position for the level instrument would be:
    a. Close to one of the two points
    b. At one-third of one of the points
    c. **At the middle point between the two points**
    d. Outside the distance between the two points
    e. At any point; it does not make a difference

27. The operation of determining the elevation of points at regular intervals along a fixed line is called:
    a. Differential leveling
    b. Linear leveling
    c. Cross-section leveling
    d. **Profile leveling**
    e. None of the above

28. The operation of determining the elevation of points at regular intervals across transverse lines perpendicular to the profile line is called:
    a. Differential leveling
    b. Linear leveling
    c. **Cross-section leveling**
    d. Profile leveling
    e. None of the above

29. Contour lines on topographic maps are imaginary lines connecting points with the same:
    a. Longitude
    b. Latitude
    c. **Elevation**
    d. All of the above
    e. None of the above

30. The contour interval on a contour map is the difference between two consecutive contour lines in terms of:
    a. Horizontal distance between the two lines
    b. Vertical distance between the two lines
    c. Perpendicular distance between the two lines
    d. Smallest distance between the two lines
    e. **Elevation**

31. The smallest closed loop (contour line) on a contour map represents:
    a. Peak of a hill
    b. Bottom of a valley
    c. Water stream

    d. **a or b**

    e. a or c

32. One of the following is *not* among the characteristics of contour lines:

    a. They do not intersect or meet or coincide only in special cases

    b. Distances between contour lines are the same when uniform slopes exist

    c. They cannot be separated into two lines with the same elevation

    d. Close contour lines indicate steep elevation differences

    e. **All of the above**

33. The contour lines have increased elevations toward the outside portion in the case of:

    a. Peaks

    b. **Valleys**

    c. Quarries

    d. Caves

    e. Mines

34. The steepest direction on a contour map is the direction that is:

    a. **Perpendicular to the contour lines**

    b. Parallel to the contour lines

    c. At 45° with the contour lines

    d. Never known

    e. None of the above

35. The central angle subtended by an arc with a length equal to the radius is the definition of:

    a. One grad

    b. One degree

    c. One minute

    d. One second

    e. **One radian**

36. An angle expressed in degrees (in the sexagesimal system) can be converted into grads (in the centesimal system) by:

    a. Multiplying the angle degrees by 0.9

    b. Multiplying the angle degrees by 57.3

    c. **Dividing the angle degrees by 0.9**

    d. Dividing the angle degrees by 57.3

    e. Dividing the angle degrees by 1.0

37. A zenith angle is measured in the vertical plane and its value can be between:

    a. 0° and 90°

    b. **0° and 180°**

    c. −90° and 90°

    d. 0° and 45°

    e. −90° and 0°

38. The acute angle that a line makes with the true meridian is called:

    a. Zenith angle

    b. Azimuth (Whole circle bearing)

    c. **True (reduced) bearing**

    d. Back bearing

    e. Back azimuth

39. The true (reduced) bearing is measured in the horizontal plane and its value can be between:

    a. **0° and 90°**

    b. 0° and 180°

    c. −90° and 90°

    d. 0° and 45°

    e. −90° and 0°

40. The true (reduced) bearing is measured in the horizontal plane and its value can be between:
    a. **0° and 90°**
    b. 0° and 180°
    c. −90° and 90°
    d. 0° and 45°
    e. −90° and 0°
41. The azimuth (whole circle bearing) is measured in the horizontal plane and can be in the range of:
    a. 0° and 90°
    b. 0° and 180°
    c. −90° and 90°
    d. **0° and 360°**
    e. −90° and 0°
42. A line with a magnetic declination (the difference between the magnetic and true meridians) of 0° is called:
    a. Meridian
    b. Isopor
    c. **Agonic line**
    d. True meridian
    e. Contour line
43. In laying out projects, the method that involves the measurement of horizontal angles is:
    a. Azimuth method
    b. Polar method
    c. Intersection method
    d. **b and c**
    e. Leveling method
44. In a traverse survey, the traverse that starts and closes at a point of known position is called:
    a. Open traverse
    b. **Closed traverse**
    c. Connecting traverse
    d. Loop traverse
    e. Regular traverse
45. The traverse that starts from a point and ends on another point of unknown position is called:
    a. **Open traverse**
    b. Closed traverse
    c. Connecting traverse
    d. Loop traverse
    e. Regular traverse
46. A special case of a closed traverse that starts and ends at the same point is called:
    a. Open traverse
    b. Irregular traverse
    c. Connecting traverse
    d. **Loop traverse**
    e. Regular traverse
47. Regular loop traverses have:
    a. Equal internal angles
    b. Equal sides
    c. **a and b**
    d. Circular shape
    e. Hexagonal shape

48. A closed traverse survey provides two independent checks on the accuracy of the field measurements that include azimuth closure and:
    a. Bearing closure
    b. Internal angle closure
    c. External angle closure
    d. **Position closure**
    e. Zenith closure

49. The procedure used to correct the coordinates of the stations of a traverse is called "balancing the traverse." There are two common rules used in balancing a traverse, the compass rule and:
    a. Azimuth rule
    b. Internal angle rule
    c. External angle rule
    d. Position rule
    e. **Transit rule**

50. One of the following methods is <u>not</u> among the methods used to compute areas:
    a. Trapezoidal rule
    b. Simpson's 1/3 rule
    c. Simpson's 3/8 rule
    d. **Average end-area method**
    e. Polar planimeter method

51. One of the following methods is <u>not</u> among the methods used to compute volumes:
    a. Spot elevations method
    b. Contours method
    c. **Polar planimeter method**
    d. Average end-area method
    e. Prismoidal method

52. The distance on the Mass Haul Diagram (MHD) at which the material is transported for free is called:
    a. Balance line
    b. Economic haul distance
    c. **Free haul distance**
    d. Over haul distance
    e. None of the above

53. The material excavated from roadway cuts that is either unsuitable or not required for making fill is called:
    a. **Waste**
    b. Excess
    c. Local borrow
    d. Important borrow
    e. Important cut

54. The borrow material excavated from sources within the right of way of the highway project is called:
    a. Waste
    b. Excess
    c. **Local borrow**
    d. Important borrow
    e. Important cut

55. The borrow material that is brought from outside the right of way of the highway project is called:
    a. Waste
    b. Excess

    c. Local borrow
    d. **Important borrow**
    e. Important cut

56. A horizontal line on the MHD that balances cut and fill quantities is called:
    a. **Balance line**
    b. Economic haul distance
    c. Free haul distance
    d. Over haul distance
    e. None of the above

57. Maximum (peak) points on the MHD indicate:
    a. Maximum cut
    b. Maximum fill
    c. Maximum change in cut
    d. Change from fill to cut
    e. **Change from cut to fill**

58. Minimum (valley) points on the MHD indicate:
    a. Maximum cut
    b. Maximum fill
    c. Maximum change in fill
    d. **Change from fill to cut**
    e. Change from cut to fill

59. A positive (+ve) slope in the MHD indicates:
    a. A cut volume
    b. A fill volume
    c. **An increase in cut volume**
    d. An increase in fill volume
    e. A maximum cut

60. A negative (−ve) slope in the MHD indicates:
    a. A cut volume
    b. A fill volume
    c. An increase in cut volume
    d. **An increase in fill volume**
    e. A maximum fill

61. The direction of hauling on the MHD of a highway project from left to right (forward hauling) occurs at:
    a. Positive slopes
    b. Negative slopes
    c. Loops below a balance line
    d. **Loops above a balance line**
    e. a or d.

62. The direction of hauling on the MHD of a highway project from right to left (backward hauling) occurs at:
    a. Positive slopes
    b. Negative slopes
    c. **Loops below a balance line**
    d. Loops above a balance line
    e. b or c

Part III

Geometric Design of Highways

8 Introduction and Standards

This chapter consists of questions that focus on important definitions, terms, and standards that are used for the geometric design of roadways. Knowledge of the different aspects of geometric design for highway facilities is crucial. The terminology used in geometric design involves terms related to the design of vertical curves, horizontal curves, superelevation, sight distance, and parking facilities. The terminology covers the factors and considerations that are taken into account in all design aspects of the aforementioned roadway facilities. The standards and requirements for the main roadway elements should also be discussed. Therefore, the questions in this chapter present a review of the basic terminology, concepts, and standards used in the geometric design of roadways. The questions are in a multiple-choice format and the answers are available at the end of this part.

8 Introduction and Standards

9 Highway Design Criteria

Chapter 9 highlights the design criteria for highway geometric design. The design parameters considered in this chapter include the design speed, width of the lane (the traveled way), width of the shoulder, and the maximum grade. The design criteria are presented for the two major categories of roadways: freeways and highways. In addition, the two types of roadways based on their functions, collectors and arterials, are considered. The location of the highway segment, whether in an urban or rural area, is also taken into consideration when selecting the design criteria. Additionally, the type of terrain (the topography of the area where the roadway is located) has a significant effect on the selection of the design criteria for a highway's geometric design. This chapter presents problems that deal with the above factors and elements.

9.1 A highway design engineer would like to determine suitable design values for the speed, lane width, shoulder width, and maximum grade for a rural freeway that is to be constructed in mountainous terrain. Provide proper design values for these four parameters.

Solution:

According to "A Policy on Geometric Design of Highways and Streets," AASHTO (2001), the recommended design speed for rural freeways is 110 km/h ($\cong$ 70 mph). However, in mountainous terrain, a design speed of 80 to 100 km/h ($\cong$ 50 to 60 mph) is recommended for rural freeways. Therefore, a design speed of 100 km/h can be selected. The functional classification of the roadway in general (and therefore, the design traffic volume) plays a role in selecting the proper design speed for a specific roadway.

The maximum grade for a rural freeway located in mountainous terrain at a design speed of 100 km/h is 6%, as shown in Table 9.1.

The recommended lane width of any roadway is within the range of 9 to 12 ft (2.7 to 3.6 m). For most high-type highways (such as freeways), the most common lane width is 12 ft (3.6 m).

The shoulder width varies from 2 ft (0.6 m) for minor rural roads to 12 ft (3.6 m) for major roads. Therefore, in the case of this rural freeway, a recommended value for the shoulder width can be 12 ft (3.6 m).

TABLE 9.1

Maximum Grades for Rural and Urban Freeways

	SI (Metric) Units Design Speed (km/h)						US Customary Units Design Speed (mph)						
	80	90	100	110	120	130	50	55	60	65	70	75	80
Type of Terrain	Grade (%)[a]						Grade (%)[a]						
Level	4	4	3	3	3	3	4	4	3	3	3	3	3
Rolling	5	5	4	4	4	4	5	5	4	4	4	4	4
Mountainous	6	6	6	5	–	–	6	6	6	5	5	–	–

[a] Grades 1% steeper than the value shown may be used for extreme cases in urban areas where development precludes the use of flatter grades and for one-way downgrades except in mountainous terrain.

Reproduced with Permission from "A Policy on Geometric Design of Highways and Streets," 2001, by AASHTO, Washington, DC, USA.

9.2   An urban freeway is to be designed and constructed in an area of level terrain. The directional design hourly volume (DDHV) for truck traffic of the freeway is 300 veh/h. Determine the recommended design values of the design speed, lane width, shoulder width, and maximum grade.

Solution:

According to "A Policy on Geometric Design of Highways and Streets," AASHTO (2001), the recommended value for the design speed of urban freeways is within the range of 80 to 100 km/h (50 to 60 mph). The quality and safety of the freeway in general, in addition to the location of the interchanges, affect the value of the design speed. For urban freeways in developing areas, a design speed of 100 km/h (60 mph) or higher can be given. Therefore, a recommended value of 100 km/h ($\cong$ 60 mph) can be selected for this freeway.

The recommended width of traffic lanes for freeways in general is 3.6 m (12 ft).

According to "A Policy on Geometric Design of Highways and Streets," AASHTO (2001), the usable paved width of the right shoulder should be at least 3 m (10 ft). When the DDHV for truck traffic is higher than 250 veh/h, the right shoulder width should be 3.6 m (12 ft). On four-lane freeways, the left shoulder width is typically 1.2 to 2.4 m (4 to 8 ft); at least 1.2 m (4 ft) of which should be paved and the remaining part stabilized. On freeways with six or more lanes, the paved width of the left shoulder should be 3 m (10 ft) and 3.6 m (12 ft) when the DDHV for truck traffic is more than 250 veh/h. Therefore, for this freeway, since the DDHV for truck traffic is 300 veh/h (more than 250 veh/h), the recommended right shoulder width is 3.6 m (12 ft). Since there is no information on the number of lanes for this freeway, a 2.4-m (8 ft) width is recommended for the left shoulder.

Finally, the recommended maximum grade values for rural and urban freeways are provided in Table 9.2.

TABLE 9.2

Maximum Grades for Rural and Urban Freeways

	SI (Metric) Units Design Speed (km/h)						US Customary Units Design Speed (mph)						
	80	90	100	110	120	130	50	55	60	65	70	75	80
Type of Terrain	Grade (%)[a]						Grade (%)[a]						
Level	4	4	3	3	3	3	4	4	3	3	3	3	3
Rolling	5	5	4	4	4	4	5	5	4	4	4	4	4
Mountainous	6	6	6	5	–	–	6	6	6	5	5	–	–

[a] Grades 1% steeper than the value shown may be used for extreme cases in urban areas where development precludes the use of flatter grades and for one-way downgrades except in mountainous terrain.

Reproduced with Permission from "A Policy on Geometric Design of Highways and Streets," 2001, by AASHTO, Washington, DC, USA.

Based on Table 9.2 and using a design speed of 100 km/h ($\cong$ 60 mph) and a "level terrain," the recommended maximum grade value is 3%.

9.3 A highway design engineer is asked to design a rural collector highway in an area of rolling terrain to carry a design traffic volume of 500 veh/day. Determine the recommended design values of the following: (a) the minimum design speed, (b) width of the traveled way, (c) width of the shoulder, and (d) the maximum grade.

Solution:

The recommended minimum design speed for a rural collector highway is given in Table 9.3.

TABLE 9.3

Minimum Design Speeds for Rural Collectors

	SI (Metric) Units Design Volume (veh/day)			US Customary Units Design Volume (veh/day)		
	0–<400	400–<2000	>2000	0–<400	400–<2000	>2000
Type of Terrain	Design Speed (km/h)			Design Speed (mph)		
Level	60	80	100	40	50	60
Rolling	50	60	80	30	40	50
Mountainous	30	50	60	20	30	40

Note: Where practical, design speeds higher than those shown should be considered.
Reproduced with Permission from "A Policy on Geometric Design of Highways and Streets," 2001, by AASHTO, Washington, DC, USA.

(a) Since the design traffic volume of the highway is 500 veh/day and the highway is located in rolling terrain, based on Table 9.3, the recommended design speed is 60 km/h ($\cong$ 40 mph).

(b) The width of the traveled way for rural collector highways is given in Table 9.4.

TABLE 9.4

Minimum Width of Traveled Way for Rural Collectors

	SI (Metric) Units Design Volume (veh/day)					US Customary Units Design Volume (veh/day)			
Design Speed (km/h)	<400	400–<1500	1500–<2000	>2000	Design Speed (mph)	<400	400–<1500	1500–<2000	>2000
	Width of Traveled Way (m)[a]					Width of Traveled Way (ft)[a]			
30	6.0[b]	6.0	6.6	7.2	20	20[b]	20	22	24
40	6.0[b]	6.0	6.6	7.2	25	20[b]	20	22	24
50	6.0[b]	6.0	6.6	7.2	30	20[b]	20	22	24
60	6.0[b]	6.6	6.6	7.2	35	20[b]	22	22	24
70	6.0	6.6	6.6	7.2	40	20	22	22	24
80	6.0	6.6	6.6	7.2	45	20	22	22	24
90	6.6	6.6	7.2	7.2	50	20	22	22	24
100	6.6	6.6	7.2	7.2	55	22	22	24	24
					60	22	22	24	24

[a] On roadways to be reconstructed, a 6.6-m (22-ft) traveled way may be retained where the alignment and safety records are satisfactory.

[b] A 5.4-m (18-ft) minimum width may be used for roadways with design volumes under 250 veh/day.

Reproduced with Permission from "A Policy on Geometric Design of Highways and Streets," 2001, by AASHTO, Washington, DC, USA.

Based on Table 9.4 and using a design traffic volume of 500 veh/day and a design speed of 60 km/h, the width of the traveled way is equal to 6.6 m ($\cong$ 22 ft).

(c) The width of the shoulder for rural collector highways is given Table 9.5.

TABLE 9.5
Minimum Width of Shoulder for Rural Collectors

	SI (Metric) Units Design Volume (veh/day)[a]					US Customary Units Design Volume (veh/day)[a]			
Design Speed (km/h)	<400	400– <1500	1500– <2000	>2000	Design Speed (mph)	<400	400– <1500	1500– <2000	>2000
	Width of Shoulder on Each Side of Road (m)					Width of Shoulder on Each Side of Road (ft)			
All Speeds	0.6	1.5[a]	1.8	2.4	All Speeds	2.0	5.0[a]	6.0	8.0

[a] Shoulder width may be reduced for design speeds greater than 50 km/h (30 mph) as long as a minimum roadway width of 9 m (30 ft) is maintained.

Reproduced with Permission from "A Policy on Geometric Design of Highways and Streets," 2001, by AASHTO, Washington, DC, USA.

Based on Table 9.5 and using a design traffic volume of 500 veh/day, the width of the shoulder for all speeds is equal to 1.5 m ($\cong$ 5 ft).

(d) The recommended maximum grade for rural collector highways is provided in Table 9.6.

TABLE 9.6
Maximum Grades for Rural Collectors

	SI (Metric) Units Design Speed (km/h)								US Customary Units Design Speed (mph)								
Type of Terrain	30	40	50	60	70	80	90	100	20	25	30	35	40	45	50	55	60
	Grade (%)								Grade (%)								
Level	7	7	7	7	7	6	6	5	7	7	7	7	7	7	6	6	5
Rolling	10	10	9	8	8	7	7	6	10	10	9	9	8	8	7	7	6
Mountainous	12	11	10	10	10	9	9	8	12	11	10	10	10	10	9	9	8

Note: Short lengths of grade in rural areas, such as grades less than 150 m (500 ft) in length, one-way downgrades, and grades on low-volume rural collectors may be up to 2% steeper than the grades shown above.

Reproduced with Permission from "A Policy on Geometric Design of Highways and Streets," 2001, by AASHTO, Washington, DC, USA.

Based on Table 9.6 and using a design speed of 60 km/h and "rolling terrain," the recommended value for the maximum grade is equal to 8%.

9.4   A collector highway is to be constructed in an urban area with level terrain. Provide recommended design values for the speed, lane width, shoulder width, and maximum grade if the highway is designed to carry a design traffic volume of 1100 veh/day.

Solution:

According to "A Policy on Geometric Design of Highways and Streets," AASHTO (2001), the recommended design speed for urban collectors is 50 km/h ($\cong$ 30 mph) or higher, depending on the available right-of-way, terrain, adjacent development, and other site factors. The type of area in this problem is given as "level terrain," and no other information is provided with regard to right-of-way and site location; therefore, a design speed of 60 km/h ($\cong$ 40 mph) is recommended.

Based on the same source, lanes can range in width from 3 to 3.6 m (10 to 12 ft). A lane width of 3.6 m (12 ft) is recommended in industrial areas except where a lack of right-of-way space exists; in these cases, a lane width of 3.3 m (11 ft) can be used. Therefore, in this problem, a 3.6 m (12 ft) is recommended for the lane width, since there is no available information on any limitation in the space for the right-of-way or the existence of industrial areas.

The shoulder width can be selected using Table 9.5 used for rural collector highways shown in Problem 9.3. Based on Table 9.5 and using a design traffic volume of 1100 veh/day, the recommended minimum value for the shoulder width is 1.5 m ($\cong$ 5 ft) for all speeds.

The maximum grade recommended values for urban collectors are provided in Table 9.7.

TABLE 9.7
Maximum Grades for Urban Collectors

Type of Terrain	SI (Metric) Units Design Speed (km/h)								US Customary Units Design Speed (mph)								
	30	40	50	60	70	80	90	100	20	25	30	35	40	45	50	55	60
	Grade (%)											Grade (%)					
Level	9	9	9	9	8	7	7	6	9	9	9	9	9	8	7	7	6
Rolling	12	12	11	10	9	8	8	7	12	12	11	10	10	9	8	8	7
Mountainous	14	13	12	12	11	10	10	9	14	13	12	12	12	11	10	10	9

Note: Short lengths of grade in urban areas, such as grades less than 150 m (500 ft) in length, one-way downgrades, and grades on low-volume urban collectors may be up to 2% steeper than the grades shown above.

Reproduced with Permission from "A Policy on Geometric Design of Highways and Streets," 2001, by AASHTO, Washington, DC, USA.

Based on Table 9.7 and using a design speed of 60 km/h and a "level terrain" as the type of the area, the recommended value of the maximum grade is 9%.

9.5   A rural arterial highway is to be designed and constructed in an area of mountain-
       ous terrain. If the design traffic volume to be carried by the highway is 1200 veh/day,
       determine the recommended design values for the minimum design speed, width of the
       traveled way, width of the shoulder, and the maximum grade.

Solution:

The design speed for rural arterials is given in Table 9.8.

TABLE 9.8
Design Speeds for Rural Arterials

	SI (Metric) Units	US Customary Units
Type of Terrain	Design Speed Range (km/h)	Design Speed Range (mph)
Level	100–120	60–75
Rolling	80–100	50–60
Mountainous	60–80	40–50

Extracted from the "A Policy on Geometric Design of Highways and Streets"
(2001), 4th Edition, American Association of State Highway and Transportation
Officials (AASHTO).

Based on Table 9.8 and using a "mountainous terrain" as the type of terrain, the recommended range of the design speed is 60 to 80 km/h ($\cong$ 40 to 50 mph).
       The width of the traveled way for rural arterial highways is provided in Table 9.9.

TABLE 9.9
Minimum Width of Traveled Way for Rural Arterials

Design Speed (km/h)	SI (Metric) Units Design Volume (veh/day)				Design Speed (mph)	US Customary Units Design Volume (veh/day)			
	<400	400– <1500	1500– <2000	>2000		<400	400– <1500	1500– <2000	>2000
	Width of Traveled Way (m)[a]					Width of Traveled Way (ft)[a]			
60	6.6	6.6	6.6	7.2	40	22	22	22	24
70	6.6	6.6	6.6	7.2	45	22	22	22	24
80	6.6	6.6	7.2	7.2	50	22	22	24	24
90	6.6	6.6	7.2	7.2	55	22	22	24	24
100	7.2	7.2	7.2	7.2	60	24	24	24	24
110	7.2	7.2	7.2	7.2	65	24	24	24	24
120	7.2	7.2	7.2	7.2	70	24	24	24	24
130	7.2	7.2	7.2	7.2	75	24	24	24	24

[a] On roadways to be reconstructed, an existing 6.6-m (22-ft) traveled way may be retained where the alignment and safety
     records are satisfactory.

Reproduced with Permission from "A Policy on Geometric Design of Highways and Streets," 2001, by AASHTO, Washington,
DC, USA.

Based on Table 9.9 and using a design traffic volume of 1200 veh/day and a design speed of 70 km/h (in the range of 60 to 80 km/h for mountainous terrain), the recommended width of traveled way is equal to 6.6 m ($\cong$ 22 ft).

On the other hand, the width of the usable shoulder for rural arterials is given in Table 9.10.

TABLE 9.10
Minimum Width of Shoulder for Rural Arterials

Design Speed (km/h)	SI (Metric) Units Design Volume (veh/day)				Design Speed (mph)	US Customary Units Design Volume (veh/day)			
	<400	400– <1500	1500– <2000	>2000		<400	400– <1500	1500– <2000	>2000
	Width of Usable Shoulder (m)[a]					Width of Usable Shoulder (ft)[a]			
All Speeds	1.2	1.8	1.8	2.4	All Speeds	4.0	6.0	6.0	8.0

[a] Usable shoulders on arterials should be paved; however, where volumes are low or a narrow section is needed to reduce construction impacts, the paved shoulder may be reduced to 0.6 m (2 ft).

Reproduced with Permission from "A Policy on Geometric Design of Highways and Streets," 2001, by AASHTO, Washington, DC, USA.

Based on Table 9.10 and using a design traffic volume of 1200 veh/day, the recommended minimum value for the shoulder width is 1.8 m ($\cong$ 6.0 ft) for all speeds.

Finally, the maximum grade for rural arterials is provided in Table 9.11.

TABLE 9.11
Maximum Grades for Rural Arterials

Type of Terrain	SI (Metric) Units Design Speed (km/h)								US Customary Units Design Speed (mph)								
	60	70	80	90	100	110	120	130	40	45	50	55	60	65	70	75	80
	Grade (%)								Grade (%)								
Level	5	5	4	4	3	3	3	3	5	5	4	4	3	3	3	3	3
Rolling	6	6	5	5	4	4	4	4	6	6	5	5	4	4	4	4	4
Mountainous	8	7	7	6	6	5	5	5	8	7	7	6	6	5	5	5	5

Reproduced with Permission from "A Policy on Geometric Design of Highways and Streets", 2001, by AASHTO, Washington, D.C., USA.

Based on Table 9.11 and using a design speed of 70 km/h ($\cong$ 45 mph) and a "mountainous terrain" for the type of terrain, the recommended maximum grade is 7%.

9.6   Determine the recommended design values for the minimum design speed, lane width, and maximum grade for an urban arterial highway that is planned to be constructed in a developing area with rolling terrain.

Solution:

According to the "A Policy on Geometric Design of Highways and Streets," AASHTO (2001), the recommended design speed for urban arterials ranges from 50 to 100 km/h (≅ 30 to 60 mph). In central business district (CBD) areas, lower speeds are recommended. Higher speeds apply in suburban and developing areas. In this problem, the area is a developing area with rolling terrain. Therefore, a design speed of 100 km/h is recommended.

According to the same source "A Policy on Geometric Design of Highways and Streets," AASHTO (2001), the recommended value for lane width varies from 3 to 3.6 m (10 to 12 ft). The lower limit value (3 m [10 ft]) is recommended in highly restricted areas with little or no truck traffic and for left-turn and combination lanes (used for traffic during peak hours and for parking during off-peak hours). On the other hand, the upper limit value (3.6 m [12 ft]) is recommended for high speed, free-flowing arterials. A lane width value of 3.3 m (11 ft) is used extensively for urban arterial streets. In conclusion, a recommended value of 3.3 m (11 ft) can be used for the lane width.

The recommended values for maximum grades for urban arterials are given in Table 9.12.

TABLE 9.12

Maximum Grades for Urban Arterials

	SI (Metric) Units Design Speed (km/h)						US Customary Units Design Speed (mph)						
	50	60	70	80	90	100	30	35	40	45	50	55	60
Type of Terrain	Grade (%)						Grade (%)						
Level	8	7	6	6	5	5	8	7	7	6	6	5	5
Rolling	9	8	7	7	6	6	9	8	8	7	7	6	6
Mountainous	11	10	9	9	8	8	11	10	10	9	9	8	8

Reproduced with Permission from "A Policy on Geometric Design of Highways and Streets", 2001, by AASHTO, Washington, D.C., USA.

Based on Table 9.12 and using a design speed of 100 km/h (≅ 60 mph) and a "rolling terrain" as the type of area, the recommended value for the maximum grade is 6%.

10 Design of Vertical Curves

Chapter 10 includes practical problems and questions related to the design of vertical curves for highways. This design involves the selection of appropriate grades and lengths for vertical curves. The type (topography) of the area, whether it is level, rolling, or mountainous, through which the highway passes has a significant effect on the design of the vertical curve. A proper design of the vertical curve must provide smooth travel (movement) for the vehicles on the curve. The vertical curve typically takes the form of a parabolic function; therefore, the properties and length of the vertical curve are based on the characteristics of a parabolic function. Two types of vertical curves are designed: crest vertical curve and sag vertical curve. In this part, practical problems related to the design of crest vertical curves as well as sag vertical curves are presented. All computations and criteria related to the design process are covered.

10.1   A +2% grade is connected with a −4% grade by a crest vertical curve at a segment of a two-lane highway. If the design speed of the highway is 60 mph (96.6 km/h), the perception–reaction time is 2.5 sec, the deceleration rate for braking (a) is 11.2 ft/sec² (3.41 m/sec²), and the stopping sight distance (SSD) is less than the length of the curve (L), determine the minimum length of the vertical curve.

Solution:

The following formulas are used to determine the stopping sight distance on the curve:

$$SSD = 1.47ut + \frac{u^2}{30\left(\dfrac{a}{g} \pm G\right)} \tag{10.1}$$

Where:
   SSD = stopping sight distance (ft)
   u = design speed (mph)
   t = perception-reaction time of drivers (typical value = 2.5 sec)
   a = acceleration/deceleration rate (typical value = 11.2 ft/sec² = 3.41 m/sec²)
   g = gravitational constant (32.18 ft/sec² = 9.81 m/sec²)
   G = grade of tangent

Since the highway has a grade (G) of +2% and another grade of −4%, the worst-case scenario for G is used to calculate the SSD. In this case, the G value of −4% is used, as shown below (see Figures 10.1 and 10.2, Figure 10.1 applies in this case):

$$SSD = 1.47ut + \frac{u^2}{30\left(\dfrac{a}{g} \pm G\right)}$$

$\Rightarrow$

$$\text{SSD} = 1.47(60)(2.5) + \frac{(60)^2}{30\left(\dfrac{11.2}{32.18} - 0.04\right)}$$

$$= 220.5 + 389.6 = 610.1 \text{ ft} \left(\cong 186 \text{ m}\right)$$

$$L_{min} = \frac{AS^2}{100\left(\sqrt{2h_1} + \sqrt{2h_2}\right)^2} \quad \text{for } S < L \tag{10.2}$$

Where:

L_{min} = minimum length of a vertical curve
A = absolute value of G_1 minus $G_2 = |G_1 - G_2|$
S = stopping sight distance
h_1 = height of driver's eye from the highway pavement surface (typical value = 3.5 ft = 1.1 m)
h_2 = height of the object from the highway pavement surface (typical value = 2.0 ft = 0.6 m)

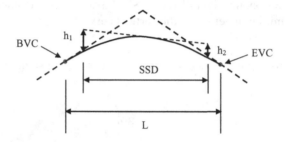

FIGURE 10.1   A crest vertical curve with a +ve grade followed by a −ve grade for Problem 10.1.

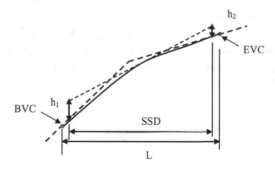

FIGURE 10.2   A crest vertical curve with a +ve grade followed by a +ve grade for Problem 10.1.

By substituting the values 3.5 ft and 2.0 ft for h_1 and h_2, respectively, the following equation is obtained:

$$L_{min} = \frac{AS^2}{2158} \quad \text{for } S < L \tag{10.3}$$

Where:

L_{min} = minimum length of vertical curve (ft)
A = absolute value of G_1 minus $G_2 = |G_1 - G_2|$
S = stopping sight distance (ft)

$$\Rightarrow$$

$$L_{min} = \frac{\left|2-(-4)\right|(610.1)^2}{2158} = 1034.8 \text{ ft} (315.4 \text{ m})$$

The MS Excel worksheet used to perform the computations in this problem in a rapid and efficient way is shown Figure 10.3.

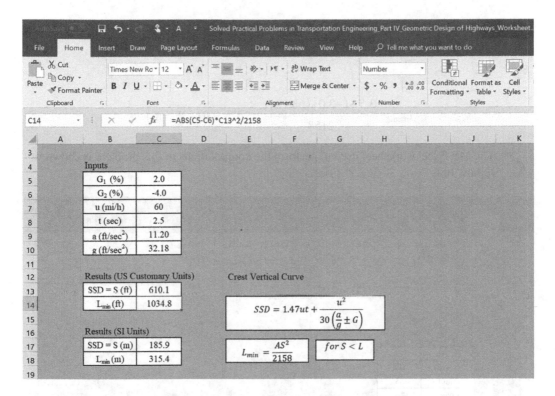

FIGURE 10.3 Image of MS Excel worksheet used for the computations of Problem 10.1

10.2 Determine the minimum length of a crest vertical curve connecting tangent grades of +3.5% and −3.0% if the design speed of the highway is 65 mph (104.6 km/h), the perception-reaction time is 2.5 sec, and the deceleration rate for braking (a) is 11.2 ft/sec² (3.41 m/sec²). Assume that the stopping sight distance (SSD) is less than the length of the curve (L).

Solution:

The stopping sight distance is first calculated on the road using the worst-case scenario (in this case, the worst grade is −3.5%):

$$SSD = 1.47ut + \frac{u^2}{30\left(\frac{a}{g} \pm G\right)}$$

$\Rightarrow$

$$SSD = 1.47(65)(2.5) + \frac{(65)^2}{30\left(\dfrac{11.2}{32.18} - 0.035\right)}$$

$$= 238.9 + 449.9 = 688.8 \text{ ft} (209.9 \text{ m})$$

$$L_{min} = \frac{AS^2}{2158} \quad \text{for } S < L$$

$\Rightarrow$

$$L_{min} = \frac{\left|3.5 - (-3.0)\right|(688.8)^2}{2158} = 1428.9 \text{ ft} (435.5 \text{ m})$$

The MS Excel worksheet used to perform the computations in this problem is shown in Figure 10.4.

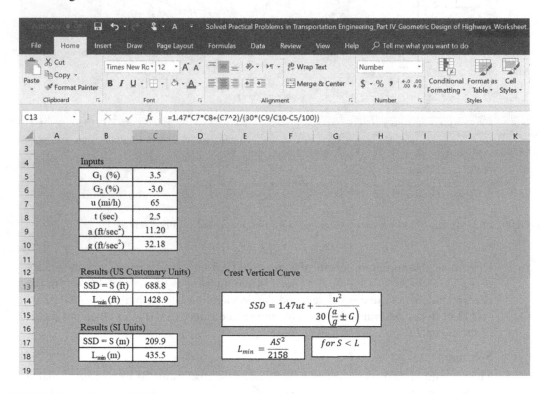

FIGURE 10.4   Image of MS Excel worksheet used for the computations of Problem 10.2.

10.3   A +2.5% and −3.5% tangent grade are connected with an existing crest vertical curve with a length of 765 ft (233.2 m). Determine the maximum allowable (safe) speed on this curve if the perception–reaction time is 2.5 sec and the deceleration rate for braking (*a*) is 11.2 ft/sec² (3.4 m/sec²). Assume that the stopping sight distance (SSD) is less than the length of the curve (*L*).

Solution:

$$L_{min} = \frac{AS^2}{2158} \quad \text{for } S < L$$

⇒

$$S = \sqrt{\frac{2158 L_{min}}{A}}$$

⇒

$$S = \sqrt{\frac{2158(765)}{|2.5-(-3.5)|}} = 524.5 \text{ ft} (159.9 \text{ m})$$

But the stopping sight distance is given by:

$$\text{SSD} = 1.47ut + \frac{u^2}{30\left(\dfrac{a}{g} \pm G\right)}$$

⇒

$$524.5 = 1.47(u)(2.5) + \frac{u^2}{30\left(\dfrac{11.2}{32.18} - 0.035\right)}$$

⇒

$$524.5 = 3.675u + 0.106482u^2$$

The MS Excel solver tool is used to solve the above quadratic equation for *u*, as shown in the procedure below:

$$\text{Left Side} = \text{SSD} \tag{10.4}$$

⇒

$$\text{Left Side} = 524.5$$

And the right side of the equation is set to:

$$\text{Right Side} = 1.47ut + \frac{u^2}{30\left(\dfrac{a}{g} \pm G\right)} \tag{10.5}$$

In this case,

$$\text{Right Side} = 3.675u + 0.106482u^2$$

An error is defined as the difference between the left side and the right side; in this case, the error is defined as follows:

$$\text{Error} = \left[\log(\text{Left Side}) - \log(\text{right side})\right]^2 \tag{10.6}$$

The objective in Excel solver is set to the "error" cell. The cell to be changed is the "u" cell. The Excel solver keeps changing the value of u (speed) through an iterative approach until the error value is zero or minimum; this is the numerical solution for the speed of the highway (u). Solving the above equation provides the following solution:

$$u = 55.0 \text{ mi/h} (88.5 \text{ km/h})$$

The MS Excel worksheet used to perform the computations in this problem is shown in Figure 10.5.

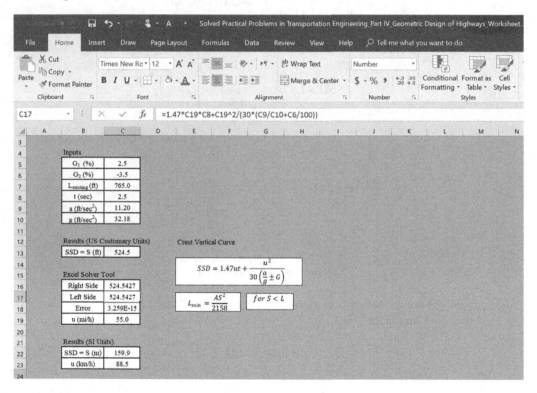

FIGURE 10.5 Image of MS Excel worksheet used for the computations of Problem 10.7.

10.4   An existing crest vertical curve for a two-lane highway connecting +4.0% and +1.5% tangent grades has a length of 150 ft (45.7 m). If the design speed of the highway at this location is 55 mph (88.5 km/h), determine if the design length of the vertical curve is appropriate for this location. Assume that the perception-reaction time is 2.5 sec and the deceleration rate for braking is 11.2 ft/sec² (3.4 m/sec²).

Solution:

$$SSD = 1.47ut + \frac{u^2}{30\left(\dfrac{a}{g} \pm G\right)}$$

The worst-case scenario for this vertical curve is when the grade is −4.0%.
⇒

$$SSD = 1.47(55)(2.5) + \frac{(55)^2}{30\left(\dfrac{11.2}{32.18} - 0.04\right)} = 529.5 \text{ ft} \left(161.4 \text{ m}\right)$$

Assume $S < L$; Therefore,

$$L_{min} = \frac{AS^2}{2158} \quad \text{for } S < L$$

⇒

$$L_{min} = \frac{|4.0 - 1.5|(529.5)^2}{2158} = 324.8 \text{ ft} \left(99.0 \text{ m}\right)$$

Since $S = 529.5$ ft $> L = 324.8$ ft, the assumption that $S < L$ is not valid. Therefore, the formula for $S > L$ should be used as shown below (Figure 10.6).

$$L_{min} = 2S - \frac{100\left(\sqrt{2h_1} + \sqrt{2h_1}\right)^2}{A} \quad \text{for } S > L \tag{10.7}$$

Where:
   L_{min} = minimum length of vertical curve
   S = stopping sight distance
   h_1 = height of driver's eye from the highway surface (typical value = 3.5 ft = 1.1 m)
   h_2 = height of the object from the highway surface (typical value = 2.0 ft = 0.6 m)
   A = absolute value of G_1 minus $G_2 = |G_1 - G_2|$

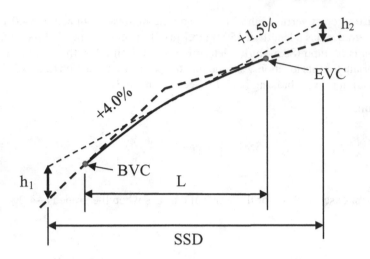

FIGURE 10.6 A crest vertical curve with a +ve grade followed by a +ve grade for Problem 10.4.

By substituting the values 3.5 ft and 2.0 ft for h_1 and h_2, respectively, the following equation is obtained:

$$L_{min} = 2S - \frac{2158}{A} \quad \text{for } S > L \tag{10.8}$$

Where:

L_{min} = minimum length of vertical curve (ft)
S = stopping sight distance (ft)
A = absolute value of G_1 minus $G_2 = |G_1 - G_2|$

$\Rightarrow$

$$L_{min} = 2(529.5) - \frac{2158}{|4.0 - 1.5|} = 195.8 \text{ ft}$$

Since the existing length of the crest vertical curve is 150 ft < 195.8 ft (the minimum required length for the curve), the existing design length is not appropriate for this location.

The MS Excel worksheet is used to conduct the computations in this problem, as shown in Figure 10.7.

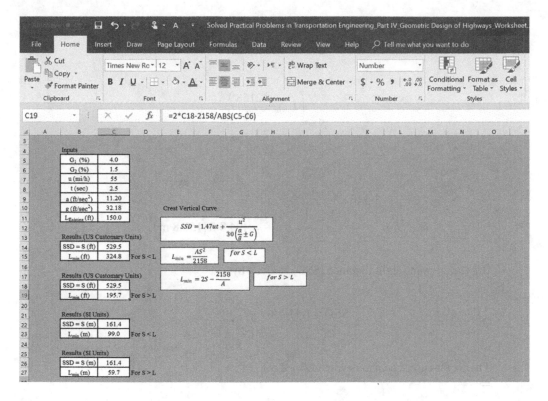

FIGURE 10.7   Image of MS Excel worksheet used for the computations of Problem 10.4.

10.5   An existing crest vertical curve with a length of 450 ft (137.2 m) is connecting two
tangent grades +1.5% and −2.5% on a highway segment. The posted speed limit on this
highway segment is 60 mph (96.6 km/h). Determine if this speed limit is appropriate
and safe for this segment if the perception–reaction time is 2.5 sec and the deceleration
rate for braking (a) is 11.2 ft/sec² (3.4 m/sec²).

Solution:

Assume $S < L$; therefore, the formula given below is used:

$$L_{min} = \frac{AS^2}{2158} \quad \text{for } S < L$$

⇒

$$S = \sqrt{\frac{2158 L_{min}}{A}}$$

⇒

$$S - \sqrt{\frac{2158(450)}{\left|1.5-(-2.5)\right|}} = 492.7 \text{ ft}(\text{m})$$

Since $S = 492.7$ ft is higher than the length of the vertical curve, $L = 450$ ft, the assumption that $S < L$ is invalid, and the formula for $S > L$ should be used as shown below:

$$L_{min} = 2S - \frac{2158}{A} \quad \text{for } S > L$$

$\Rightarrow$

$$S = \frac{\left(L_{min} + \dfrac{2158}{A}\right)}{2}$$

$\Rightarrow$

$$S = \frac{\left(450 + \dfrac{2158}{|1.5 - (-2.5)|}\right)}{2} = 494.8 \text{ ft}\left(150.8 \text{ m}\right)$$

But the stopping sight distance (SSD) is given by:

$$SSD = 1.47ut + \frac{u^2}{30\left(\dfrac{a}{g} \pm G\right)}$$

The worst-case scenario is when $G = -2.5\%$.

$\Rightarrow$

$$494.8 = 1.47(u)(2.5) + \frac{u^2}{30\left(\dfrac{11.2}{32.18} - 0.025\right)}$$

$\Rightarrow$

$$494.8 = 3.675u + 0.103186u^2$$

Using the MS Excel solver tool to solve the above quadratic equation for u, it implies that:

$$u = 53.7 \text{ mi/h}\left(86.4 \text{ km/h}\right)$$

Since the safest speed according to the existing length of the vertical crest curve is 53.7 mph < the posted speed limit on the highway segment = 60 mph; therefore, the posted speed limit is not safe for this highway segment. It should be reduced to 53.7 mph (or 50 mph).

The MS Excel worksheet used to perform the computations in this problem is shown in Figure 10.8.

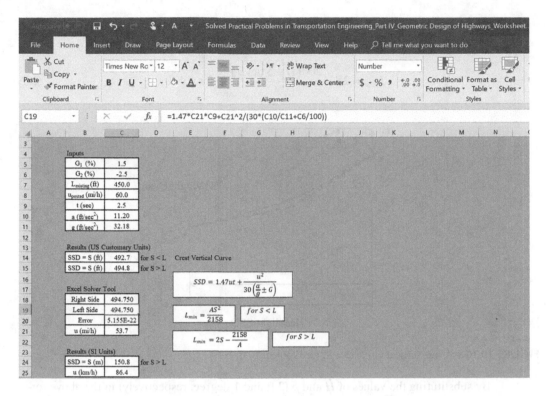

FIGURE 10.8 Image of MS Excel worksheet used for the computations of Problem 10.5.

10.6 Determine the length of a vertical sag curve connecting two tangent grades of −3% and +4% on a highway segment (as shown in Figure 10.9) with a design speed of 50 mph (80.5 km/h) that satisfies the stopping sight distance, comfort, and overall appearance criteria for designing vertical sag curves. Make appropriate assumptions in your solution.

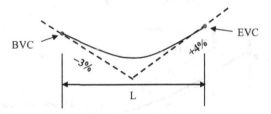

FIGURE 10.9 A sag vertical curve with a −ve grade followed by a +ve grade for Problem 10.6.

Solution:

The criteria that control the design of a sag vertical curve are:

(1) Stopping sight distance provided by the headlight.
(2) Comfort of driving on the curve.
(3) Overall appearance of the curve.
(4) Control of drainage at the lowest point of the curve.

The length of the sag vertical curve is determined based on the first three criteria as shown in the procedure below:

(1) Length of sag vertical curve based on the SSD criterion:
The headlight stopping sight distance criterion (Figures 10.10 and 10.11):

$$L_{min} = 2S - \frac{200(H + S\tan\beta)}{A} \quad \text{for } S > L \qquad (10.9)$$

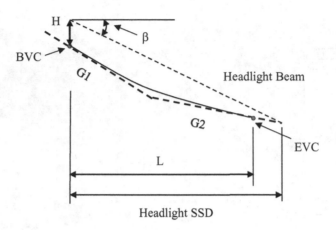

FIGURE 10.10 A sag vertical curve $(S > L)$ for Problem 10.6.

By substituting the values of H and β (2 ft and 1 degree, respectively) in the above formula, the following formula is obtained:

$$L_{min} = 2S - \frac{400 + 3.5S}{A} \quad \text{for } S > L \qquad (10.10)$$

$$L_{min} = \frac{AS^2}{200(H + S\tan\beta)} \quad \text{for } S < L \qquad (10.11)$$

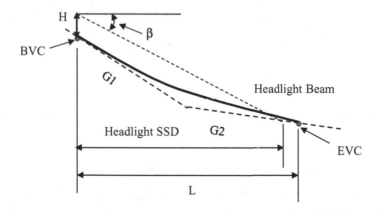

FIGURE 10.11 A sag vertical curve $(S < L)$ for Problem 10.6.

In a similar manner, substituting the values of H and β (2 ft and 1 degree, respectively) in the above formula provides the following formula:

$$L_{min} = \frac{AS^2}{400 + 3.5S} \quad \text{for } S < L \tag{10.12}$$

Where:

L_{min} = minimum length of sag vertical curve (ft)

S = stopping sight distance

H = height of vehicle's headlight from the surface of the highway pavement (typical value = 2.0 ft = 0.6 m)

β = angle of upward inclination of headlight beam from horizontal line (typical value = 1 degree)

A = absolute value of G_1 minus $G_2 = |G_1 - G_2|$

- Assume $S > L$:

$$SSD = 1.47ut + \frac{u^2}{30\left(\dfrac{a}{g} \pm G\right)}$$

The worst-case scenario is when $G = -4\%$.

$t = 2.5$ sec

$a = 11.2$ ft/sec^2

$g = 32.18$ ft/sec^2

$\Rightarrow$

$$SSD = 1.47(50)(2.5) + \frac{(50)^2}{30\left(\dfrac{11.2}{32.18} - 0.04\right)} = 454.3 \text{ ft} (138.5 \text{ m})$$

$$L_{min} = 2S - \frac{400 + 3.5S}{A} \quad \text{for } S > L$$

$\Rightarrow$

$$L_{min} = 2(454.3) - \frac{400 + 3.5(454.3)}{|-3-4|} = 624.3 \text{ ft} (190.3 \text{ m})$$

But S is not higher than L; therefore, the assumption that $S > L$ is not valid.

- For $S < L$:

$$L_{min} = \frac{AS^2}{400 + 3.5S} \quad \text{for } S < L$$

$\Rightarrow$

$$L_{min} = \frac{|-3-4|(454.3)^2}{400 + 3.5(454.3)} = 725.9 \text{ ft} (221.3 \text{ m})$$

(2) Length of sag vertical curve based on comfort of driving on the curve:

$$L_{min} = \frac{Au^2}{46.5}$$ (10.13)

Where:
L_{min} = minimum length of sag vertical curve (ft)
u = design speed (mph)

$\Rightarrow$

$$L_{min} = \frac{|-3-4|(50)^2}{46.5} = 376.3 \text{ ft} (114.7 \text{ m})$$

(3) Length of sag vertical curve based on overall appearance:

$$L_{min} = 100A$$ (10.14)

$\Rightarrow$

$$L_{min} = 100 \times |-3-4| = 700 \text{ ft} (213.4 \text{ m})$$

It is noticed from the above computations that the minimum length that satisfies all three criteria (SSD, comfort, and overall appearance) is 700 ft (213.4 m).

The computations to determine the minimum length of the sag vertical curve using the three different criteria in this problem are performed using the MS Excel worksheet shown in Figure 10.12.

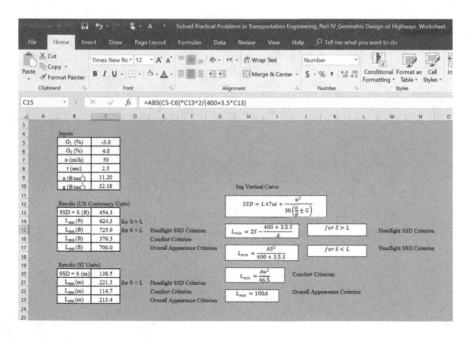

FIGURE 10.12   Image of MS Excel worksheet used for the computations of Problem 10.6.

10.7   Design a vertical sag curve connecting two tangent grades of −3.5% and −1.0% on a
highway segment with a design speed of 55 mph (88.5 km/h) that satisfies the headlight
SSD criterion. Assume that the perception–reaction time is 2.5 sec and the deceleration
rate for braking (a) is 11.2 ft/sec² (3.4 m/sec²).

Solution:

The headlight SSD criterion provides a minimum length for the sag vertical curve, as
shown in the formulas below:

$$L_{\min} = 2S - \frac{400 + 3.5S}{A} \quad \text{for } S > L$$

$$L_{\min} = \frac{AS^2}{400 + 3.5S} \quad \text{for } S < L$$

- Assume $S > L$ (Figure 10.13):

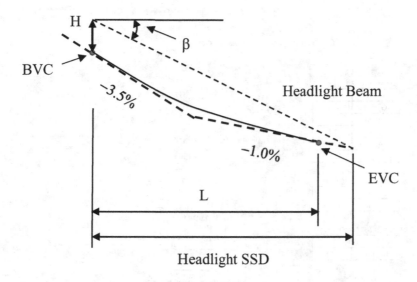

FIGURE 10.13   A sag vertical curve $(S > L)$ for Problem 10.7.

$$SSD = 1.47ut + \frac{u^2}{30\left(\dfrac{a}{g} \pm G\right)}$$

The worst-case scenario is when $G = -3.5\%$.
$t = 2.5$ sec
$a = 11.2$ ft/sec²
$g = 32.18$ ft/sec²

$\Rightarrow$

$$SSD = 1.47(55)(2.5) + \frac{(55)^2}{30\left(\dfrac{11.2}{32.18} - 0.035\right)} = 524.2 \text{ ft} (159.8 \text{ m})$$

$$L_{min} = 2S - \frac{400 + 3.5S}{A} \quad \text{for } S > L$$

$$\Rightarrow$$

$$L_{min} = 2(524.2) - \frac{400 + 3.5(524.2)}{|-3.5 - (-1.0)|} = 154.5 \text{ ft} (47.1 \text{ m})$$

Since $S = 524.2$ ft $> L = 154.5$ ft, the assumption that $S > L$ is valid, and the minimum length of the sag vertical curve that satisfies the headlight SSD is equal to 154.5 ft (47.1 m). The computations of this problem are performed using the MS Excel worksheet shown in Figure 10.14.

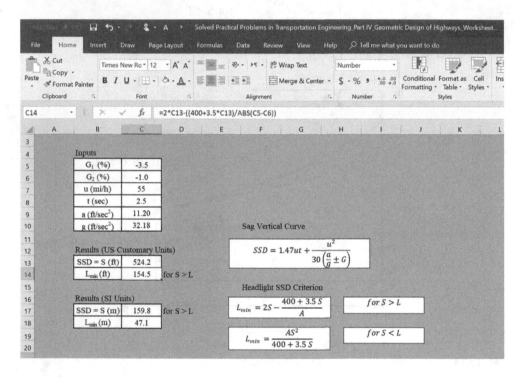

FIGURE 10.14 Image of MS Excel worksheet used for the computations of Problem 10.7.

10.8   An existing vertical sag curve (as shown in Figure 10.15) with a length of 280 ft (85.3
        m) is connecting two tangent grades of −1.0% and +2.5% on a highway segment with a
        speed limit of 50 mph (80.5 km/h). A highway engineer would like to determine if the
        combination of the existing vertical curve length and speed satisfy the headlight SSD
        criterion. What recommendations can the highway engineer make so that the sag curve
        will satisfy the criterion? Assume that the perception–reaction time is 2.5 sec and the
        deceleration rate for braking (a) is 11.2 ft/sec^2 (3.4 m/sec^2).

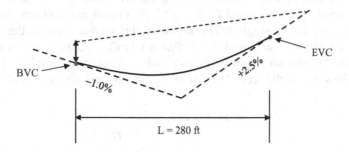

FIGURE 10.15   A sag vertical curve with a −ve grade followed by a +ve grade for Problem 10.8.

Solution:

The following formulas provide the minimum length required for a sag vertical curve
to satisfy the headlight SSD criterion:

$$L_{min} = 2S - \frac{400 + 3.5S}{A} \quad \text{for } S > L$$

$$L_{min} = \frac{AS^2}{400 + 3.5S} \quad \text{for } S < L$$

- Assume $S > L$:

$$SSD = 1.47ut + \frac{u^2}{30\left(\dfrac{a}{g} \pm G\right)}$$

The worst-case scenario is when $G = -2.5\%$.
$t = 2.5$ sec
$a = 11.2$ ft/sec^2
$g = 32.18$ ft/sec^2

$\Rightarrow$

$$SSD = 1.47(50)(2.5) + \frac{(50)^2}{30\left(\dfrac{11.2}{32.18} - 0.025\right)} = 441.7 \text{ ft} (134.6 \text{ m})$$

$$L_{min} = 2S - \frac{400 + 3.5S}{A} \quad \text{for } S > L$$

$\Rightarrow$

$$L_{min} = 2(441.7) - \frac{400 + 3.5(441.7)}{\left|-1.0 - (2.5)\right|} = 327.4 \text{ ft} (99.8 \text{ m})$$

Since $S=441.7$ ft $> L=327.4$ ft, the assumption that $S>L$ is valid, and the minimum required length of the sag vertical curve that satisfies the headlight SSD is 327.4 ft (99.8 m).

Since the existing length of the sag vertical length is 280 ft, it is not adequate for this highway segment with a speed limit of 50 mph. One of two recommendations can be made to satisfy the headlight SSD criterion on this vertical curve: either (1) increasing the existing length to at least 327.4 ft (99.8 m), or (2) reducing the speed limit so that the required minimum length would be equal to the existing length of the sag vertical curve. The speed in the second recommendation can be determined as follows:

$$SSD = 1.47ut + \frac{u^2}{30\left(\dfrac{a}{g} \pm G\right)}$$

The worst-case scenario is when $G=-2.5\%$.

$\Rightarrow$

$$SSD = 1.47(u)(2.5) + \frac{u^2}{30\left(\dfrac{11.2}{32.18} - 0.025\right)}$$

$\Rightarrow$

$$SSD = 3.675u + 0.103186u^2$$

But the minimum required length of the sag vertical curve is given by:

$$L_{min} = 2S - \frac{400 + 3.5S}{A} \quad \text{for } S > L$$

$\Rightarrow$

$$280 = 2(3.674u + 0.103186u^2) - \frac{400 + 3.5(3.674u + 0.103186u^2)}{\left|-1.0 - (2.5)\right|}$$

Or:

$$394.3 = 3.674u + 0.103186u^2$$

Solving the above equation using MS Excel solver provides the following value for the required speed:

$$u_{recommended} = 46.5 \text{ mph} (74.9 \text{ km/h})$$

The MS Excel worksheet used to perform the computations in this problem is shown in Figure 10.16.

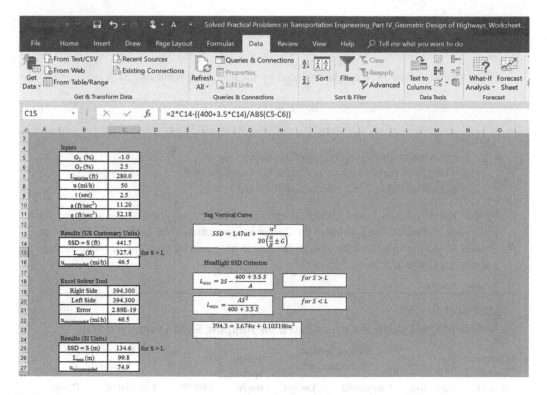

FIGURE 10.16 Image of MS Excel worksheet used for the computations of Problem 10.8.

10.9 Determine the minimum length of a crest vertical curve that connects a +2% grade with a −3% grade on a segment of a two-lane highway (Figure 10.17) using *K* factors if the design speed of the highway is 45 mph (72.4 km/h).

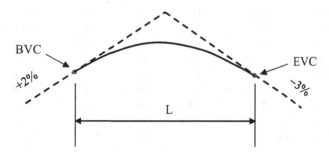

FIGURE 10.17 A crest vertical curve with a +ve grade followed by a −ve grade for Problem 10.9.

Solution:

The formulas used to determine the minimum length of a crest vertical curve are shown below:

$$L_{min} = 2S - \frac{2158}{A} \quad \text{for } S > L$$

$$L_{min} = \frac{AS^2}{2158} \quad \text{for } S < L$$

The formula when $S < L$ can be replaced by the following formula:

$$L = KA \qquad\qquad (10.15)$$

Where:
    L = minimum length of crest vertical curve
    K = length of vertical curve per percent of change of A
    A = absolute value of G_1 minus $G_2 = |G_1 - G_2|$

Since the K value is based on the design speed value that governs the stopping sight distance, the use of the K value in the design is appropriate (see Table 10.1). For the case when $S > L$, the values of the minimum length obtained by the typical formula shown above are not practical.

TABLE 10.1
K Values of Crest Vertical Curves Based on Stopping Sight Distance (SSD)

Design Speed (km/h)	SSD (m)	Rate of Vertical Curvature, K*		Design Speed (mph)	SSD (ft)	Rate of Vertical Curvature, K*	
		Calculated	Design			Calculated	Design
20	20	0.6	1	15	80	3.0	15
30	35	1.9	2	20	115	6.1	20
40	50	3.8	4	25	155	11.1	25
50	65	6.4	7	30	200	18.5	30
60	85	11.0	11	35	250	29.0	35
70	105	16.8	17	40	305	43.1	40
80	130	25.7	26	45	360	60.1	45
90	160	38.9	39	50	425	83.7	50
100	185	52.0	52	55	495	113.5	55
110	220	73.6	74	60	570	150.6	60
120	250	95.0	95	65	645	192.8	65
130	285	123.4	124	70	730	246.9	70
				75	820	311.6	75
				80	910	383.7	80

* Rate of vertical curvature, K, is the length of curve per percent algebraic difference intersecting grades (A), K = L/A.

Reproduced with Permission from "A Policy on Geometric Design of Highways and Streets," 2001, by AASHTO, Washington, DC, USA.

At a speed = 45 mph, $K = 61$. Using the formula that is based on the K factor, the minimum length of the vertical curve is determined as shown below:

$$L = KA$$

$\Rightarrow$

$$L = 61 \times \left| 2 - (-3) \right| = 305 \text{ ft} \left(93.0 \text{ m} \right)$$

The MS Excel worksheet shown in the image below is used to perform the computations of this problem (Figure 10.18).

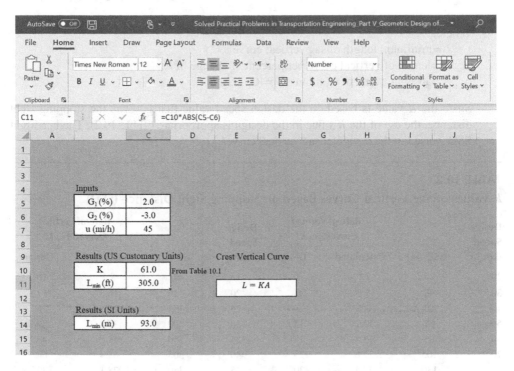

FIGURE 10.18 Image of MS Excel worksheet used for the computations of Problem 10.9.

10.10 Determine the minimum length of a sag vertical curve that connects a −4% grade with a +3% grade on a segment of a two-lane highway (Figure 10.19) using K factors if the design speed of the highway is 50 mph (80.5 km/h).

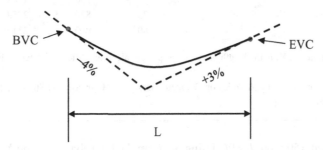

FIGURE 10.19 A sag vertical curve with a −ve grade followed by a +ve grade for Problem 10.10.

Solution:

The formulas used to determine the minimum length of a crest vertical curve are shown below:

$$L_{min} = 2S - \frac{400 + 3.5S}{A} \quad \text{for } S > L$$

$$L_{min} = \frac{AS^2}{400 + 3.5S} \quad \text{for } S < L$$

The formula when $S < L$ can be replaced by the same formula used in the design of crest vertical curves that is shown below:

$$L = KA$$

Where:
  L = minimum length of sag vertical curve
  K = length of vertical curve per percent of change of A
  A = absolute value of G_1 minus $G_2 = |G_1 - G_2|$

Since the K value is based on the design speed value that governs the stopping sight distance, the use of the K value in the design is appropriate (Table 10.2). The headlight SSD is used in the design of sag vertical curves.

TABLE 10.2
K Values of Sag Vertical Curves Based on Stopping Sight Distance (SSD)

Design Speed (km/h)	SSD (m)	Rate of Vertical Curvature, K^* Calculated	Design	Design Speed (mph)	SSD (ft)	Rate of Vertical Curvature, K^* Calculated	Design
20	20	2.1	3	15	80	9.4	10
30	35	5.1	6	20	115	16.5	17
40	50	8.5	9	25	155	25.5	26
50	65	12.2	13	30	200	36.4	37
60	85	17.3	18	35	250	49.0	49
70	105	22.6	23	40	305	63.4	64
80	130	29.4	30	45	360	78.1	79
90	160	37.6	38	50	425	95.7	96
100	185	44.6	45	55	495	114.9	115
110	220	54.4	55	60	570	135.7	136
120	250	62.8	63	65	645	156.5	157
130	285	72.7	73	70	730	180.3	181
				75	820	205.6	206
				80	910	231.0	231

* Rate of vertical curvature, K, is the length of curve per percent algebraic difference intersecting grades (A), K = L/A.

Reproduced with Permission from "A Policy on Geometric Design of Highways and Streets", 2001, by AASHTO, Washington, DC, USA.

At a speed = 50 mph, $K = 96$. Using the formula that is based on the K factor, the minimum length of the vertical curve is determined as shown below:

$$L = KA$$

$\Rightarrow$

$$L = 96 \times |-4(-3)| = 672 \text{ ft} (204.8 \text{ m})$$

The MS Excel worksheet shown in Figure 10.20 is used to perform the computations of this problem.

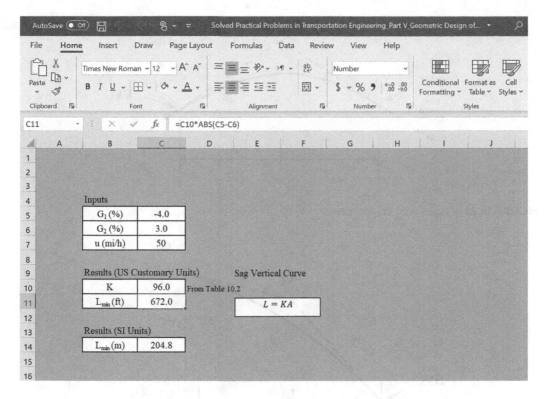

FIGURE 10.20 Image of MS Excel worksheet used for the computations of Problem 10.10.

10.11 For the crest curve in Problem 10.9 above, if the tangent grades are intersecting at a station of 125 + 50 and at an elevation of 142.0 ft (43.3 m), determine the stations and elevations of the beginning point of the vertical curve (BVC) and the end point of the vertical curve (EVC) if a length for the vertical curve (L) of 310 ft (94.5 m) is used (Figure 10.21).

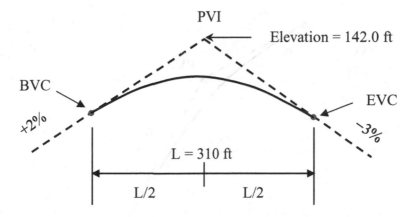

FIGURE 10.21 A crest vertical curve with a +ve grade followed by a −ve grade for Problem 10.11.

Solution:

The length of the vertical curve (L) is 310 ft.

There are three types of crest vertical curves in general. These three types are illustrated in Figures 10.22 through 10.24:

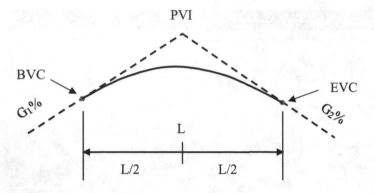

FIGURE 10.22   Type I crest vertical curve.

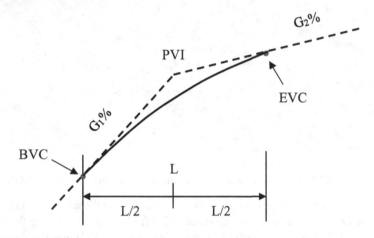

FIGURE 10.23   Type II crest vertical curve.

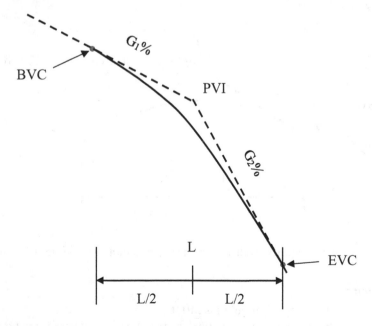

FIGURE 10.24   Type III crest vertical curve.

In all three types, the station of BVC and station of EVC are given by the same formulas shown below:

$$\text{Station of BVC} = \text{Station of PVI} - \frac{L}{2} \qquad (10.16)$$

For this problem,
⇒

$$\text{Station of BVC} = (125 + 50) - \frac{310}{2} = 123 + 95$$

$$\text{Station of EVC} = \text{Station of BVC} + L \qquad (10.17)$$

⇒

$$\text{Station of EVC} = (123 + 95) + 310 = 127 + 05$$

The elevation of BVC and elevation of EVC are determined using the following formulas:

For crest vertical curves as in Type I and Type II above, the formula used to determine the elevation of BVC is the same and described below:

$$\text{Elevation of BVC} = \text{Elevation of PVI} - \left|\frac{G_1}{100}\right|\left(\frac{L}{2}\right) \qquad (10.18)$$

For crest vertical curves of Type III, the formula used to determine the elevation of BVC is shown below:

$$\text{Elevation of BVC} = \text{Elevation of PVI} + \left|\frac{G_1}{100}\right|\left(\frac{L}{2}\right) \qquad (10.19)$$

Where:
   BVC = beginning point of vertical curve
   PVI = point of vertical intersection
   L = length of vertical curve
   G_1 and G_2 = tangent grades (%)

For this problem, since the crest vertical curve is of Type I:
⇒

$$\text{Elevation of BVC} = 142.0 - \left|\frac{2}{100}\right|\left(\frac{310}{2}\right) = 138.9 \text{ ft} (42.3 \text{ m})$$

The elevation of EVC for crest vertical curves of Type I and Type III is determined using the formula shown in the following expression:

$$\text{Elevation of EVC} = \text{Elevation of PVI} - \left|\frac{G_2}{100}\right|\left(\frac{L}{2}\right) \qquad (10.20)$$

However, the elevation of EVC for crest vertical curves of Type II is determined using the following formula:

$$\text{Elevation of EVC} = \text{Elevation of PVI} + \left|\frac{G_2}{100}\right|\left(\frac{L}{2}\right) \qquad (10.21)$$

Where:

BVC = beginning point of vertical curve
PVI = point of vertical intersection
L = length of vertical curve
G_1 and G_2 = tangent grades (%)

For this problem, since the crest vertical curve is of Type I,

⇒

$$\text{Elevation of EVC} = 142.0 - \left|\frac{-3}{100}\right|\left(\frac{310}{2}\right) = 137.4 \text{ ft}\left(41.9 \text{ m}\right)$$

The MS Excel worksheet used to perform the computations in this problem is shown in Figure 10.25.

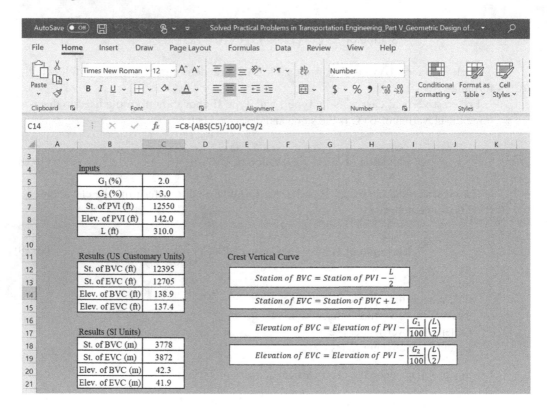

FIGURE 10.25   Image of MS Excel worksheet used for determining the stations and elevations of BVC and EVC of the crest vertical curve for Problem 10.11.

10.12   Using Problem 10.11, compute the elevations of points at 20-ft ($\cong$ 6-m) intervals on the vertical curve (Figure 10.26).

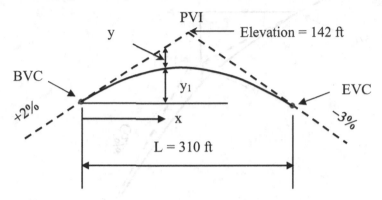

FIGURE 10.26   A crest vertical curve with a +ve grade followed by a −ve grade for Problem 10.12.

Solution:

The elevations of points on a crest vertical curve can be determined using the procedure described below. In this procedure, the three types of crest vertical curves are considered (Figures 10.27 through 10.29).

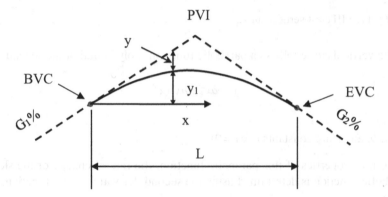

FIGURE 10.27   Type I crest vertical curve.

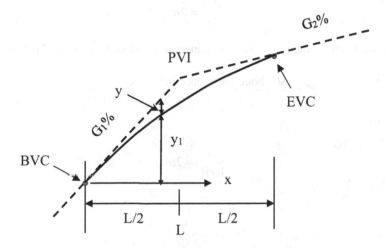

FIGURE 10.28   Type II crest vertical curve.

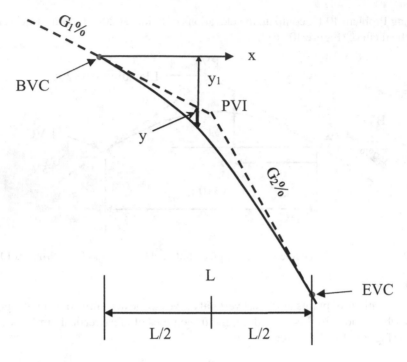

FIGURE 10.29 Type III crest vertical curve.

The vertical curve follows a parabolic function (polynomial of the second degree):

$$y = ax^2 + bx + c \tag{10.22}$$

Where:
    a, b, and c are constants ($b = c = 0$).

Using the properties of this parabolic function, the rate of change of the slope of the parabolic function is determined using the second derivative of the function, as shown below:

$$\frac{d^2y}{dx^2} = 2a \tag{10.23}$$

However, from geometry, the rate of change of the slope (s) can also be determined as:

$$s'' = \frac{\text{Algebraic Difference in Grades}}{L} = \frac{A}{100L} \tag{10.24}$$

Therefore,

$$\frac{A}{100L} = 2a \tag{10.25}$$

$\Rightarrow$

$$a = \frac{A}{200L} \tag{10.26}$$

$$y = \left(\frac{A}{200L}\right)x^2 \tag{10.27}$$

Where:

y = vertical distance from tangent to vertical curve (tangent offset), as shown in the figures above

A = absolute value of G_1 minus $G_2 = |G_1 - G_2|$

L = length of crest vertical curve

x = horizontal distance from BVC

The vertical distance from the horizontal line to the vertical curve (y_1 in the figures above) for all three types of crest vertical curves (Type I, Type II, and Type III) is equal to:

$$y_1 = \frac{G_1 x}{100} - y \tag{10.28}$$

$\Rightarrow$

$$y_1 = \frac{G_1 x}{100} - \left(\frac{A}{200L}\right)x^2 \tag{10.29}$$

Where:

y_1 = vertical distance from the horizontal line to vertical curve as shown in the figures above (+ve for Type I and Type II crest curves and −ve for Type III crest curve)

A = absolute value of G_1 minus $G_2 = |G_1 - G_2|$

L = length of crest vertical curve

x = horizontal distance from BVC

Consequently, for a crest vertical curve of any type, the elevation of any point on the curve is given by the following formula:

$$\text{Elevation of any point} = \text{Elevation of BVC} + y_1 \text{ at that point} \tag{10.30}$$

Or:

$$\text{Elevation of any point} = \text{Elevation of BVC} + \frac{G_1 x}{100} - \left(\frac{A}{200L}\right)x^2 \tag{10.31}$$

Note that the term ($G_1 x/100$) for Type I and Type II crest vertical curves will be +ve since G_1 is +ve. On the other hand, ($G_1 x/100$) for Type III will be −ve since G_1 is −ve. That is why the above two formulas are valid for all three types of crest curves and the elevation in Type III for any point on the curve will be lower than the elevation of the BVC due to the fact that y_1 will be −ve (the term $G_1 x/100$ is −ve because G_1 is −ve and the term $Ax^2/200L$ has already a −ve sign in the formula).

To determine the elevations of points on the crest vertical curve at 20-ft intervals, the above formulas are used.

Sample Calculation:

At point 4:

The station is 124 + 60, as a 20-ft distance is added to the previous station to get the new station.

$$x = \text{Station of point 4} - \text{Station of BVC}$$

$\Rightarrow$

$$x = (124 + 60) - (123 + 95) = 65 \text{ ft}$$

y_1 at $x = 65$ ft is calculated as follows:

$$y_1 = \frac{G_1 x}{100} - \left(\frac{A}{200L}\right)x^2$$

$\Rightarrow$

$$y_1 = \frac{2(65)}{100} - \left(\frac{|2-(-3)|}{200(310)}\right)(65)^2 = 0.959 \text{ ft}$$

Therefore, the elevation of this point (i.e., at $x = 65$ ft) is computed, as shown below:

$$\text{Elevation of point } 4 = \text{Elevation of BVC} + y_1$$

$\Rightarrow$

$$\text{Elevation of point } 4 = 138.90 + 0.959 = 139.9 \text{ ft}(42.6 \text{ m})$$

Table 10.3 shows the results of all points.

TABLE 10.3

Elevations of Points on the Crest Vertical Curve at 20-ft Intervals for Problem 10.12

Point	Station (ft)	x (ft)	Elevation of Point (ft)
BVC	123 + 95	0	138.9
1	124 + 00	5	139.00
2	124 + 20	25	139.35
3	124 + 40	45	139.64
4	124 + 60	65	139.86
5	124 + 80	85	140.02
6	125 + 00	105	140.11
7	125 + 20	125	140.14
8	125 + 40	145	140.10
9	125 + 60	165	140.00
10	125 + 80	185	139.84
11	126 + 00	205	139.61
12	126 + 20	225	139.32
13	126 + 40	245	138.96
14	126 + 60	265	138.54
15	126 + 80	285	138.05
16	127 + 00	305	137.50
EVC	127 + 05	310	137.35

Figures 10.30 and 10.31 show the MS Excel worksheet used to perform the computations of this problem.

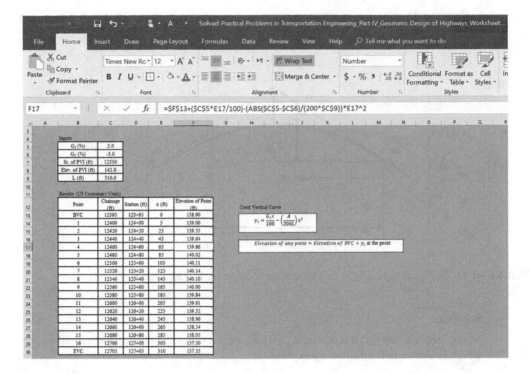

FIGURE 10.30 Image of MS Excel worksheet used for determining the elevations of points on the crest vertical curve for Problem 10.12 (US Customary units).

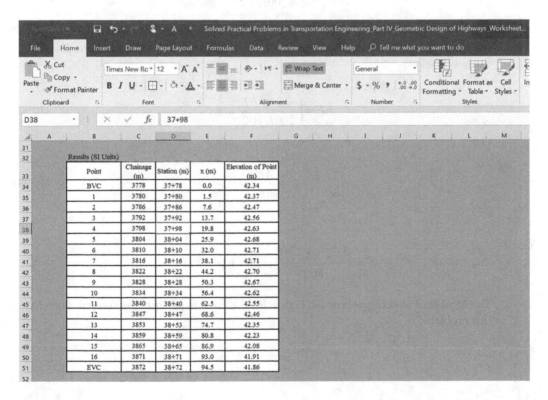

FIGURE 10.31 Image of MS Excel worksheet used for determining the elevations of points on the crest vertical curve for Problem 10.12 (SI units).

10.13 In Problem 10.12, determine the external distance (E) between the PVI and the vertical curve. See Figure 10.32.

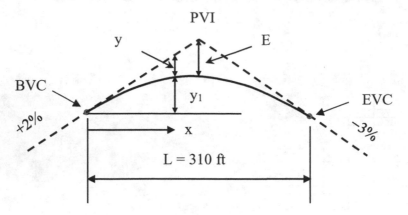

FIGURE 10.32 External distance (E) on a crest vertical curve for Problem 10.13.

Solution:

At any given x, the vertical distance y is determined using the formula described above, which is shown in the following expression:

$$y = \left(\frac{A}{200L}\right)x^2$$

The external distance (E) is the vertical distance when $x = L/2$. Therefore, substituting the value of $x = L/2$ in the above formula yields the expression to determine the E value, as shown below:

$$E = \left(\frac{A}{200L}\right)\left(\frac{L}{2}\right)^2 \tag{10.32}$$

Or:

$$E = \frac{AL}{800} \tag{10.33}$$

$\Rightarrow$

$$E = \frac{\left|2 - (-3)\right| \times 310}{800} = 1.94 \text{ ft} \left(0.59 \text{ m}\right)$$

The MS Excel worksheet shown in Figure 10.33 is used to perform the computations for this problem:

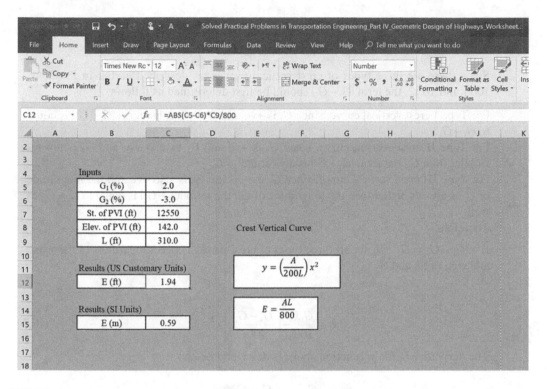

FIGURE 10.33 Image of MS Excel worksheet used for determining the external distance (E) of the crest vertical curve for Problem 10.13.

10.14 Using Problem 10.12, determine the location and elevation of the highest point on the crest vertical curve (Figure 10.34).

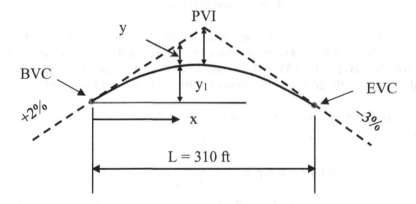

FIGURE 10.34 Highest point on a crest vertical curve for Problem 10.14.

Solution:

The highest point of the crest vertical curve occurs when the derivative of the y_1 expression is equal to zero.

$$y_1 = \frac{G_1 x}{100} - \left(\frac{A}{200L}\right)x^2$$

But:

$$A = |G_1 - G_2|$$

Since there are three types of crest vertical curves, as described earlier in Problems 10.11 and 10.12, the term $|G_1 - G_2|$ should be first determined as follows:

For Type I crest vertical curve, since G_1 is +ve and G_2 is −ve, the term $|G_1 - G_2|$ is equal to $G_1 - G_2$.

For Type II crest vertical curve, since G_1 is +ve and G_2 is also +ve and G_2 is always smaller than G_1, the term $|G_1 - G_2|$ is also equal to $G_1 - G_2$.

For Type III crest vertical curve, since G_1 is −ve and G_2 is also −ve and the magnitude of G_2 is always higher than the magnitude of G_1, the term $|G_1 - G_2|$ is equal to $G_1 - G_2$ as well.

Therefore:

$A = |G_1 - G_2| = G_1 - G_2$ for all three types of crest vertical curves regardless of the value and sign of G_1 and G_2.

⇒

$$y_1 = \frac{G_1 x}{100} - \left(\frac{G_1 - G_2}{200L} \right) x^2 \tag{10.34}$$

The derivative of the y_1 function above is described below:

$$\frac{dy_1}{dx} = \frac{G_1}{100} - 2 \left(\frac{G_1 - G_2}{200L} \right) x = 0 \tag{10.35}$$

Solving the above equation for x yields:

$$x \text{ at the highest point} = \frac{LG_1}{G_1 - G_2} \tag{10.36}$$

Note that the location of the highest point according to the above formula will be after the EVC for Type II crest vertical curves and before the BVC for Type III crest vertical curves. In this case, the EVC will be considered the highest point for Type II crest vertical curves and the BVC will be considered the highest point for Type III crest vertical curves.

For this problem, the crest vertical curve is of Type I,

⇒

$$x \text{ at the highest point} = \frac{310(2)}{2 - (-3)} = 124.0 \text{ ft} (37.8 \text{ m})$$

This is the location of the highest point from the beginning point of the vertical curve (BVC).

To determine the elevation of the highest point, the x value at the highest point is substituted in the following formula used to determine the elevation of any point on the crest vertical curve:

$$\text{Elevation of any point} = \text{Elevation of BVC} + y_1 \text{ at that point}$$

Where:

$$y_1 = \frac{G_1 x}{100} - \left(\frac{A}{200L} \right) x^2$$

Therefore, the elevation of the highest point takes the following expression:

$$\text{Elev. of highest point} = \text{Elev. of BVC} + \frac{G_1^2 L}{100(G_1 - G_2)} - \left(\frac{G_1 - G_2}{200L}\right)\left(\frac{LG_1}{G_1 - G_2}\right)^2 \quad (10.37)$$

⇒

$$\text{Elev. of highest point} = \text{Elev. of BVC} + \frac{G_1^2 L}{100(G_1 - G_2)} - \frac{G_1^2 L}{200(G_1 - G_2)} \quad (10.38)$$

⇒

$$\text{Elev. of highest point} = \text{Elev. of BVC} + \frac{G_1^2 L}{200(G_1 - G_2)} \quad (10.39)$$

This applies for all three types of crest vertical curves taking into consideration that the highest point in the case of Type II crest vertical curves is the EVC point and for Type III crest vertical, the highest point is the BVC point.

For this problem, since the crest vertical curve is of Type I,

⇒

$$\text{Elev. of highest point} = 138.9 + \frac{(2)^2 (310)}{200(2 - (-3))} = 140.1 \text{ ft} (42.7 \text{ m})$$

Figure 10.35 is an image of the MS Excel worksheet used to perform the computations of this problem:

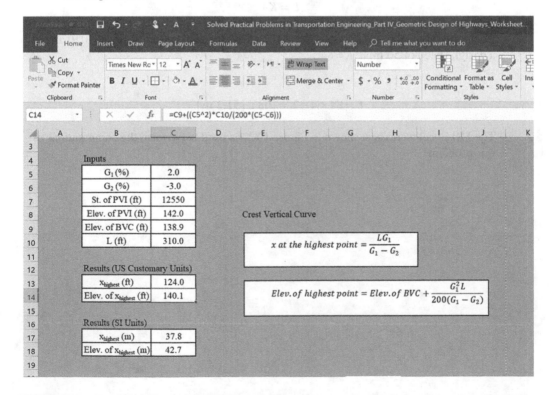

FIGURE 10.35 Image of MS Excel worksheet used for determining the location and elevation of the highest point of the crest vertical curve for Problem 10.14.

10.15   Determine the minimum length of a crest vertical curve that connects a +4% grade with
        a +2% grade on a segment of two-lane highway (Figure 10.36) using K factors if the
        design speed of the highway is 50 mph (80.5 km/h).

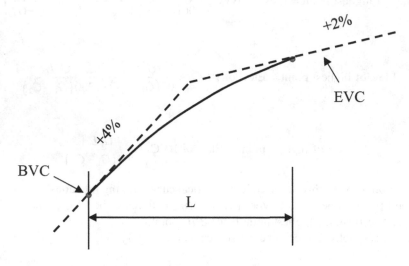

FIGURE 10.36   A crest vertical curve with a +ve grade followed by a +ve grade for Problem 10.15.

Solution:

The K-factor formula used to determine the minimum length of a crest vertical curve is
shown below:

$$L = KA$$

Using the same table for K values of crest vertical curves (used in Problem 10.9) and at
a design speed of 50 mph, $K = 84$.
⇒

$$L = 84 \times |4 - 2| = 168 \text{ ft} (51.2 \text{ m})$$

The MS Excel worksheet shown in Figure 10.37 is used to perform the computations of
this problem.

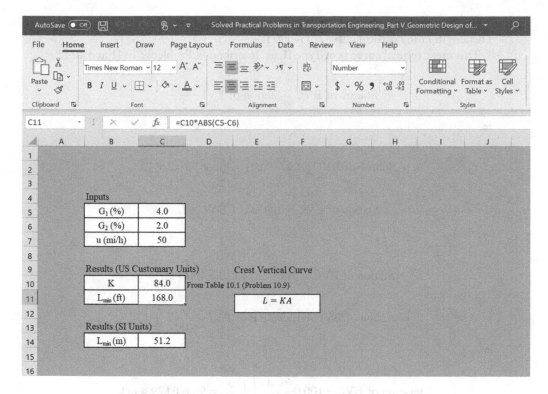

FIGURE 10.37 Image of MS Excel worksheet used for the computations of Problem 10.15.

10.16 For the crest curve in Problem 10.15, if the tangent grades are intersecting at a station of 100 + 10 and at an elevation of 100.0 ft (30.5 m), determine the stations and elevations of the beginning point of the vertical curve (BVC) and the end point of the vertical curve (EVC). Use a length of 170 ft for the vertical curve (Figure 10.38).

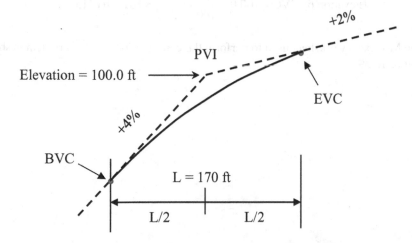

FIGURE 10.38 Stations and elevations of BVC and EVC for a crest vertical curve for Problem 10.22.

Solution:

The length of the vertical curve (L) is 170 ft.

This crest vertical curve is of Type II, as described in Problems 10.11 and 10.12.

$$\text{Station of BVC} = \text{Station of PVI} - \frac{L}{2}$$

$\Rightarrow$

$$\text{Station of BVC} = (100+10) - \frac{170}{2} = 99+25$$

$$\text{Station of EVC} = \text{Station of BVC} + L$$

$\Rightarrow$

$$\text{Station of EVC} = (99+25) + 170 = 100+95$$

$$\text{Elevation of BVC} = \text{Elevation of PVI} - \left|\frac{G_1}{100}\right|\left(\frac{L}{2}\right)$$

$\Rightarrow$

$$\text{Elevation of BVC} = 100.0 - \left|\frac{4}{100}\right|\left(\frac{170}{2}\right) = 96.6 \text{ ft} (29.4 \text{ m})$$

$$\text{Elevation of EVC} = \text{Elevation of PVI} + \left|\frac{G_2}{100}\right|\left(\frac{L}{2}\right)$$

$\Rightarrow$

$$\text{Elevation of EVC} = 100.0 + \left|\frac{2}{100}\right|\left(\frac{170}{2}\right) = 101.7 \text{ ft} (31.0 \text{ m})$$

The MS Excel worksheet used to perform the computations in this problem is shown in Figure 10.39.

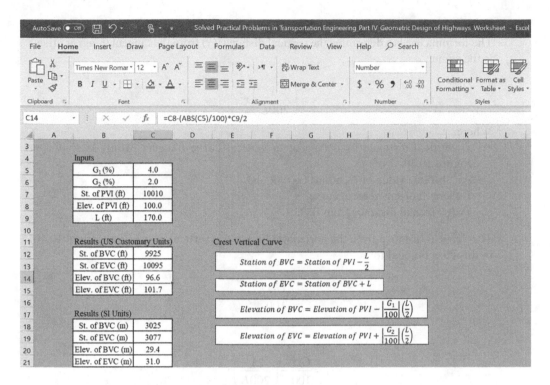

FIGURE 10.39 Image of MS Excel worksheet used for determining the stations and elevations of BVC and EVC of the crest vertical curve for Problem 10.16.

10.17 In the above problem, compute the elevations of points at 20-ft ($\cong$ 6-m) intervals on the vertical curve (Figure 10.40).

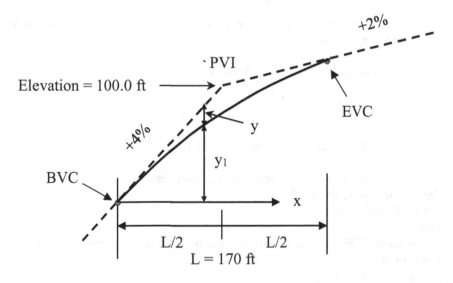

FIGURE 10.40 Elevations of points on a crest vertical curve for Problem 10.23.

Solution:

The formula used to determine the vertical distance (y) from the tangent (tangent offset) is the same:

$$y = \left(\frac{A}{200L}\right)x^2$$

Where:

y = vertical distance from tangent to vertical curve (tangent offset) as shown in Figure 10.40

A = absolute value of G_1 minus $G_2 = |G_1 - G_2|$

L = length of sag vertical curve

x = horizontal distance from BVC

The vertical distance from the horizontal line to the vertical curve (y_1 in Figure 10.40) in this case is equal to:

$$y_1 = \frac{G_1 x}{100} - y$$

$\Rightarrow$

$$y_1 = \frac{G_1 x}{100} - \left(\frac{A}{200L}\right)x^2$$

Where:

y_1 = vertical distance from horizontal line to vertical curve as shown in Figure 10.40

A = absolute value of G_1 minus $G_2 = |G_1 - G_2|$

L = length of crest vertical curve

x = horizontal distance from BVC

Consequently, for a crest vertical curve, the elevation of any point on the curve is given by the following formula:

Elevation of any point = Elevation of BVC + y_1 at that point

Or:

$$\text{Elevation of any point} = \text{Elevation of BVC} + \frac{G_1 x}{100} - \left(\frac{A}{200L}\right)x^2$$

To determine the elevations of points on the crest vertical curve at 20-ft intervals, the above formulas are used.

Sample Calculation:

At point 2:

The station is 99 + 50, as a 20-ft distance is added to the previous station to get the new station.

$$x = \text{Station of point 2} - \text{Station of BVC}$$

$\Rightarrow$

$$x = (99 + 50) - (99 + 25) = 25 \text{ ft}$$

y_1 at $x = 25$ ft is calculated as follows:

$$y_1 = \frac{G_1 x}{100} - \left(\frac{A}{200L} \right) x^2$$

$\Rightarrow$

$$y_1 = \frac{4(25)}{100} - \left(\frac{|4-2|}{200(170)} \right)(25)^2 = 0.963 \text{ ft}$$

Therefore, the elevation of this point (i.e., at $x = 25$ ft) is computed, as shown below:

$$\text{Elevation of point } 2 = \text{Elevation of BVC} + y_1$$

$\Rightarrow$

$$\text{Elevation of point } 2 = 96.60 + 0.963 = 97.6 \text{ ft} (29.7 \text{ m})$$

Table 10.4 shows the results of all points.

TABLE 10.4

Elevations of Points on the Crest Vertical Curve at 20-ft Intervals for Problem 10.17

Point	Station (ft)	x (ft)	Elevation of Point (ft)
BVC	99 + 25	0	96.60
1	99 + 30	5	96.80
2	99 + 50	25	97.56
3	99 + 70	45	98.28
4	99 + 90	65	98.95
5	100 + 10	85	99.58
6	100 + 30	105	100.15
7	100 + 50	125	100.68
8	100 + 70	145	101.16
9	100 + 90	165	101.60
EVC	100 + 95	170	101.70

Figures 10.41 and 10.42 show the MS Excel worksheet used to perform the computations of this problem:

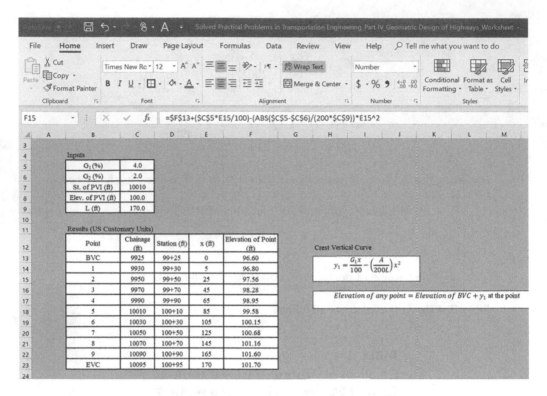

FIGURE 10.41 Image of MS Excel worksheet used for determining the elevations of points on the crest vertical curve for Problem 10.17 (US Customary units).

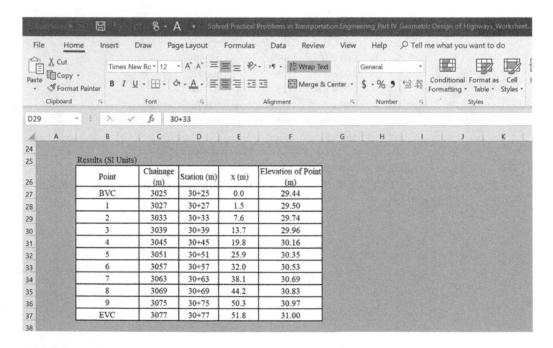

FIGURE 10.42 Image of MS Excel worksheet used for determining the elevations of points on the crest vertical curve for Problem 10.17 (SI units).

10.18   In Problem 10.16, determine the external distance (E) between the PVI and the vertical curve. See Figure 10.43.

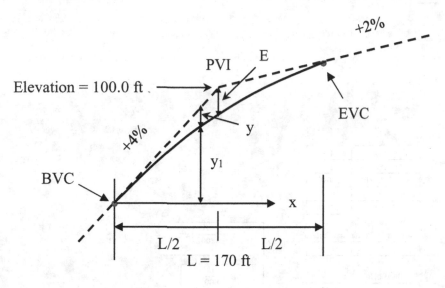

FIGURE 10.43   External distance (E) of a crest vertical curve for Problem 10.18.

Solution:

The external distance (E) is given by the expression shown in the formula below:

$$E = \left(\frac{A}{200L}\right)\left(\frac{L}{2}\right)^2$$

Or:

$$E = \frac{AL}{800}$$

⇒

$$E = \frac{|4-2|\times 170}{800} = 0.42 \text{ ft}\left(0.13 \text{ m}\right)$$

The MS Excel worksheet shown in Figure 10.44 is used to perform the computations for this problem:

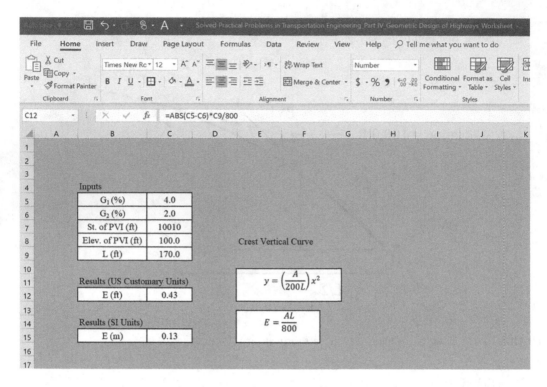

FIGURE 10.44 Image of MS Excel worksheet used for determining the external distance (*E*) of the crest vertical curve for Problem 10.18.

10.19 Using Problem 10.16, determine the location and elevation of the highest point on the crest vertical curve (Figure 10.45).

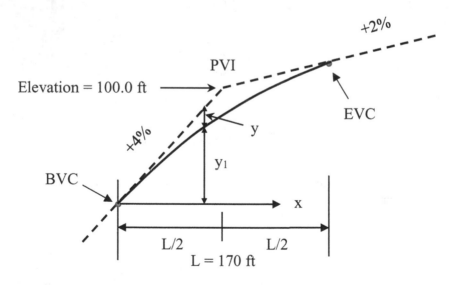

FIGURE 10.45 Highest point on a crest vertical curve for Problem 10.19.

Solution:

The location of the highest point of a crest vertical curve is determined using the following formula:

$$x \text{ at the highest point} = \frac{LG_1}{G_1 - G_2}$$

⇒

$$x \text{ at the highest point} = \frac{170(4)}{4-2} = 340.0 \text{ ft} \left(103.6 \text{ m}\right)$$

This is the location of the highest point from the beginning point of the vertical curve (BVC).

The elevation of the highest point takes the expression shown in the formula below:

$$\text{Elev. of highest point} = \text{Elev. of BVC} + \frac{G_1^2 L}{200(G_1 - G_2)}$$

⇒

$$\text{Elev. of highest point} = 96.60 + \frac{(4)^2 (170)}{200(4-2)} = 103.4 \text{ ft} \left(31.5 \text{ m}\right)$$

Note that the location of the highest point for this crest vertical curve (Type II) occurs beyond the EVC point ($x=340.0$ ft). The location of the EVC is 170.0 ft from the BVC. The elevation of the highest point computed using the formula above represents the elevation of a point outside the length of the vertical curve. Therefore, the highest point on the crest vertical curve in this case is the EVC point. Consequently, the elevation of the highest point is equal to the elevation of the EVC = 101.7 ft (31.0 m).

10.20 A sag vertical curve connects two tangent grades of −3.5% and +2.5% on a highway with a design speed of 60 mph (96.6 km/h), as shown in Figure 10.46. If the station and elevation of the PVI point are 112 + 80 and 165.0 ft (50.3 m), respectively, determine the stations and elevations of the beginning point of the vertical curve (BVC) and the end point of the vertical curve (EVC).

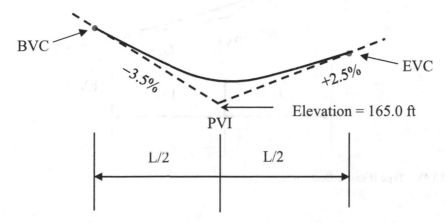

FIGURE 10.46 Stations and elevations of BVC and EVC for a sag vertical curve for Problem 10.20.

Solution:

The length of the vertical curve (L) is determined using the following formula:

$$L = KA$$

At a design speed = 60 mph, the K value is 136 (Table 10.2 in Problem 10.10 for sag vertical curves):

$\Rightarrow$

$$L = 136 \times \left| -3.5(-2.5) \right| = 816 \text{ ft} \left(248.7 \text{ m} \right)$$

The formulas used to determine the stations of the BVC and EVC points for sag vertical curves are the same as those used for crest vertical curves. There are three types for sag vertical curves as described in Figures 10.47 through 10.49.

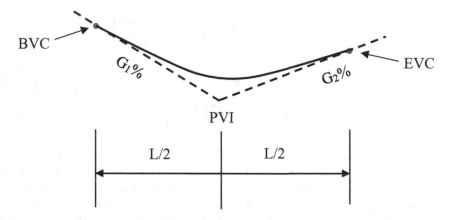

FIGURE 10.47   Type I sag vertical curve.

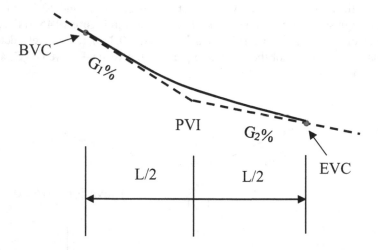

FIGURE 10.48   Type II sag vertical curve.

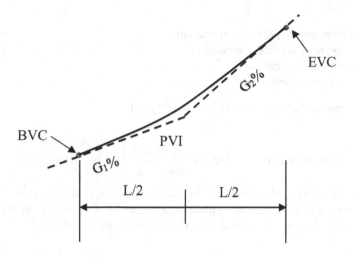

FIGURE 10.49 Type III sag vertical curve.

Therefore:

$$\text{Station of BVC} = \text{Station of PVI} - \frac{L}{2}$$

$\Rightarrow$

$$\text{Station of BVC} = \left(112 + 80\right) - \frac{816}{2} = 108 + 72$$

$$\text{Station of EVC} = \text{Station of BVC} + L$$

$\Rightarrow$

$$\text{Station of EVC} = \left(108 + 72\right) + 816 = 116 + 88$$

The formulas used to determine the elevations of the BVC and EVC points for sag vertical curves are based on the type of sag vertical curve:

For Type I and Type II sag vertical curves, the elevation of BVC is given by the following formula, which is the same formula used to determine the elevation of BVC for Type III crest vertical curves described earlier in Problems 10.11 and 10.12:

$$\text{Elevation of BVC} = \text{Elevation of PVI} + \left|\frac{G_1}{100}\right|\left(\frac{L}{2}\right)$$

On the other hand, for Type III sag vertical curves, the elevation of BVC is given by the formula below, which is the same formula used for Type I and Type II crest vertical curves described earlier in Problems 10.11 and 10.12:

$$\text{Elevation of BVC} = \text{Elevation of PVI} - \left|\frac{G_1}{100}\right|\left(\frac{L}{2}\right)$$

Where:

BVC = beginning point of vertical curve
PVI = point of vertical intersection
L = length of vertical curve
G_1 and G_2 = tangent grades (%)

In this problem, the sag vertical curve is of Type I,
$\Rightarrow$

$$\text{Elevation of BVC} = 165.0 + \left|\frac{-3.5}{100}\right|\left(\frac{816}{2}\right) = 179.3\ \text{ft}\left(54.6\ \text{m}\right)$$

For Type I and Type III sag vertical curves, the elevation of EVC is given by the following formula, which is the same formula used to determine the elevation of EVC for Type II crest vertical curves described earlier in Problems 10.11 and 10.12:

$$\text{Elevation of EVC} = \text{Elevation of PVI} + \left|\frac{G_2}{100}\right|\left(\frac{L}{2}\right)$$

And for Type II sag vertical curves, the elevation of EVC is given by the formula below, which is the same formula used for Type I and Type III crest vertical curves described earlier in Problems 10.11 and 10.12:

$$\text{Elevation of EVC} = \text{Elevation of PVI} - \left|\frac{G_2}{100}\right|\left(\frac{L}{2}\right)$$

Where:

BVC = beginning point of vertical curve
PVI = point of vertical intersection
L = length of vertical curve
G_1 and G_2 = tangent grades (%)

In this problem, the sag vertical curve is of Type I,
$\Rightarrow$

$$\text{Elevation of EVC} = 165.0 + \left|\frac{2.5}{100}\right|\left(\frac{816}{2}\right) = 175.2\ \text{ft}\left(53.4\ \text{m}\right)$$

The MS Excel worksheet used to perform the computations of this problem is shown in Figure 10.50.

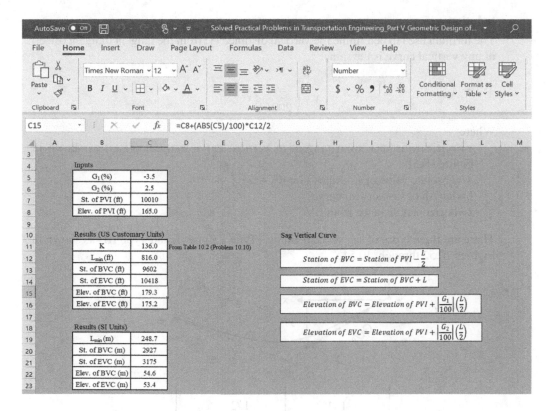

FIGURE 10.50 Image of MS Excel worksheet used for determining the stations and elevations of BVC and EVC of the sag vertical curve for Problem 10.20.

10.21 Using Problem 10.20, compute the elevations of points at 50-ft ($\cong$ 15-m) intervals on the vertical curve (Figure 10.51).

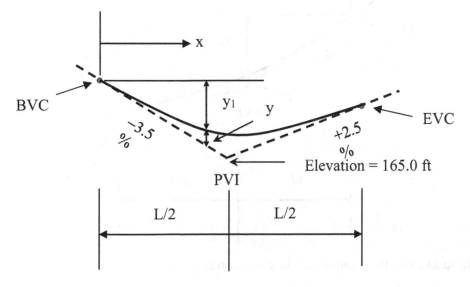

FIGURE 10.51 Elevations of points on a sag vertical curve for Problem 10.21.

Solution:

The formula used to determine the vertical distance (y) from the tangent (tangent offset) of any vertical curve is the same:

$$y = \left(\frac{A}{200L} \right) x^2$$

Where:

y = vertical distance from tangent to vertical curve (tangent offset) as shown in Figure 10.51

A = absolute value of G_1 minus $G_2 = |G_1 - G_2|$

L = length of sag vertical curve

x = horizontal distance from BVC

There are three types of sag vertical curves (Type I, Type II, and Type III) as shown in Figures 10.52 through 10.54.

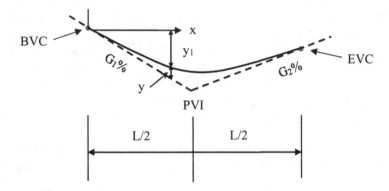

FIGURE 10.52   Type I sag vertical curve for Problem 10.21.

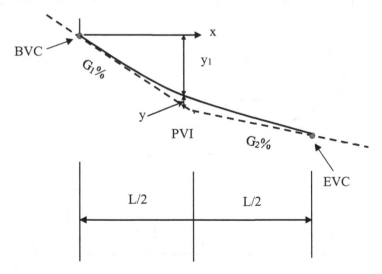

FIGURE 10.53   Type II sag vertical curve for Problem 10.21.

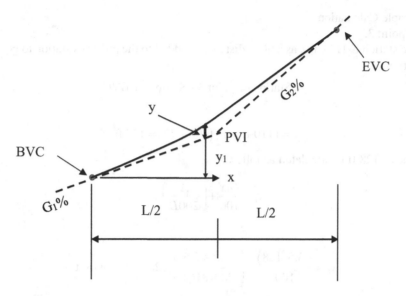

FIGURE 10.54 Type III sag vertical curve for Problem 10.21.

For Types I , Type II, and Type III sag vertical curves, the vertical distance from the horizontal line to the vertical curve (y_1 in Figures 10.52 through 10.54) is equal to:

$$y_1 = \frac{G_1 x}{100} + y \tag{10.40}$$

$\Rightarrow$

$$y_1 = \frac{G_1 x}{100} + \left(\frac{A}{200L}\right) x^2 \tag{10.41}$$

Where:
y_1 = vertical distance from horizontal line to vertical curve as shown in the figures above (−ve for Type I and Type II sag curves and +ve for Type III sag curve)
A = absolute value of G_1 minus $G_2 = |G_1 - G_2|$
L = length of crest vertical curve
x = horizontal distance from BVC

Consequently, for a sag vertical curve of any type, the elevation of any point on the curve is given by the following formula:

Elevation of any point = Elevation of BVC + y_1 at that point

Or:

$$\text{Elevation of any point} = \text{Elevation of BVC} + \frac{G_1 x}{100} + \left(\frac{A}{200L}\right) x^2 \tag{10.42}$$

Note that the term ($G_1 x/100$) for Type I and Type II sag vertical curves will be −ve since G_1 is −ve. On the other hand, ($G_1 x/100$) for Type III will be +ve since G_1 is +ve. That is why the above two formulas are valid for all three types of sag curves and the elevation in Type III for any point on the curve will be higher than the elevation of the BVC due to the fact that y_1 will be +ve (the term $G_1 x/100$ is +ve because G_1 is +ve and the term $Ax^2/200L$ has already a +ve sign in the formula).

To determine the elevations of points on the sag vertical curve at 50-ft intervals, the above formulas are used.

Sample Calculation:

At point 3:

The station is $110 + 00$, as a 50-ft distance is added to the previous station to get the new station.

$$x = \text{Station of point 3} - \text{Station of BVC}$$

$\Rightarrow$

$$x = (110 + 00) - (108 + 72) = 128 \text{ ft}$$

y_1 at $x = 128$ ft is calculated as follows:

$$y_1 = \frac{G_1 x}{100} + \left(\frac{A}{200L} \right) x^2$$

$\Rightarrow$

$$y_1 = \frac{-3.5(128)}{100} + \left(\frac{|-3.5 - 2.5|}{200(816)} \right)(128)^2 = -3.878 \text{ ft}$$

Therefore, the elevation of this point (i.e., at $x = 128$ ft) is computed, as shown below:

$$\text{Elevation of point 3} = \text{Elevation of BVC} + y_1$$

$\Rightarrow$

$$\text{Elevation of point 3} = 179.3 + (-3.878) = 175.4 \text{ ft} (53.5 \text{ m})$$

Table 10.5 shows the results of all points.

TABLE 10.5

Elevations of Points on the Sag Vertical Curve at 50-ft Intervals for Problem 10.21

Point	Station (ft)	x (ft)	Elevation of Point (ft)
BVC	108 + 72	0	179.28
1	109 + 00	28	178.33
2	109 + 50	78	176.77
3	110 + 00	128	175.40
4	110 + 50	178	174.21
5	111 + 00	228	173.21
6	111 + 50	278	172.39
7	112 + 00	328	171.76
8	112 + 50	378	171.30
9	113 + 00	428	171.03
10	113 + 50	478	170.95
11	114 + 00	528	171.05
12	114 + 50	578	171.33
13	115 + 00	628	171.80
14	115 + 50	678	172.45
15	116 + 00	728	173.28
16	116 + 50	778	174.30
EVC	116 + 88	816	175.20

Figures 10.55 and 10.56 show the MS Excel worksheet used to perform the computations of this problem.

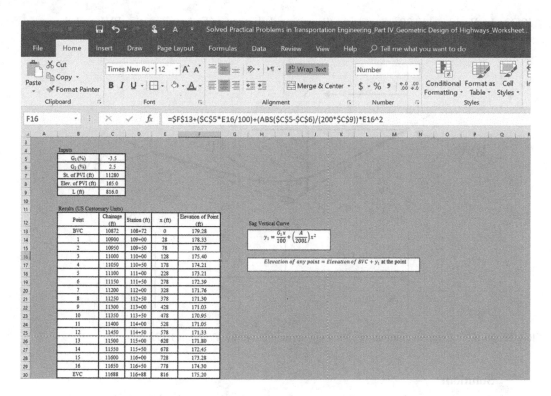

FIGURE 10.55 Image of MS Excel worksheet used for determining the elevations of points on the crest vertical curve for Problem 10.21 (US Customary units).

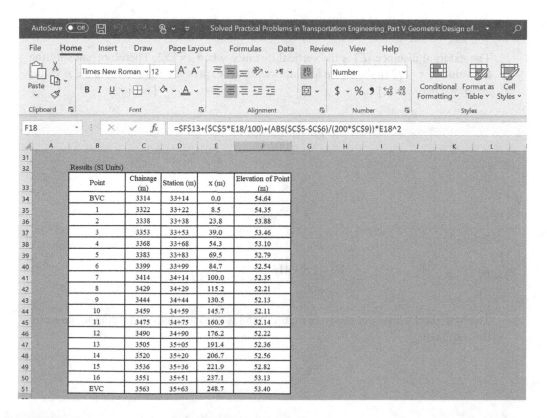

FIGURE 10.56 Image of MS Excel worksheet used for determining the elevations of points on the crest vertical curve for Problem 10.21 (SI units).

10.22 In Problem 10.20, determine the external distance (E) between the PVI and the sag
       vertical curve. See Figure 10.57.

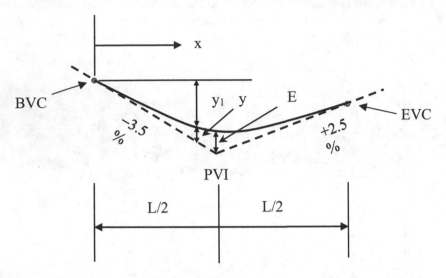

FIGURE 10.57 External distance (E) for a sag vertical curve for Problem 10.22.

Solution:

At any given x, the vertical distance y is determined using the formula described above,
which is shown in the following expression:

$$y = \left(\frac{A}{200L}\right)x^2$$

The external distance (E) is the vertical distance when $x=L/2$. Therefore, substituting
the value of $x=L/2$ in the above formula yields the expression to determine the E value,
as shown below:

$$E = \left(\frac{A}{200L}\right)\left(\frac{L}{2}\right)^2$$

Or:

$$E = \frac{AL}{800}$$

$\Rightarrow$

$$E = \frac{|-3.5-2.5|\times 816}{800} = 6.12 \text{ ft}\left(1.87 \text{ m}\right)$$

The MS Excel worksheet shown in Figure 10.58 is used to perform the computations for
this problem.

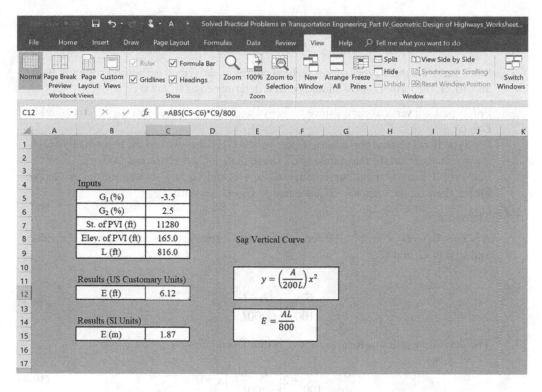

FIGURE 10.58 Image of MS Excel worksheet used for determining the external distance (e) of the sag vertical curve for Problem 10.22.

10.23 For the same problem (Problem 10.20), determine the location and elevation of the lowest point on the sag vertical curve (Figure 10.59).

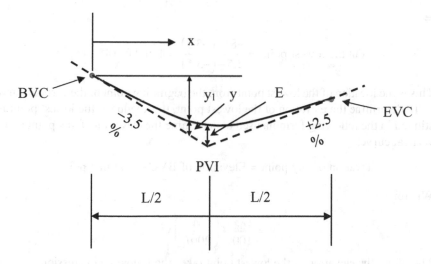

FIGURE 10.59 Lowest point on a sag vertical curve for Problem 10.23.

Solution:

The lowest point of the sag vertical curve occurs when the derivative of the y_1 expression is equal to zero.

$$y_1 = \frac{G_1 x}{100} + \left(\frac{A}{200L}\right)x^2$$

But:

$$A = |G_1 - G_2|$$

Since there are three types of sag vertical curves as described earlier in Problems 10.20 and 10.21, the term $|G_1 - G_2|$ should be first determined as follows:

For Type I sag vertical curve, since G_1 is −ve and G_2 is +ve, the term $|G_1 - G_2|$ is equal to $G_2 - G_1$.
For Type II sag vertical curve, since G_1 is −ve and G_2 is also −ve and the magnitude of G_2 is always smaller than the magnitude of G_1, the term $|G_1 - G_2|$ is also equal to $G_2 - G_1$.
For Type III sag vertical curve, since G_1 is +ve and G_2 is also +ve and G_2 is always higher than G_1, the term $|G_1 - G_2|$ is equal to $G_2 - G_1$ as well.

Therefore:

$A = |G_1 - G_2| = G_2 - G_1$ for all three types of sag vertical curves regardless of the value and sign of G_1 and G_2.
$\Rightarrow$

$$y_1 = \frac{G_1 x}{100} + \left(\frac{G_2 - G_1}{200L} \right) x^2 \tag{10.43}$$

The derivative of the y_1 function above is described below:

$$\frac{dy_1}{dx} = \frac{G_1}{100} + 2 \left(\frac{G_2 - G_1}{200L} \right) x = 0 \tag{10.44}$$

Solving the above equation for x yields:

$$x \text{ at the lowest point} = \frac{-LG_1}{G_2 - G_1} \tag{10.45}$$

$\Rightarrow$

$$x \text{ at the lowest point} = \frac{-816(-3.5)}{2.5 - (-3.5)} = 476.0 \text{ ft} (145.1 \text{ m})$$

This is the location of the lowest point from the beginning point of the vertical curve (BVC).

To determine the elevation of the lowest point, the x value at the lowest point is substituted in the following formula used to determine the elevation of any point on the sag vertical curve:

Elevation of any point = Elevation of BVC + y_1 at that point

Where:

$$y_1 = \frac{G_1 x}{100} + \left(\frac{A}{200L} \right) x^2$$

Therefore, the elevation of the lowest point takes the following expression:

$$\text{Elev. of lowest point} = \text{Elev. of BVC} - \frac{G_1^2 L}{100 (G_2 - G_1)}$$

$$+ \left(\frac{G_2 - G_1}{200L} \right) \left(\frac{-LG_1}{G_2 - G_1} \right)^2 \tag{10.46}$$

$\Rightarrow$

$$\text{Elev. of lowest point} = \text{Elev. of BVC} - \frac{G_1^2 L}{100(G_2 - G_1)} + \frac{G_1^2 L}{200(G_2 - G_1)} \qquad (10.47)$$

$\Rightarrow$

$$\text{Elev. of lowest point} = \text{Elev. of BVC} - \frac{G_1^2 L}{200(G_2 - G_1)} \qquad (10.48)$$

$\Rightarrow$

$$\text{Elev. of lowest point} = 179.3 - \frac{(-3.5)^2 (816)}{200(2.5 - (-3.5))} = 171.0 \text{ ft} (52.1 \text{ m})$$

Figure 10.60 is an image of the MS Excel worksheet used to perform the computations of this problem.

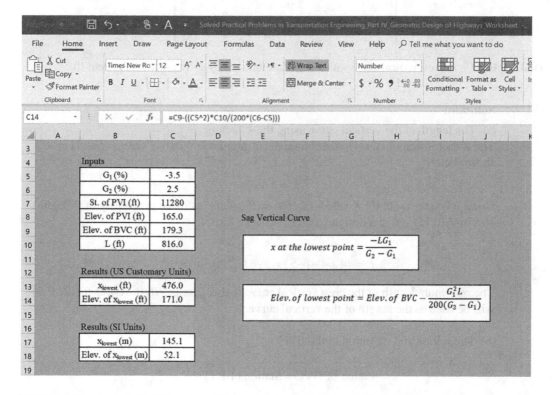

FIGURE 10.60 Image of MS Excel worksheet used for determining the location and elevation of the lowest point of the sag vertical curve for Problem 10.23.

10.24 Two tangent grades of +4% and −5% are connected by a crest vertical curve, as shown in (Figure 10.61). If the PVI is located at a station of 20 + 50 and at an elevation of 850.0 ft and the design speed is 40 mph, determine the following:

(a) The length of the vertical curve using the K factors.
(b) The station and elevation of the BVC.
(c) The station and elevation of the EVC.
(d) The station and elevation of the highest point on the curve.

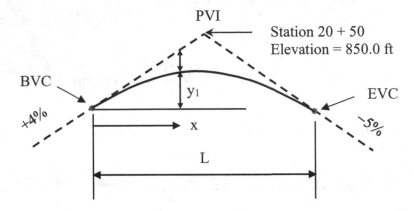

FIGURE 10.61   A Crest vertical curve with a +ve grade followed by a −ve grade for Problem 10.24.

Solution:

(a) The length of the vertical curve using the K factors:
The K-factor formula used to determine the minimum length of a crest vertical curve is shown below:

$$L = KA$$

Using the same table for K values of crest vertical curves (used previously in Problems 10.9 and 10.15) and at a design speed of 40 mph, $K=44$.
⇒

$$L = 44 \times \left| 4 - (-5) \right| = 396 \text{ ft} \left(120.7 \text{ m} \right)$$

The minimum length of the crest vertical curve in this case is 396 ft. A 400-ft (121.9 m) will be used as the length of the vertical curve.

(b) The station and elevation of the BVC:

$$\text{Station of BVC} = \text{Station of PVI} - \frac{L}{2}$$

⇒

$$\text{Station of BVC} = \left(20 + 50 \right) - \frac{400}{2} = 18 + 50$$

For Type I crest vertical curves, the formula used to determine the elevation of BVC is shown below:

$$\text{Elevation of BVC} = \text{Elevation of PVI} - \left| \frac{G_1}{100} \right| \left(\frac{L}{2} \right)$$

$\Rightarrow$

$$\text{Elevation of BVC} = 850.0 - \left|\frac{4}{100}\right|\left(\frac{400}{2}\right) = 842.0 \text{ ft}\,(256.6 \text{ m})$$

(c) The station and elevation of the EVC:

$$\text{Station of EVC} = \text{Station of BVC} + L$$

$\Rightarrow$

$$\text{Station of EVC} = (18+50) + 400 = 22 + 50$$

The elevation of EVC for crest vertical curves of Type I is determined using the formula shown in the following expression:

$$\text{Elevation of EVC} = \text{Elevation of PVI} - \left|\frac{G_2}{100}\right|\left(\frac{L}{2}\right)$$

$\Rightarrow$

$$\text{Elevation of EVC} = 850.0 - \left|\frac{-5}{100}\right|\left(\frac{400}{2}\right) = 840.0 \text{ ft}\,(256.0 \text{ m})$$

(d) The station and elevation of the highest point on the curve:
The station of the highest point for Type I crest vertical curves is given by the following formula:

$$x \text{ at the highest point} = \frac{LG_1}{G_1 - G_2}$$

$\Rightarrow$

$$x \text{ at the highest point} = \frac{400(4)}{4-(-5)} = 177.8 \text{ ft}\,(54.2 \text{ m})$$

This is the location of the highest point from the beginning point of the vertical curve (BVC).
The elevation of the highest point for Type I crest vertical curves takes the following expression:

$$\text{Elev. of highest point} = \text{Elev. of BVC} + \frac{G_1^2 L}{200(G_1 - G_2)}$$

$\Rightarrow$

$$\text{Elev. of highest point} = 842.0 + \frac{(4)^2(400)}{200(4-(-5))} = 845.6 \text{ ft}\,(257.7 \text{ m})$$

Figure 10.62 and 10.63 are images of the MS Excel worksheet used to perform the computations of this problem.

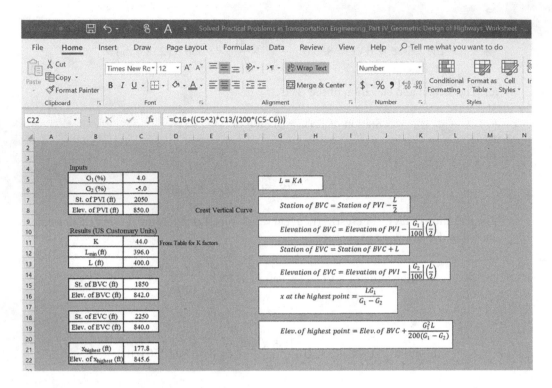

FIGURE 10.62 Image of MS Excel worksheet used for the computations of Problem 10.24 (US Customary units).

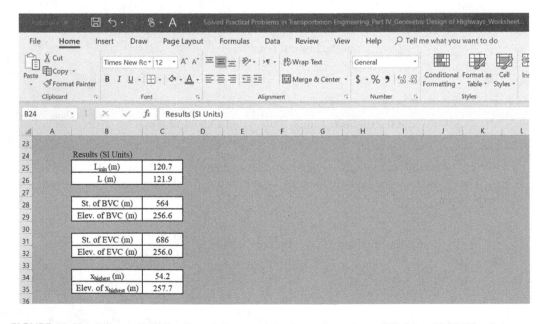

FIGURE 10.63 Image of MS Excel worksheet used for the computations of Problem 10.24 (SI units).

10.25 For Problem 10.24 above, determine the elevation of a point on the vertical curve at a station 19 + 25.

Solution:

The elevation of any point on a crest vertical curve of any type is given by the following formula:

$$\text{Elevation of any point} = \text{Elevation of BVC} + y_1 \text{ at that point}$$

Or:

$$\text{Elevation of any point} = \text{Elevation of BVC} + \frac{G_1 x}{100} - \left(\frac{A}{200L}\right)x^2$$

For this point at a station 19 + 25, the x value is computed first as follows:

$$x = \text{station of this point} - \text{station of BVC}$$

$$\Rightarrow$$

$$x = (19+25) - (18+50) = 75 \text{ ft} (22.9 \text{ m})$$

$$\Rightarrow$$

$$\text{Elevation of the point} = 842.0 + \frac{4(75)}{100} - \left(\frac{|4-(-5)|}{200(400)}\right)(75)^2$$

$$= 844.4 \text{ ft} (257.4 \text{ m})$$

The MS Excel worksheet used to perform the computations in this problem is hown in Figure 10.64.

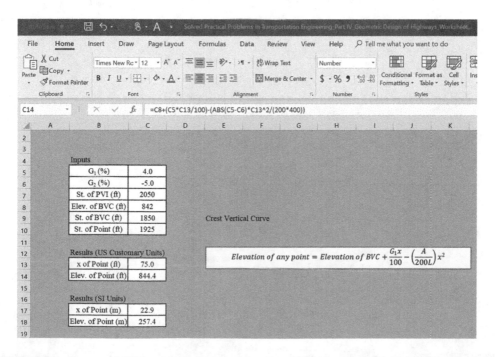

FIGURE 10.64 Image of MS Excel worksheet used for determining the location and elevation of a point on the crest vertical curve for Problem 10.25.

11 Design of Horizontal Curves and Superelevation

Chapter 11 highlights the engineering concepts and principles used in the design of horizontal curves. The types of horizontal curves considered in this chapter include simple circular curves, composite curves with two symmetric spirals, and compound curves. The design criteria as well as the engineering formulas used in the design are introduced. Computations required to set out the horizontal curves, including simple circular curves and compound curves, are also detailed. The method of deflection angles and chord lengths is used to set out simple horizontal and compound curves. Superelevation and stopping sight distance for simple horizontal curves are addressed when needed in each section of this chapter. The need for spiral curves in simple horizontal curves is assessed and the sightline offset (middle ordinate) computations are emphasized as well.

11.1   A simple horizontal curve with a degree of curvature of 5° and an intersection angle (Δ) of 40° is shown in Figure 11.1. Determine the radius and the length of the horizontal curve.

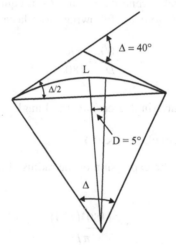

FIGURE 11.1   A simple horizontal curve with $D = 5°$ for Problem 11.1.

Solution:

The degree of curvature (D) of a simple horizontal (circular) curve is defined as the central angle (in radians) that is subtended by an arc length of 100 ft as shown in Figure 11.2. Therefore, the central angle, θ (in radians) can be written, as shown in the expression below:

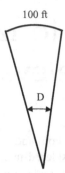

100 ft

D

FIGURE 11.2 A central angle subtended by an arc length of 100 ft.

$$\theta = \frac{\pi D}{180} \qquad (11.1)$$

Where:

θ = central angle (in radians)

D = degree of curvature (in degrees) = central angle (in degrees) opposite to an arc length of 100 ft

The arc length that subtends a central angle of 180° is equal to half of the circumference of the circle, i.e., πR. Therefore, the following formula can be written:

$$\frac{100}{D} = \frac{\pi R}{180} \qquad (11.2)$$

Where:

D = degree of curvature (in degrees) = central angle (in degrees) opposite to an arc length of 100 ft

R = radius of simple circular curve (ft)

From the above formula, the expression of the radius of a simple horizontal curve can be derived:

$$R = \frac{(180)(100)}{\pi D} \qquad (11.3)$$

Or:

$$R = \frac{5729.6}{D} \qquad (11.4)$$

The length of the arc is equal to the radius of the circular curve multiplied by the central angle (in radians):

$$\text{Arc Length} = R\theta \qquad (11.5)$$

And since $\theta = \frac{\pi D}{180}$, therefore, the expression of the arc length can be re-written as:

$$\text{Arc Length} = \frac{R\pi D}{180} \qquad (11.6)$$

For a simple circular curve, the central angle is the intersection angle (Δ). Consequently, the length of the arc that subtends this angle is the length of the horizontal curve and can be written based on the above formula as:

$$L = \frac{R\pi\Delta}{180}$$ (11.7)

Where:
L = length of simple horizontal (circular) curve
R = radius of horizontal curve
Δ = intersection angle (in degrees)

$\Rightarrow$

$$R = \frac{(180)(100)}{\pi(5)} = 1146 \text{ ft} (349.3 \text{ m})$$

$$L = \frac{(1146)(\pi)(40)}{180} = 800 \text{ ft} (243.8 \text{ m})$$

Figure 11.3 shows the MS Excel worksheet used to do the computations of this problem.

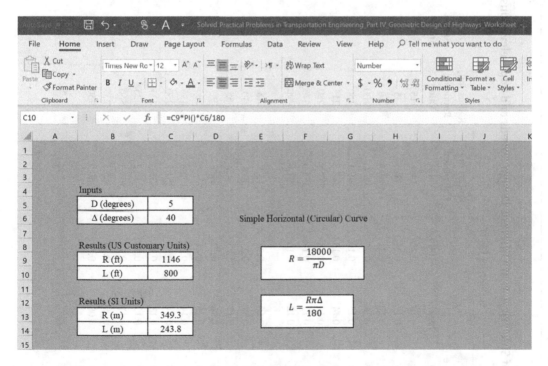

FIGURE 11.3 Image of MS Excel worksheet used for the computations of Problem 11.1.

11.2 Determine the radius and length of a simple horizontal (circular) curve with an intersection angle of 45° if the curve is within a highway segment with a design speed of 50 mph (80.5 km/h). Use a superelevation of 6.0%.

Solution:

At a superelevation of 6.0% and a design speed of 50 mph, the limiting value of the side friction (f) is 0.14 using Table 11.1.

TABLE 11.1
Maximum Value of Side Friction at Different Design Speeds and Superelevations

SI (Metric) Units

Design Speed (km/h)	Max. e (%)	Max. f	(e/100 + f)	Radius (m), Calculated Value	Radius (m), Rounded Value
20	4.0	0.18	0.22	14.3	15
30	4.0	0.17	0.21	33.7	35
40	4.0	0.17	0.21	60.0	60
50	4.0	0.16	0.20	98.4	100
60	4.0	0.15	0.19	149.1	150
70	4.0	0.14	0.18	214.2	215
80	4.0	0.14	0.18	279.8	280
90	4.0	0.13	0.17	375.0	375
100	4.0	0.12	0.16	491.9	490
20	6.0	0.18	0.24	13.1	15
30	6.0	0.17	0.23	30.8	30
40	6.0	0.17	0.23	54.7	55
50	6.0	0.16	0.22	89.4	90
60	6.0	0.15	0.21	134.9	135
70	6.0	0.14	0.20	192.8	195
80	6.0	0.14	0.20	251.8	250
90	6.0	0.13	0.19	335.5	335
100	6.0	0.12	0.18	437.2	435
110	6.0	0.11	0.17	560.2	560
120	6.0	0.09	0.15	755.5	755
130	6.0	0.08	0.14	950.0	950

US Customary Units

Design Speed (mph)	Max. e (%)	Max. f	(e/100 + f)	Radius (ft), Calculated Value	Radius (ft), Rounded Value
15	4.0	0.175	0.215	70.0	70
20	4.0	0.170	0.210	127.4	125
25	4.0	0.165	0.205	203.9	205
30	4.0	0.160	0.200	301.0	300
35	4.0	0.155	0.195	420.2	420
40	4.0	0.150	0.190	563.3	565
45	4.0	0.145	0.185	732.2	730
50	4.0	0.140	0.180	929.0	930
55	4.0	0.130	0.170	1190.2	1190
60	4.0	0.120	0.160	1505.0	1505
15	6.0	0.175	0.235	64.0	65
20	6.0	0.170	0.230	116.3	115
25	6.0	0.165	0.225	185.8	185
30	6.0	0.160	0.220	273.6	275
35	6.0	0.155	0.215	381.1	380
40	6.0	0.150	0.210	509.6	510
45	6.0	0.145	0.205	660.7	660
50	6.0	0.140	0.200	836.1	835
55	6.0	0.130	0.190	1065.0	1065
60	6.0	0.120	0.180	1337.8	1340
65	6.0	0.110	0.170	1662.4	1660
70	6.0	0.100	0.160	2048.5	2050
75	6.0	0.090	0.150	2508.4	2510
80	6.0	0.080	0.140	3057.8	3060

(Continued)

TABLE 11.1 (CONTINUED)
Maximum Value of Side Friction at Different Design Speeds and Superelevations

SI (Metric) Units

Design Speed (km/h)	Max. e (%)	Max. f	(e/100 + f)	Radius (m), Calculated Value	Radius (m), Rounded Value
20	8.0	0.18	0.28	12.1	10
30	8.0	0.17	0.25	28.3	30
40	8.0	0.17	0.25	50.4	50
50	8.0	0.16	0.24	82.0	80
60	8.0	0.15	0.23	123.2	125
70	8.0	0.14	0.22	175.3	175
80	8.0	0.14	0.22	228.9	230
90	8.0	0.13	0.21	303.6	305
100	8.0	0.12	0.20	393.5	395
110	8.0	0.11	0.19	501.2	500
120	8.0	0.09	0.17	666.6	665
130	8.0	0.08	0.18	831.3	830
20	10.0	0.18	0.28	11.2	10
30	10.0	0.17	0.27	26.2	25
40	10.0	0.17	0.27	46.6	45
50	10.0	0.16	0.26	75.7	75
60	10.0	0.15	0.25	113.3	115
70	10.0	0.14	0.24	160.7	160
80	10.0	0.14	0.24	209.9	210
90	10.0	0.13	0.23	277.2	275
100	10.0	0.12	0.22	357.7	360
110	10.0	0.11	0.21	453.5	455

US Customary Units

Design Speed (mph)	Max. e (%)	Max. f	(e/100 + f)	Radius (ft), Calculated Value	Radius (ft), Rounded Value
15	8.0	0.175	0.255	59.0	60
20	8.0	0.170	0.250	107.0	105
25	8.0	0.165	0.245	170.8	170
30	8.0	0.160	0.240	250.8	250
35	8.0	0.155	0.235	348.7	350
40	8.0	0.150	0.230	465.3	465
45	8.0	0.145	0.225	502.0	500
50	8.0	0.140	0.220	760.1	760
55	8.0	0.130	0.210	963.5	965
60	8.0	0.120	0.200	1204.0	1205
65	8.0	0.110	0.190	1487.4	1485
70	8.0	0.100	0.180	1820.9	1820
75	8.0	0.090	0.170	2213.3	2215
80	8.0	0.080	0.160	2675.6	2675
15	10.0	0.175	0.275	54.7	55
20	10.0	0.170	0.270	99.1	100
25	10.0	0.165	0.265	157.8	160
30	10.0	0.160	0.280	231.5	230
35	10.0	0.155	0.255	321.3	320
40	10.0	0.150	0.250	428.1	430
45	10.0	0.145	0.245	552.9	555
50	10.0	0.140	0.240	696.8	695
55	10.0	0.130	0.230	879.7	880
60	10.0	0.120	0.220	1094.6	1095

(Continued)

TABLE 11.1 (CONTINUED)
Maximum Value of Side Friction at Different Design Speeds and Superelevations

SI (Metric) Units

Design Speed (km/h)	Max. e (%)	Max. f	(e/100+f)	Radius (m), Calculated Value	Radius (m), Rounded Value
120	10.0	0.09	0.19	596.5	595
130	10.0	0.08	0.18	738.9	740
20	12.0	0.18	0.30	10.5	10
30	12.0	0.17	0.29	24.4	15
40	12.0	0.17	0.29	43.4	45
50	12.0	0.16	0.28	70.3	70
60	12.0	0.15	0.27	104.9	105
70	12.0	0.14	0.26	148.3	150
80	12.0	0.14	0.26	193.7	195
90	12.0	0.13	0.25	255.0	255
100	12.0	0.12	0.24	327.9	330
110	12.0	0.11	0.23	414.0	415
120	12.0	0.09	0.21	539.7	540
130	12.0	0.08	0.20	665.0	665

US Customary Units

Design Speed (mph)	Max. e (%)	Max. f	(e/100+f)	Radius (ft), Calculated Value	Radius (ft), Rounded Value
65	10.0	0.110	0.210	1345.8	1345
70	10.0	0.100	0.200	1838.8	1840
75	10.0	0.090	0.190	1980.3	1980
80	10.0	0.080	0.180	2378.3	2380
15	12.0	0.175	0.295	51.0	50
20	12.0	0.170	0.290	92.3	90
25	12.0	0.165	0.285	146.7	145
30	12.0	0.160	0.280	215.0	215
35	12.0	0.155	0.275	298.0	300
40	12.0	0.150	0.270	396.4	395
45	12.0	0.145	0.265	511.1	510
50	12.0	0.140	0.260	643.2	645
55	12.0	0.130	0.250	809.4	810
60	12.0	0.120	0.240	1003.4	1005
65	12.0	0.110	0.230	1228.7	1230
70	12.0	0.100	0.220	1489.8	1490
75	12.0	0.090	0.210	1791.7	1790
80	12.0	0.080	0.200	2140.5	2140

Reproduced with Permission from "A Policy on Geometric Design of Highways and Streets," 2001, by AASHTO, Washington, DC, USA.

The formula used to determine the radius of a simple horizontal (circular) curve is shown below:

$$R = \frac{u^2}{g(e+f)} \tag{11.8}$$

Where:
R=radius of a simple horizontal (circular) curve
g=acceleration of gravity or gravitational constant (9.81 m/sec² or 32.2 ft/sec²)
u=design speed
e=rate of superelevation (as a fraction)
f=coefficient of side friction

When the units of the design speed (u) are mph, the conversion factor from mph to ft/sec is 1.467, and g=32.2 ft/sec². Therefore, a constant of (1.467)²/32.2=1/14.97 is used in the formula. The new expression of the formula becomes:

$$R = \frac{u^2}{14.97(e+f)} \tag{11.9}$$

Where:
R=radius of simple horizontal (circular) curve (ft)
g=acceleration of gravity or gravitational constant (32.2 ft/sec²)
u=design speed (mph)
e=rate of superelevation as a fraction (ft/ft)
f=coefficient of side friction

$\Rightarrow$

$$R = \frac{(50)^2}{14.97(0.06+0.14)} = 835 \text{ ft} (\cong 255 \text{ m})$$

The length of the horizontal curve is determined using the following formula:

$$L = \frac{R\pi \Delta}{180}$$

$\Rightarrow$

$$L = \frac{(835)(\pi)(45)}{180} = 655.8 \text{ ft} (\cong 200 \text{ m})$$

The MS Excel worksheet used to perform the computations of this problem is shown in Figure 11.4.

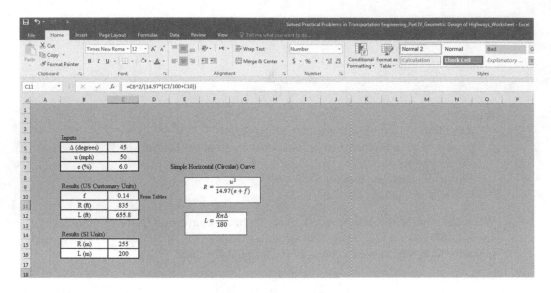

FIGURE 11.4　Image of MS Excel worksheet used for the computations of Problem 11.2.

11.3　A simple horizontal (circular) curve is to be designed for a rural highway with the following design data:
- Design speed = 60 mph (96.6 km/h)
- Maximum superelevation rate = 6.0%
- Intersection angle (Δ) = 30°
- Station of point of intersection (PI) = 10 + 40

Determine the radius of the horizontal curve, the tangent length, the station of the beginning point of the curve (BC), and the station of the end point of the curve (EC).

Solution:

The radius of the simple horizontal curve is determined using the following formula:

$$R = \frac{u^2}{14.97\left(e + f\right)}$$

At a design speed of 60 mph and a superelevation of 6.0%, the coefficient of side friction according to Table 11.1 (used in Problem 11.2) is equal to 0.12, therefore:

$$R = \frac{\left(60\right)^2}{14.97\left(0.06 + 0.12\right)} = 1336 \text{ ft}\left(407.2 \text{ m}\right)$$

The tangent length can be computed using the formula shown below:

$$T = R \tan\left(\frac{\Delta}{2}\right) \tag{11.10}$$

Where:
　T = length of simple horizontal curve tangent
　R = radius of simple horizontal curve
　Δ = intersection angle

$\Rightarrow$

$$T = 1336 \tan\left(\frac{30°}{2}\right) = 358.0 \text{ ft} (109.1 \text{ m})$$

The station of the BC point and the station of the EC point are determined using the following formulas, respectively:

$$\text{Station of BC Point} = \text{Station of } PI - T \qquad (11.11)$$

$$\text{Station of EC Point} = \text{Station of } BC + L \qquad (11.12)$$

Where:
  BC = beginning point of simple horizontal curve
  PI = point of intersection of simple horizontal curve
  T = tangent length
  EC = end point of simple horizontal curve
  L = length of simple horizontal curve

$$\text{Station of BC Point} = (10 + 40) - 358 = 6 + 82$$

Before determining the station of the EC point, the length of the simple horizontal curve is computed as follows:

$$L = \frac{R\pi\,\Delta}{180}$$

$\Rightarrow$

$$L = \frac{(1336)(\pi)(30)}{180} = 699.5 \text{ ft} \cong 700 \text{ ft} (\cong 213.2 \text{ m})$$

$\Rightarrow$

$$\text{Station of EC Point} = (6 + 82) + 700 = 13 + 82$$

The computations of this problem are done using the MS Excel worksheet shown in Figure 11.5.

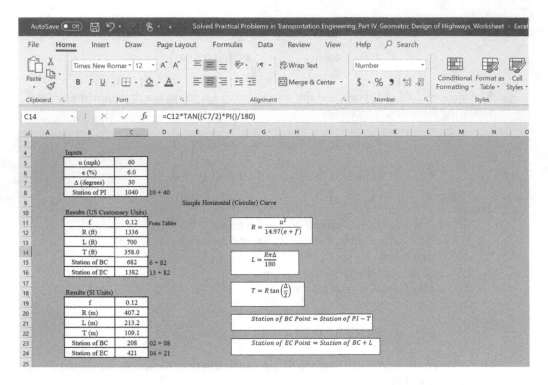

FIGURE 11.5 Image of MS Excel worksheet used for the computations of Problem 11.3.

11.4   A highway design engineer intends to design a simple horizontal (circular) curve to a proposed highway with two tangents, as shown in Figure 11.6. If the middle (inside) ordinate (M) has to be 35 ft (10.7 m) due to an obstruction near the line between the BC and EC points, determine the appropriate radius and length of the horizontal curve. If the point of intersection (PI) is at a station of $20+32$, what are the stations of the BC and EC points?

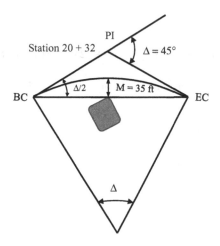

FIGURE 11.6 Designing a simple horizontal curve with the existence of obstruction for Problem 11.4.

Solution:

The middle (inside) ordinate (M) is given by the formula shown below:

$$M = R\left(1 - \cos\frac{\Delta}{2}\right) \tag{11.13}$$

Where:
T = length of simple horizontal curve tangent
R = radius of simple horizontal curve
Δ = intersection angle

⇒

$$35 = R\left(1 - \cos\frac{45°}{2}\right)$$

Solving the above equation for R
⇒

$$R = 459.8 \text{ ft} (140.1 \text{ m})$$

$$L = \frac{R\pi\,\Delta}{180}$$

⇒

$$L = \frac{(459.8)(\pi)(45)}{180} = 361.1 \text{ ft} (110.1 \text{ m})$$

Station of BC Point = Station of PI − T

But:

$$T = R\tan\left(\frac{\Delta}{2}\right)$$

$$T = 460\tan\left(\frac{45°}{2}\right) = 190.5 \text{ ft} (58.1 \text{ m})$$

⇒

Station of BC Point $= (20 + 32) - 190.5 = 18 + 41.5 \cong 18 + 42$

Station of EC Point = Station of BC + L

⇒

Station of EC Point $= (18 + 41.5) + 361.3 \cong 22 + 03$

The MS Excel worksheet used to perform the computations of this problem is illustrated in Figure 11.7.

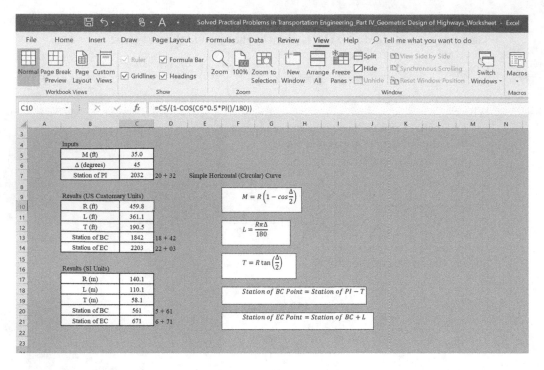

FIGURE 11.7   Image of MS Excel worksheet used for the computations of Problem 11.4.

11.5   A simple horizontal (circular) curve is to be designed and constructed on a highway segment located in a mountainous area. If the point of intersection of the two tangents of the proposed curve has to be at the edge of the nearest mountain, as shown in Figure 11.8, and the external distance from that point to the proposed curve is 100 ft (30.5 m), determine an appropriate value for the radius and the length of the horizontal curve.

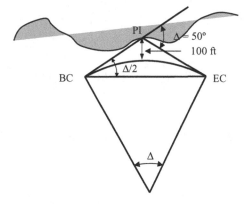

FIGURE 11.8   Designing a simple horizontal curve in a mountainous area for Problem 11.5.

Solution:

The external distance between the point of intersection of a horizontal curve and the curve is given by the formula:

$$E = R\left(\sec\frac{\Delta}{2} - 1\right)$$

(11.14)

Where:

E = external distance from PI and simple horizontal curve

R = radius of simple horizontal curve

Δ = intersection angle

$\Rightarrow$

$$100 = R\left(\sec\frac{50°}{2} - 1\right)$$

Solving the above equation for $R \Rightarrow$

$$R = 967.3 \text{ ft} \left(294.8 \text{ m}\right)$$

$$L = \frac{R\pi\Delta}{180}$$

$\Rightarrow$

$$L = \frac{(967.3)(\pi)(50)}{180} = 844.1 \text{ ft} \left(257.3 \text{ m}\right)$$

The MS Excel worksheet used to perform the computations in this problem is shown in Figure 11.9.

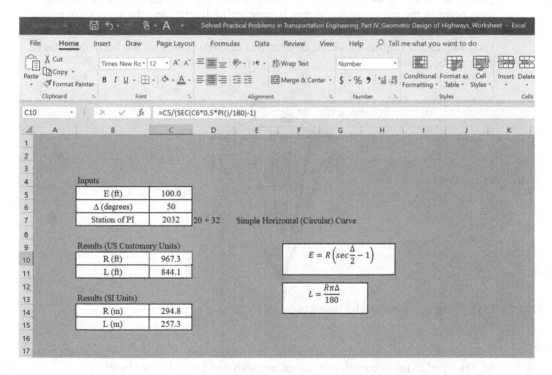

FIGURE 11.9 Image of MS Excel worksheet used for the computations of Problem 11.5.

11.6   A simple horizontal (circular) curve with an intersection angle (Δ) of 25° is designed
        within a highway segment having a design speed of 60 mph (96.6 km/h) and a supereleva-
        tion rate of 6.0%. Setting out of the horizontal curve is to be done. Determine the deflec-
        tion angles and the chord lengths at 100-ft stations form the beginning point of the curve
        (BC) if the point of intersection (PI) is located at a station 100 + 40 (see Figure 11.10).

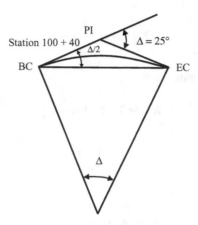

FIGURE 11.10   A simple horizontal curve with an intersection angle of 25° for Problem 11.6.

Solution:

The process of setting out simple horizontal curves includes locating successive
points on the curve using angles (called deflection angles) measured from the beginning point of the curve (BC) and chords connecting the BC and the successive points
on the curve (as shown in Figure 11.11). Therefore, the deflection angles and the corresponding chord lengths should be determined at these points on the horizontal curve
as described below:

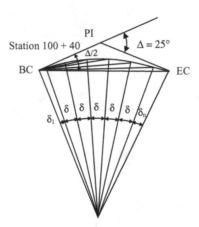

FIGURE 11.11   Setting out a simple horizontal curve for Problem 11.6.

The deflection angle to any point on the curve (measured from the tangent at the point
BC) is equal to the accumulative central angle to that point divided by 2:

$$\left(\text{Deflection Angle} \right)_i = \sum_{i=1}^{n} \delta_i \qquad (11.15)$$

$$\text{nth Deflection Angle} = \frac{\delta_1}{2} + (n-1)\frac{\delta}{2} + \frac{\delta_n}{2} = \frac{\Delta}{2} \tag{11.16}$$

Where:

δ_i=central angle between successive points on the horizontal curve

δ=central angle between successive points excluding the first and the nth angles (middle central angles as shown in the figure above)

n=number of points to be staked out on the horizontal curve including the last point (EC)

Note that the first central angle and the last central angle are typically smaller than the other middle central angles because the distance from BC to the first point and the distance from the EC to the $(n-1)$ point are smaller than a complete station.

The length of the arc that is subtended by the central angle between two successive points on the curve is given by the following formula:

$$l_i = \delta_i \frac{R\pi}{180} \tag{11.17}$$

Since the length of the simple horizontal curve is given by the following formula:

$$L = \frac{R\pi\,\Delta}{180}$$

Therefore,

$$\frac{l_i}{\delta_i} = \frac{L}{\Delta} \tag{11.18}$$

Where:

l_i=length of arc i subtended by any central angle δ_i

δ_i=central angle i

R=radius of simple horizontal curve

L=length of simple horizontal curve

Δ=intersection angle (central angle) of simple horizontal curve

The arc lengths between successive points on the horizontal curve are used to compute the chord lengths corresponding to these arc lengths. To locate the points on the horizontal curve, the chord lengths are required in addition to the deflection angles. The chord length is computed as follows:

$$C_i = 2R \sin\frac{\delta_i}{2} \tag{11.19}$$

Where:

C_i=chord length corresponding to central angle δ_i

δ_i=central angle i

R=radius of simple horizontal curve

The radius of the simple horizontal curve is determined using the following formula:

$$R = \frac{u^2}{14.97(e+f)}$$

At a design speed of 60 mph and a superelevation of 6.0%, the coefficient of side friction according to Table 11.1 (used in Problem 11.2) is equal to 0.12, therefore:

$$R = \frac{(60)^2}{14.97(0.06+0.12)} = 1336 \text{ ft}(407.2 \text{ m})$$

The length of the horizontal curve is determined using the following formula:

$$L = \frac{R\pi\,\Delta}{180}$$

$\Rightarrow$

$$L = \frac{(1336)(\pi)(25)}{180} = 582.9 \text{ ft}(\cong 177.7 \text{ m})$$

The tangent length is determined using the formula shown below:

$$T = R\tan\left(\frac{\Delta}{2}\right)$$

$$T = 1336\tan\left(\frac{25°}{2}\right) = 296.2 \text{ ft}(90.3 \text{ m})$$

Station of BC Point = Station of PI $- T$

$\Rightarrow$

Station of BC Point $= (100+40) - 296.2 = 97+43.8$

Station of EC Point = Station of BC $+ L$

$\Rightarrow$

Station of EC Point $= (97+43.8) + 582.9 = 103+26.7$

Sample Calculation:
The length of the first arc is determined as follows:

$$l_1 = (98+00) - (97+43.8) = 56.2 \text{ ft}(17.1 \text{ m})$$

$$\frac{l_i}{\delta_i} = \frac{L}{\Delta}$$

$\Rightarrow$

$$\frac{l_1}{\delta_1} = \frac{L}{\Delta}$$

$\Rightarrow$

$$\frac{56.2}{\delta_1} = \frac{582.9}{25°}$$

⇒

$$\delta_1 = 2.41°$$

The chord length is computed using the formula below:

$$C_i = 2R\sin\frac{\delta_i}{2}$$

⇒

$$C_1 = 2R\sin\frac{\delta_1}{2}$$

⇒

$$C_1 = 2(1336)\sin\frac{2.41°}{2} = 56.2\,\text{ft}(17.1\,\text{m})$$

In a similar manner, the other chord lengths and deflection angles are computed, as shown above in the sample calculation. The results are summarized in Tables 11.2 (US Customary units) and 11.3 (SI units):

TABLE 11.2

Calculations of Arc Length, Deflection Angle, and Chord Length of Simple Horizontal Curve (US Customary Units) for Problem 11.6

Station	Arc Length, l (ft)	Deflection Angle, δ (degrees)	Chord Length (ft)
BC: 97+43.8	0.0	0.00	0.0
98+00	56.2	2.41	56.2
99+00	156.2	6.70	156.1
100+00	256.2	10.99	255.8
101+00	356.2	15.28	355.1
102+00	456.2	19.56	454.0
103+00	556.2	23.85	552.2
EC: 103+26.7	582.9	25.00	578.3

TABLE 11.3

Calculations of Arc Length, Deflection Angle, and Chord Length of Simple Horizontal Curve (SI Units) for Problem 11.6

Station	Arc Length, l (m)	Deflection Angle, δ (degrees)	Chord Length (m)
BC: 29+70	0.0	0.00	0.0
29+87	17.1	0.73	17.1
30+18	47.6	2.04	47.6
30+48	78.1	3.35	78.0
30+78	108.6	4.66	108.2
31+09	139.0	5.96	138.4
31+39	169.5	7.27	168.3
EC: 31+48	177.7	7.62	176.3

Figures 11.12 and 11.13 show the MS Excel worksheet used to perform the computations of this problem.

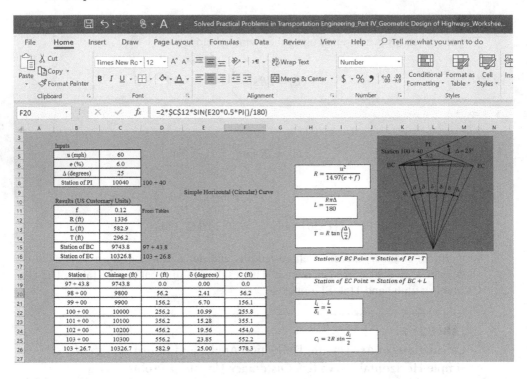

FIGURE 11.12 Image of MS Excel worksheet used for setting out a simple horizontal curve for Problem 11.6 (US Customary units).

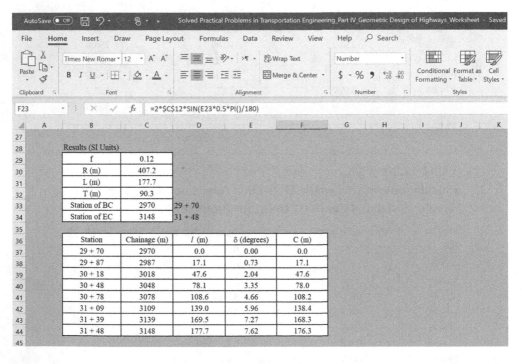

FIGURE 11.13 Image of MS Excel worksheet used for setting out a simple horizontal curve for Problem 11.6 (SI units).

11.7 A study is being conducted on a specific highway segment with a high number of car accidents. The horizontal alignment of the highway segment includes a simple horizontal (circular) curve with a radius of 800 ft and a superelevation of 6%. It is believed that this problem is caused by a defect in the geometric design of the horizontal curve under existing conditions within this segment. If the speed limit posted on the highway segment is 65 mph, provide recommendations on this highway segment to reduce/minimize the number of accidents on the highway.

Solution:

For a speed of 65 mph, and a superelevation rate of 6%, the coefficient of side friction recommended by the AASHTO "A Policy on Geometric Design of Highways and Streets," 2001 (Table 11.1 in Problem 11.2) is equal to 0.11.

$$R = \frac{u^2}{g(e+f)}$$

$\Rightarrow$

$$R = \frac{u^2}{14.97(e+f)}$$

$\Rightarrow$

$$R = \frac{(65)^2}{14.97(0.06+0.11)} = 1660 \text{ ft} \left(\cong 506 \text{ m}\right)$$

The radius of the existing horizontal curve is 800 ft. Thus, either the radius should be increased from 800 ft to 1660 ft to meet the specifications and the minimum requirement based on the speed and the other factors, or the speed limit posted on the highway should be reduced to fulfill this requirement. In the second recommendation, the reduced speed limit that should be posted on the highway is calculated as follows:

$$R = \frac{u^2}{14.97(e+f)}$$

$\Rightarrow$

$$800 = \frac{u^2}{14.97(0.06+0.11)}$$

$\Rightarrow$

Solving the above equation for speed (u) provides:

$$u = 45.1 \text{ mph} \left(72.6 \text{ km/h}\right)$$

The MS Excel worksheet that is used to perform the computations of this problem is shown in Figure 11.14:

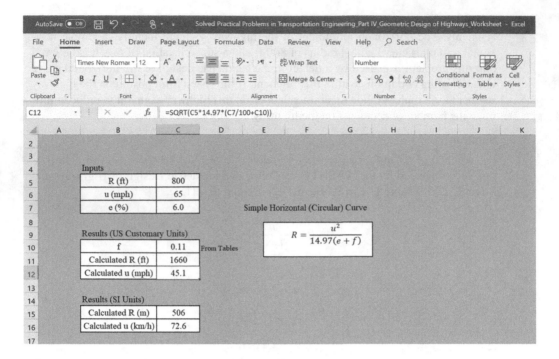

FIGURE 11.14 Image of MS Excel worksheet used for the computations of Problem 11.7.

11.8   Two tangents of a proposed roadway to be constructed in mountainous terrain have reduced bearings of S 60° 55′ W and S 35° 15′ E as shown in Figure 11.15. A highway design engineer intends to connect these two tangents with the best alternative for a simple horizontal (circular) curve. If the middle ordinate of the curve has to be 135.0 ft (41.15 m) due to right-of-way limitations, determine the intersection angle of the curve, the central angle of the curve, the radius of the curve, and the length of the curve.

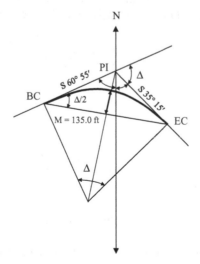

FIGURE 11.15 Designing a simple horizontal curve with right-of-way limitations for Problem 11.8.

Solution:

Using the reduced bearings of the two tangents and the geometry in the diagram above, the intersection angle of the horizontal curve is determined, as shown below:

$$\text{Intersection Angle}(\Delta) = 180 - (60°55' + 35°15') = 83°50'$$

$$\text{Central Angle} = \text{Intersection Angle} = \Delta = 83°50'$$

The middle ordinate (*M*) of a simple horizontal (circular) curve is given by the following formula:

$$M = R\left(1 - \cos\frac{\Delta}{2}\right)$$

$\Rightarrow$

$$135 = R\left(1 - \cos\frac{83°50'}{2}\right)$$

Solving for *R* (the radius of the horizontal curve) in the above equation provides:

$$R = 527.6 \text{ ft}\,(160.8 \text{ m})$$

The length of the horizontal curve is determined using the following formula:

$$L = \frac{R\pi\Delta}{180}$$

$\Rightarrow$

$$L = \frac{(527.6)(\pi)(83.83°)}{180°} = 771.9 \text{ ft}\,(235.3 \text{ m})$$

. Figure 11.16 is an image for the Excel worksheet used to conduct the computations in this problem.

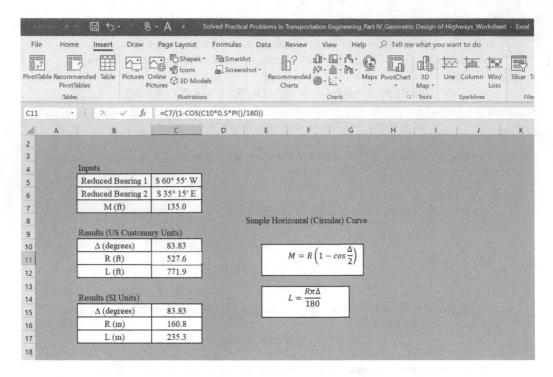

FIGURE 11.16 Image of MS Excel worksheet used for the computations of Problem 11.8.

11.9   For the problem above (Problem 11.8), what would be the appropriate speed limit that the highway design engineer would adopt for this proposed roadway if the superelevation rate is 8.0% and the coefficient of side friction is 0.14 using the radius calculated above?

Solution:

The relationship between the radius of the simple horizontal curve and the design speed of a highway is given by the following formula:

$$R = \frac{u^2}{14.97(e+f)}$$

⇒

$$527.6 = \frac{u^2}{14.97(0.08+0.14)}$$

⇒

$$u = 41.7 \, \text{mph} \, (67.1 \, \text{km/h})$$

An appropriate value for the posted speed limit on this roadway would be about 10 mph lower than the design speed. Therefore,

$$\text{Posted Speed Limit} = 30 \, \text{mph} \, (48.3 \, \text{km/h})$$

An image of the MS Excel worksheet used to perform the computations of this problem is shown in Figure 11.17.

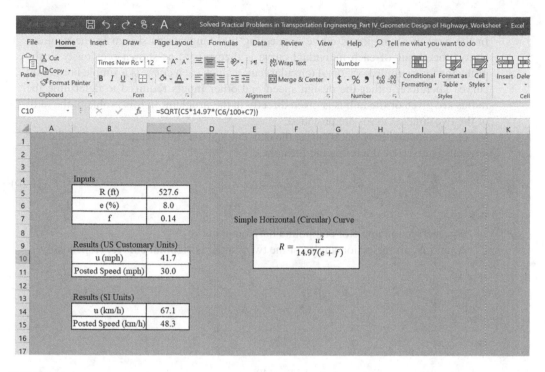

FIGURE 11.17   Image of MS Excel worksheet used for the computations of Problem 11.9.

11.10   Determine the values of all required parameters needed to set out a compound curve
that connects two tangents intersecting at an angle of 85° as shown in Figure 11.18 if the
radii of the two simple horizontal curves of the compound curve are 400 ft (121.9 m)
and 800 ft (243.8 m), respectively, the deflection angle of the first curve is 45°, and the
station of the point of compound curve (PCC) is 105+20.

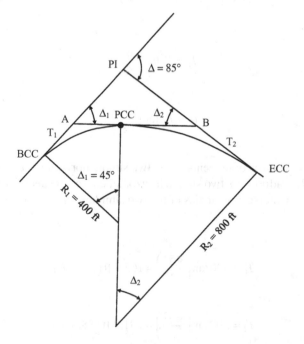

FIGURE 11.18   A compound curve for Problem 11.10.

Solution:

In the case of a compound curve connecting two tangents as shown in the diagram above, the radius of each individual curve is determined as shown previously in the problems for a simple horizontal (circular) curve since each curve is a simple horizontal curve. Therefore, either the design speed and other inputs required to determine the radius are given or the radius is given directly in the problem. In this problem, the radii of the two simple curves are provided.

For a compound horizontal curve, the deflection angle (intersection angle) of the two tangents is given by the following formula:

$$\Delta = \Delta_1 + \Delta_2 \tag{11.20}$$

Where:

Δ = intersection angle of compound horizontal curve

Δ_1 and Δ_2 = intersection angles of the two simple horizontal curves, respectively.

$\Rightarrow$

$$85° = 45° + \Delta_2$$

$\Rightarrow$

$$\Delta_2 = 40°$$

The tangent of each simple horizontal curve is computed, as shown previously in the problems for simple horizontal curves. In other words, the formula is as shown in the expression below:

$$T = R\tan\left(\frac{\Delta}{2}\right)$$

Therefore,

$$T_1 = R_1 \tan\left(\frac{\Delta_1}{2}\right) \tag{11.21}$$

$$T_2 = R_2 \tan\left(\frac{\Delta_2}{2}\right) \tag{11.22}$$

Where:

T_1 and T_2 = lengths of tangents of the two simple horizontal curves, respectively

R_1 and R_2 = radii of the two simple horizontal curves, respectively

Δ_1 and Δ_2 = intersection angles of the two simple horizontal curves, respectively

$\Rightarrow$

$$T_1 = 400\tan\left(\frac{45°}{2}\right) = 165.7 \text{ ft} \left(50.5 \text{ m}\right)$$

$$T_2 = 800\tan\left(\frac{40°}{2}\right) = 291.2 \text{ ft} \left(88.8 \text{ m}\right)$$

The lengths of the tangents of the compound curve are determined, as shown in the procedure below:

Using the geometry of the above diagram (Figure 11.18), the following formulas can be written:

$$T_{cc1} = T_1 + \overline{\text{A PI}} \tag{11.23}$$

$$T_{cc2} = T_2 + \overline{\text{B PI}} \tag{11.24}$$

Where:

T_{cc1} = length of the first tangent of a compound curve
T_{cc2} = length of the second tangent of the compound curve
T_1 and T_2 = lengths of tangents of the two simple horizontal curves, respectively
$\overline{\text{A PI}}$ = distance between point A and point of intersection (PI) of the compound curve
$\overline{\text{B PI}}$ = distance between point B and point of intersection of the compound curve

To determine the distance from point A and PI and the distance from point B and PI, the sin law for the triangle A B PI will be used, as shown below:

$$\frac{\overline{\text{A PI}}}{\sin \Delta_2} = \frac{\overline{\text{B PI}}}{\sin \Delta_1} = \frac{\overline{\text{A B}}}{\sin\left(180° - \Delta\right)} = \frac{\overline{\text{A B}}}{\sin \Delta} \tag{11.25}$$

Where:

$\overline{\text{A PI}}$ = distance between point A and point of intersection of the compound curve
$\overline{\text{B PI}}$ = distance between point B and point of intersection of the compound curve
Δ_1 and Δ_2 = intersection angles of the two simple horizontal curves, respectively
$\overline{\text{A B}}$ = distance between point A and point B

But the distance between point A and point B can be determined as below:

$$\overline{\text{A B}} = \overline{\text{APCC}} + \overline{\text{BPCC}} = T_1 + T_2$$

Where:

T_1 and T_2 = lengths of tangents of the two simple horizontal curves, respectively
$\overline{\text{A PCC}}$ = distance between point A and PCC = T_1
$\overline{\text{B PCC}}$ = distance between point B and PCC = T_2

$\Rightarrow$

$$\overline{\text{A B}} = 165.7 + 291.2 = 456.9 \text{ ft} \left(139.3 \text{ m}\right)$$

And therefore,

$$\frac{\overline{\text{A PI}}}{\sin 40°} = \frac{\overline{\text{B PI}}}{\sin 45°} = \frac{456.9}{\sin\left(180° - 85°\right)}$$

$$\overline{\text{A PI}} = 294.8 \text{ ft} \left(89.9 \text{ m}\right)$$

$$\overline{\text{B PI}} = 324.3 \text{ ft} \left(98.8 \text{ m}\right)$$

Therefore, the lengths of the two tangents of the compound curve can now be computed as below:

⇒

$$T_{cc1} = 165.7 + 294.8 = 460.5 \text{ ft} \left(140.4 \text{ m}\right)$$

$$T_{cc2} = 291.2 + 324.3 = 615.5 \text{ ft} \left(187.6 \text{ m}\right)$$

In addition, the length of each simple horizontal curve is determined using the formula used earlier for simple horizontal curves in the problems solved previously. This formula is given below:

$$L = \frac{R\pi \Delta}{180}$$

Consequently:

$$L_1 = \frac{R_1\pi \Delta_1}{180} \tag{11.26}$$

$$L_2 = \frac{R_2\pi \Delta_2}{180} \tag{11.27}$$

Where:
L_1 and L_2 = lengths of the two simple horizontal curves, respectively
R_1 and R_2 = radii of the two simple horizontal curves, respectively
Δ_1 and Δ_2 = intersection angles of the two simple horizontal curves, respectively

⇒

$$L_1 = \frac{400\left(\pi\right)\left(45\right)}{180} = 314.2 \text{ ft} \left(95.8 \text{ m}\right)$$

$$L_2 = \frac{800\left(\pi\right)\left(40\right)}{180} = 558.5 \text{ ft} \left(170.2 \text{ m}\right)$$

The stations of the beginning point of the compound curve (BCC) and the end point of the compound curve (ECC) are determined using the following formulas:

$$\text{Station of BCC Point} = \text{Station of PCC} - L_1 \tag{11.28}$$

$$\text{Station of ECC Point} = \text{Station of PCC} + L_2 \tag{11.29}$$

Where:
BCC = beginning point of compound curve
PCC = point of compound curve
ECC = end point of compound curve
L_1 and L_2 = lengths of the two simple horizontal curves, respectively

⇒

$$\text{Station of BCC Point} = \left(105 + 20\right) - 314.2 = 102 + 05.8$$

$\Rightarrow$
$$\text{Station of ECC Point} = (105 + 20) + 558.5 = 110 + 78.5$$

The length of the arc that is subtended by the central angle between two successive points on the compound curve is determined using the same formula used before for simple horizontal curves in the problems solved previously; it is given below:

$$l_i = \delta_i \frac{R\pi}{180}$$

Therefore, for arcs on the first simple horizontal curve of the compound curve, the length of the arc is given by the formula below:

$$l_i = \delta_i \frac{R_1\pi}{180} \tag{11.30}$$

Since the length of the first simple horizontal curve is given by the following formula:

$$L_1 = \frac{R_1\pi\,\Delta_1}{180}$$

Therefore,

$$\frac{l_i}{\delta_i} = \frac{L_1}{\Delta_1} \tag{11.31}$$

Where:
l_i = length of arc i subtended by any central angle δ_i
δ_i = central angle i
L_1 = length of the first simple horizontal curve
R_1 = radius of the first simple horizontal curve
Δ_1 = intersection angle of the first simple horizontal curve

And for arcs on the second simple horizontal curve of the compound curve, the length of the arc is given by the formula below:

$$l_j = \delta_j \frac{R_2\pi}{180} \tag{11.32}$$

Since the length of the second simple horizontal curve is given by the following formula:

$$L_2 = \frac{R_2\pi\,\Delta_2}{180}$$

Therefore,

$$\frac{l_j}{\delta_j} = \frac{L_2}{\Delta_2} \tag{11.33}$$

Where:
l_j = length of arc j subtended by any central angle δ_j
δ_j = central angle j
L_2 = length of the second simple horizontal curve
R_2 = radius of the second simple horizontal curve
Δ_2 = intersection angle of the second simple horizontal curve

The arc lengths between successive points on the compound curve are used to compute the chord lengths corresponding to these arc lengths. To locate the points on the compound curve (on the two simple horizontal curves of the compound curve), the chord lengths and the deflection angles are required. The chord length is computed using the typical formula used before for simple horizontal curves in earlier problems as shown below:

$$C_i = 2R\sin\frac{\delta_i}{2}$$

Therefore, to determine the chord lengths on the first simple horizontal curve of the compound curve, the following formula is used:

$$C_i = 2R_1\sin\frac{\delta_i}{2} \tag{11.34}$$

Where:
  C_i = chord length corresponding to central angle δ_i
  δ_i = central angle i
  R_1 = radius of the first simple horizontal curve

And to determine the chord lengths on the second simple horizontal curve of the compound curve, the following formula is used:

$$C_j = 2R_2\sin\frac{\delta_j}{2} \tag{11.35}$$

Where:
  C_j = chord length corresponding to central angle δ_j
  δ_j = central angle j
  R_2 = radius of the second simple horizontal curve

Sample Calculation:
First Simple Horizontal Curve:
The length of the first arc on the first simple horizontal curve is determined as follows:

$$l_1 = \text{next whole station} - \text{station of BCC}$$

$\Rightarrow$

$$l_1 = (103+00)-(102+05.8) = 94.2 \text{ ft}\,(28.7 \text{ m})$$

$$\frac{l_i}{\delta_i} = \frac{L_1}{\Delta_1}$$

$\Rightarrow$

$$\frac{l_1}{\delta_1} = \frac{L_1}{\Delta_1}$$

$\Rightarrow$

$$\frac{94.2}{\delta_1} = \frac{314.2}{45°}$$

$\Rightarrow$

$$\delta_1 = 13.49°$$

The chord length is computed using the formula below:

$$C_i = 2R_1 \sin\frac{\delta_i}{2}$$

$\Rightarrow$

$$C_1 = 2R_1 \sin\frac{\delta_1}{2}$$

$\Rightarrow$

$$C_1 = 2(400)\sin\frac{13.49°}{2} = 94.0 \text{ ft} (28.6 \text{ m})$$

Second Simple Horizontal Curve:

The length of the first arc on the second simple horizontal curve is determined as follows:

$$l_1 = \text{next whole station} - \text{station of PCC}$$

$\Rightarrow$

$$l_1 = (106 + 00) - (105 + 20) = 80.0 \text{ ft} (24.4 \text{ m})$$

$$\frac{l_j}{\delta_j} = \frac{L_2}{\Delta_2}$$

$\Rightarrow$

$$\frac{l_1}{\delta_1} = \frac{L_2}{\Delta_2}$$

$\Rightarrow$

$$\frac{80.0}{\delta_1} = \frac{558.5}{40°}$$

$\Rightarrow$

$$\delta_1 = 5.73°$$

The chord length is computed using the formula below:

$$C_j = 2R_2 \sin\frac{\delta_j}{2}$$

$\Rightarrow$

$$C_1 = 2R_2 \sin\frac{\delta_1}{2}$$

$\Rightarrow$

$$C_1 = 2(800)\sin\frac{5.73°}{2} = 80.0 \text{ ft} (24.4 \text{ m})$$

In a similar manner, the other chord lengths and deflection angles on the two simple horizontal curves of the compound curve are computed, as shown above in the sample calculation. The results are summarized in Tables 11.4 (US Customary units) and 11.5 (SI units).

TABLE 11.4

Calculations of Arc Length, Deflection Angle, and Chord Length of Compound Curve (US Customary Units) for Problem 11.19

Station	Arc Length, l (ft)	Deflection Angle, δ (degrees)	Chord Length (ft)
First Simple Horizontal Curve ($R_1 = 400$ ft)			
BCC: 102+05.8	0.0	0.00	0.0
103+00	94.2	13.49	94.0
104+00	194.2	27.82	192.3
105+00	294.2	42.14	287.6
PCC: 105+20	314.2	45.01	306.2
Second Simple Horizontal Curve ($R_2 = 800$ ft)			
PCC: 105+20	0.0	0.00	0.0
106+00	80.0	5.73	80.0
107+00	180.0	12.89	179.6
108+00	280.0	20.05	278.6
109+00	380.0	27.22	376.4
110+00	480.0	34.38	472.8
ECC: 110+78.5	558.5	40.00	547.2

TABLE 11.5

Calculations of Arc Length, Deflection Angle, and Chord Length of Compound Curve (SI Units) for Problem 11.19

Station	Arc Length, l (m)	Deflection Angle, δ (degrees)	Chord Length (m)
First Simple Horizontal Curve ($R_1 = 121.9$ m)			
BCC: 31+10.7	0.0	0.00	0.0
31+39.4	28.7	13.49	28.6
31+69.9	59.2	27.82	58.6
32+00.4	89.7	42.14	87.7
PCC: 32+06.5	95.8	45.01	93.3
Second Simple Horizontal Curve ($R_2 = 243.8$ m)			
PCC: 32+06.5	0.0	0.00	0.0
32+30.9	24.4	5.73	24.4
32+61.4	54.9	12.89	54.7
32+91.8	85.3	20.05	84.9
33+22.3	115.8	27.22	114.7
33+52.8	146.3	34.38	144.1
ECC: 33+76.7	170.2	40.00	166.8

All the computations for this problem are performed using the MS Excel worksheet shown in Figures 11.19 through 11.21.

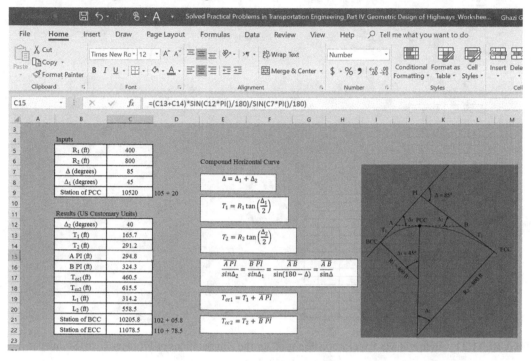

FIGURE 11.19 Image of MS Excel worksheet (Part 1) used for the computations of Problem 11.10 (US Customary units).

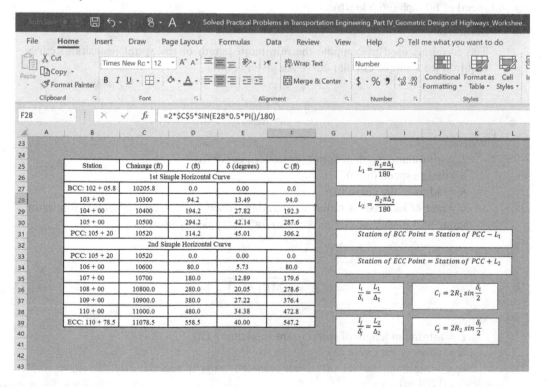

FIGURE 11.20 Image of MS Excel worksheet (Part 2) used for the computations of Problem 11.10 (US Customary units).

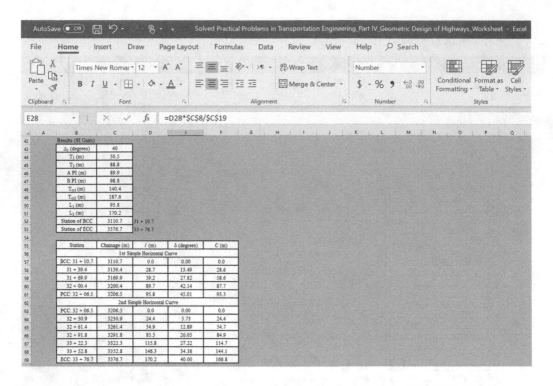

FIGURE 11.21   Image of MS Excel worksheet used for the computations of Problem 11.10 (SI units).

11.11   Determine the length of the transition (spiral) curve for a simple horizontal (circular) curve having a radius of 400 ft (121.9 m) on a two-lane highway segment with a design speed of 60 mph (96.6 km/h).

Solution:

The formulas used to compute the minimum length of the spiral (transition) curve required for a simple circular curve are given below:

$$L_s = \frac{3.155u^3}{RC} \qquad (11.36)$$

Where:
   L_s = minimum length of spiral (transition) curve (ft)
   u = design speed (mph)
   R = radius of the curve (ft)
   C = rate of increase of radial acceleration (ft/sec²/sec). A recommended minimum value of 4 ft/sec²/sec (1.2 m/sec²/sec) for C is typically used.

Or:

$$L_s = \sqrt{24R\left(p_{\min}\right)} \qquad (11.37)$$

Where:
   L_s = minimum length of spiral (transition) curve (ft)
   R = radius of the curve (ft)
   $p_{\min}$ = minimum lateral offset between the tangent and the circular curve (0.66 ft = 0.201 m)

By assuming a value of $C = 4$ ft/sec²/sec, the length of the spiral curve is computed as follows:
⇒

$$L_s = \frac{3.155(60)^3}{400(4)} = 425.9 \text{ ft} (129.8 \text{ m})$$

Using the other formula in Equation 11.37, the minimum spiral length is calculated as:

$$L_s = \sqrt{24(400)(0.66)} = 79.6 \text{ ft} (24.3 \text{ m})$$

Therefore, the larger of the two values, which is equal to 425.9 ft (129.8 m), is the value recommended for the length of the spiral curve for the horizontal curve on this highway segment.

11.12 Compute the distance required to transition a normal crown section (with a cross slope of 2%) to a full superelevation section of 8% cross slope on a two-lane highway if the design speed on the highway is 60 mph (96.6 km/h) and the lane width is 12 ft (3.66 m).

Solution:

This distance is called the superelevation transition length, which is composed of two components: the superelevation runoff and the tangent runout. The superelevation run-off is simply the distance required to change the cross slope of the outside lane pavement from zero to full superelevation. On the other hand, the length of the tangent runout is the distance needed to accomplish a change of the cross slope of the outside lane pavement from normal (2.0%) to 0.0% or from 0.0% to normal (2.0%).

$$\text{Superelevation Transition Length} = L_r + L_t \qquad (11.38)$$

Where:
L_r = length of superelevation runoff
L_t = length of tangent runout

The formula shown in the equation below is used to calculate the superelevation runoff:

$$L_r = \frac{wn_1 e_d}{\Delta}(b_w) \qquad (11.39)$$

Where:
L_r = minimum length of superelevation runoff
w = width of one lane (ft) (the typical width is 12 ft)
n_1 = number of rotated lanes
e_d = design superelevation rate (%)
b_w = adjustment factor for the number of lanes rotated, determined from Table 11.6
Δ = maximum relative gradient (%) determined from Table 11.7
The lane width $(w) = 12$ ft (given)
The number of lanes rotated $(n_1) = 1$ since the highway is a two-lane highway
The adjustment factor (b_w) is determined from Table 11.6, $b_w = 1.00$ for the number of lanes rotated = 1

TABLE 11.6
Adjustment Factor for Number of Lanes Rotated

SI Units			US Customary Units		
Number of Lanes Rotated (n_1)	Adjustment Factor (b_w)	Length Increase Relative to One-Lane Rotated $(n_1 b_w)$	Number of Lanes Rotated (n_1)	Adjustment Factor (b_w)	Length Increase Relative to One-Lane Rotated $(n_1 b_w)$
1.0	1.00	1.00	1.0	1.00	1.00
1.5	0.83	1.25	1.5	0.83	1.25
2.0	0.75	1.50	2.0	0.75	1.50
2.5	0.70	1.75	2.5	0.70	1.75
3.0	0.67	2.00	3.0	0.67	2.00
3.5	0.64	2.25	3.5	0.64	2.25

Reproduced with Permission from "A Policy on Geometric Design of Highways and Streets," 2001, by AASHTO, Washington, DC, USA.

TABLE 11.7
Maximum Relative Gradient (Δ) at Different Design Speeds

SI Units			US Customary Units		
Design Speed (km/h)	Maximum Relative Gradient, Δ (%)	Equivalent Maximum Relative Slope	Design Speed (mph)	Maximum Relative Gradient, Δ (%)	Equivalent Maximum Relative Slope
20	0.80	1:125	15	0.78	1:128
30	0.75	1:133	20	0.74	1:135
40	0.70	1:143	25	0.70	1:143
50	0.65	1:150	30	0.66	1:152
60	0.60	1:167	35	0.62	1:161
70	0.55	1:182	40	0.58	1:172
80	0.50	1:200	45	0.54	1:185
90	0.47	1:213	50	0.50	1:200
100	0.44	1:227	55	0.47	1:213
110	0.41	1:244	60	0.45	1:222
120	0.38	1:263	65	0.43	1:233
130	0.35	1:286	70	0.40	1:250
			75	0.38	1:263
			80	0.35	1:286

Reproduced with Permission from "A Policy on Geometric Design of Highways and Streets," 2001, by AASHTO, Washington, DC, USA.

The maximum relative gradient (Δ) is determined from Table 11.7, $\Delta = 0.45$ for a design speed $= 60$ mph.

$\Rightarrow$

$$L_r = \frac{(12)(1)(8)}{0.45}(1) = 213.3 \text{ ft} (65.0 \text{ m})$$

The length of the tangent runout is determined using the formula shown below:

$$L_t = \frac{e_{NC}}{e_d}(L_r) \tag{11.40}$$

Where:

L_t = length of tangent runout
e_{NC} = normal cross slope rate (%)
e_d = design superelevation rate (%)
L_r = minimum length of superelevation runoff

$\Rightarrow$

$$L_t = \frac{2}{8}(213.3) = 53.3 \text{ ft} (16.2 \text{ m})$$

$\Rightarrow$

The distance needed to transition the normal crown section to an 8%-full superelevation section on this two-lane highway segment is:

$$\text{Superelevation Transition Length} = 213.3 + 53.3 = 266.6 \text{ ft} (81.3 \text{ m})$$

11.13   Determine the distance needed to transition a normal crown section (with a cross slope of 2%) to a full superelevation section of 6% cross slope on a four-lane highway segment if the design speed on the highway is 65 mph (104.6 km/h) and the lane width is 12 ft (3.66 m).

Solution:

The superelevation transition length is basically two components: the superelevation runoff and the tangent runout. The superelevation runoff is computed using the formula shown below:

$$L_r = \frac{wn_1 e_d}{\Delta}(b_w)$$

The lane width (w) = 12 ft (given).
The number of lanes rotated (n_1) = 2 since the highway is a four-lane highway.
The adjustment factor (b_w) is determined from Table 11.6 in Problem 11.12, $b_w = 0.75$ for the number of lanes rotated = 2.
The maximum relative gradient (Δ) is also determined from Table 11.7 in Problem 11.12, $\Delta = 0.43$ for a design speed = 65 mph.

$\Rightarrow$

$$L_r = \frac{(12)(2)(6)}{0.43}(0.75) = 251.2 \text{ ft} (76.6 \text{ m})$$

The tangent runout is determined using the formula shown below:

$$L_t = \frac{e_{NC}}{e_d}(L_r)$$

$\Rightarrow$

$$L_t = \frac{2}{6}(251.2) = 83.7 \text{ ft}(25.5 \text{ m})$$

Therefore:

$$\text{Superelevation Transition Length} = 251.2 + 83.7 = 334.9 \text{ ft}(102.1 \text{ m})$$

11.14   A composite curve (two symmetric spiral curves on both sides of a circular curve) is shown in Figure 11.22 with the following data:
   - Intersection angle (Δ) = 50°
   - Radius of the circular curve = 1400 ft (426.7 m)
   - Design speed = 60 mph (96.6 km/h)
   - Superelevation rate = 6.0%
   - Station of PI = 110 + 00

It is required to set out the spiral (transition) curve at 20-ft intervals.

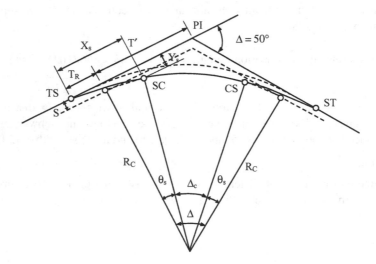

FIGURE 11.22   A composite curve with two symmetric spirals for Problem 11.14.

Solution:

The length of the spiral curve is determined using the following formula:

$$L_s = \frac{3.155u^3}{RC}$$

$C = 4$ ft/sec^2/sec (1.2 m/sec^2/sec)
$R = R_c = 1400$ ft (426.7 m)
$u = 60$ mph (96.6 km/h)

$\Rightarrow$

$$L_s = \frac{3.155(60)^3}{1400(4)} = 121.7 \text{ ft} (37.1 \text{ m})$$

Or:

$$L_s = \sqrt{24R(p_{min})}$$

$R = 1400$ ft (426.7 m)
$p_{min} = 0.66$ ft (0.201 m)
$\Rightarrow$

$$L_s = \sqrt{24(1400)(0.66)} = 148.9 \text{ ft} (45.4 \text{ m})$$

Therefore, the minimum length of the spiral curve is the larger of the two values above.
$\Rightarrow$

$$L_s = 148.9 \text{ ft} (45.4 \text{ m})$$

The angle θ_s in the diagram is determined using the formula below:

$$\theta_s = \frac{L_s}{2R_c} \tag{11.41}$$

Where:
$\quad \theta_s$ = central angle of spiral curve
$\quad L_s$ = length of spiral curve
$\quad R_c$ = radius of circular curve

$\Rightarrow$

$$\theta_s = \frac{148.9}{2(1400)} = 0.053 \text{ rad} (3.047°)$$

The coordinates X and Y of any point on the spiral curve can be determined by the following series, which are given as functions of the length (L):

$$X = L - \frac{L^5}{40A^4} + \frac{L^9}{3456A^8} + \cdots \tag{11.42}$$

$$Y = \frac{L^3}{6A^2} - \frac{L^7}{336A^6} + \frac{L^{11}}{42240A^{10}} + \cdots \tag{11.43}$$

Where:
$\quad X$ and $Y = X$ and Y coordinates of any point on the spiral curve
$\quad L$ = length on the spiral curve
$\quad A$ = constant given by the following formula:

$$A^2 = R_c L_s = RL \tag{11.44}$$

Where:
  R_c = radius of circular curve
  L_s = length of spiral curve
  R = radius of spiral curve (R changes from 0 to $1/R_c$)
  L = length on the spiral curve

At the point SC (the end of first spiral and beginning of the circular curve), $X = X_s$ and $Y = Y_s$; therefore:

$$X_s = L_s - \frac{L_s^5}{40A^4} + \frac{L_s^9}{3456A^8} + \dots$$

$$Y_s = \frac{L_s^3}{6A^2} - \frac{L_s^7}{336A^6} + \frac{L_s^{11}}{42240A^{10}} + \dots$$

But:

$$A^2 = R_c L_s$$

$\Rightarrow$

$$A = \sqrt{1400(148.9)} = 456.6$$

Using three terms of the series above used for calculating X_s provides the following solution for X_s:

$$X_s = 148.9 - \frac{(148.9)^5}{40(456.6)^4} + \frac{(148.9)^9}{3456(456.6)^8} = 148.86 \text{ ft} (45.4 \text{ m})$$

And using three terms of the series above used for calculating Y_s provides the following solution for Y_s:

$$Y_s = \frac{(148.9)^3}{6(456.6)^2} - \frac{(148.9)^5}{336(456.6)^6} + \frac{(148.9)^9}{42240(456.6)^{10}}$$

$$= 2.639 \text{ ft} (0.804 \text{ m})$$

The shift (S) of the circular curve that is shown on the diagram above is given by the following formula:

$$S = Y_s - R_c (1 - \cos\theta_s) \tag{11.45}$$

Where:
  S = shift of circular curve from its original position
  Y_s = Y-coordinate of the point SC on the spiral curve
  R_c = radius of circular curve
  θ_s = central angle of spiral curve

$\Rightarrow$

$$S = 2.639 - 1400(1 - \cos(3.047°)) = 0.66 \text{ ft} (0.20 \text{ m})$$

The distance (T_R) on the diagram is determined using the following formula:

$$T_R = X_s - R_c \sin \theta_s \tag{11.46}$$

Where:
T_R = the distance along the extended tangent from the TS to the point where the circular curve is tangent to the offset (shifted) tangents
X_s = X-coordinate of the point SC on the spiral curve
R_c = radius of circular curve
θ_s = central angle of spiral curve

⇒

$$T_R = 148.86 - 1400 \sin(3.047°) = 74.44 \text{ ft} (22.69 \text{ m})$$

The distance (T') along the extended tangent from PI to the point where the circular curve is tangent to the offset (shifted) tangents is determined using the following formula:

$$T' = (R_c + S) \tan\left(\frac{\Delta}{2}\right) \tag{11.47}$$

Where:
T' = the distance along the extended tangent from PI to the point where the circular curve is tangent to the offset (shifted) tangents
R_c = radius of circular curve
S = shift of circular curve from its original position
Δ = central angle of the circular curve

⇒

$$T' = (1400 + 0.66) \tan\left(\frac{50°}{2}\right) = 653.14 \text{ ft} (199.08 \text{ m})$$

The length of the circular curve of the composite curve is computed using the formula shown below:

$$L_c = R_c (\Delta - 2\theta_s) \tag{11.48}$$

Or:

$$L_c = R_c \Delta - L_s \tag{11.49}$$

Since:

$$\theta_s = \frac{L_s}{2R_c}$$

Where:
L_c = length of the circular curve of the composite curve
R_c = radius of circular curve
Δ = central angle of the circular curve (in radians)
θ_s = central angle of the spiral curve (in radians)
L_s = length of spiral curve

⇒

$$L_c = 1400 \left(50 \times \frac{\pi}{180} - 2(0.053) \right) = 1073.33 \text{ ft} (327.15 \text{ m})$$

The station of the point TS on the extended tangent and at the beginning of the first spiral curve is determined as follows:

$$\text{Station of TS} = \text{Station of PI} - (T' + T_R) \tag{11.50}$$

⇒

$$\text{Station of TS} = (110 + 00) - (653.14 + 74.44) = 102 + 72.4$$

The station of the point SC at the end of the first spiral curve and beginning of the circular curve is calculated using the following formula:

$$\text{Station of SC} = \text{Station of TS} + L_s \tag{11.51}$$

⇒

$$\text{Station of SC} = (102 + 72.42) + 148.9 = 104 + 21.3$$

In addition, the station of the point CS at the end of the circular curve and beginning of the second spiral curve is determined using the following formula:

$$\text{Station of CS} = \text{Station of SC} + L_c \tag{11.52}$$

⇒

$$\text{Station of CS} = (104 + 21.3) + 1073.33 = 114 + 94.6$$

And finally, the station of the point ST at the end of the second spiral curve on the extended tangent is determined as follows:

$$\text{Station of ST} = \text{Station of CS} + L_s \tag{11.53}$$

⇒

$$\text{Station of ST} = (114 + 94.6) + 148.9 = 116 + 43.5$$

The spiral angle (θ) at any length on the spiral curve can be determined using the following formula:

$$\theta_i = \frac{L_i}{2R_i} \tag{11.54}$$

Where:
    θ_i = spiral angle i
    L_i = length i on the spiral curve
    R_i = radius of spiral curve (R changes from 0 to $1/R_c$)

But:

$$A^2 = R_c L_s = RL$$

$\Rightarrow$

$$R = \frac{R_c L_s}{L} \tag{11.55}$$

Substituting this equation in the above equation provides:

$$\theta_i = \frac{L_i^2}{2R_c L_s} \tag{11.56}$$

The deflection angle at any point on the spiral curve (δ) is given by the following expression:

$$\delta_i = \tan^{-1}\left(\frac{Y_i}{X_i}\right) \tag{11.57}$$

The length of the chord (C) at any point on the spiral curve is given by the following formula:

$$C_i = \sqrt{X_i^2 + Y_i^2} \tag{11.58}$$

Where:
δ_i=deflection angle at any point i on the spiral curve
X_i and Y_i=X and Y coordinates of any point i on the spiral curve
C_i=length of the chord at any point i on the spiral curve

Sample Calculation:
At station $102 + 80$

$$L = (102 + 80) - (102 + 72.4) = 7.6 \text{ ft}$$

$$X = L - \frac{L^5}{40A^4} + \frac{L^9}{3456A^8} + \dots$$

Using three terms of the above series to determine X provides the following value for X:

$$X = 7.6 - \frac{(7.6)^5}{40(456.6)^4} + \frac{(7.6)^9}{3456(456.6)^8} = 7.60 \text{ ft}$$

$$Y = \frac{L^3}{6A^2} - \frac{L^7}{336A^6} + \frac{L^{11}}{42240A^{10}} + \dots$$

Using three terms of the above series to determine Y provides the following value for Y:

$$Y = \frac{(7.6)^3}{6(456.6)} - \frac{(7.6)^7}{336(456.6)^6} + \frac{(7.6)^{11}}{42240(456.6)^{10}} = 0.0004 \text{ ft}$$

$$\theta_i = \frac{L_i^2}{2R_cL_s}$$

$\Rightarrow$

$$\theta = \frac{(7.6)^2}{2(1400)(148.9)} = 0.00014 \text{ radian}$$

$$\delta_i = \tan^{-1}\left(\frac{Y_i}{X_i}\right)$$

$\Rightarrow$

$$\delta = \tan^{-1}\left(\frac{0.0004}{7.60}\right) = 0.00005 \text{ radian}$$

$$C_i = \sqrt{X_i^2 + Y_i^2}$$

$\Rightarrow$

$$C = \sqrt{(7.60)^2 + (0.0004)^2} = 7.60 \text{ ft}$$

The calculations for setting out the spiral curve of the composite curve in this problem are summarized in Table 11.8.

TABLE 11.8

Calculations of Deflection Angle and Chord Length on Spiral (Transition) Curve of a Composite Curve (US Customary Units) for Problem 11.14

Station	L (ft)	X (ft)	Y (ft)	Spiral Angle (θ) Radians	Degrees	Deflection Angle (δ) Radians	Degrees	Chord Length, C (ft)
TS: 102+72.4	0.0	0.000	0.0000	0.00000	0° 00′ 00″	0.00000	0° 00′ 00″	0.00
102+80	7.6	7.600	0.0004	0.00014	0° 00′ 29″	0.00005	0° 00′ 10″	7.60
103+00	27.6	27.600	0.0168	0.00183	0° 06′ 17″	0.00061	0° 02′ 06″	27.60
103+20	47.6	47.600	0.0862	0.00543	0° 18′ 41″	0.00181	0° 06′ 14″	47.60
103+40	67.6	67.599	0.2470	0.01096	0° 37′ 41″	0.00365	0° 12′ 34″	67.60
103+60	87.6	87.597	0.5374	0.01841	1° 03′ 16″	0.00613	0° 21′ 05″	87.60
103+80	107.6	107.592	0.9958	0.02777	1° 35′ 28″	0.00926	0° 31′ 49″	107.60
104+00	127.6	127.581	1.6607	0.03905	2° 14′ 15″	0.01302	0° 44′ 45″	127.59
104+20	147.6	147.560	2.5701	0.05225	2° 59′ 38″	0.01742	0° 59′ 52″	147.58
SC: 104+21.3	148.9	148.858	2.6386	0.05318	3° 02′ 49″	0.01772	1° 00′ 56″	148.88

11.15 Check whether a transition (spiral) curve is needed for a horizontal curve designed for a two-lane highway segment with a design speed of 65 mph and superelevation rate of 6%.

Solution:

At a design speed of 65 mph and superelevation of 6%, the coefficient of side friction is equal to 0.110 (Table 11.1 in Problem 11.2). The radius of the horizontal curve on this highway segment is calculated using the following formula:

$$R = \frac{u^2}{14.97(e+f)}$$

$\Rightarrow$

$$R = \frac{(65)^2}{14.97(0.06+0.110)} = 1660.2 \text{ ft} (506.0 \text{ m})$$

Based on Table 11.9 (the warrant to use a spiral curve), at a design speed of 65 mph, the maximum radius for the horizontal curve for use of a spiral curve is 2138 ft. Since the radius of the horizontal curve determined above is 1660.1 ft < 2138 ft; therefore, a spiral curve is needed for this horizontal curve at this highway segment.

TABLE 11.9
Maximum Radius for the Use of a Spiral Curve Transition

SI Units		US Customary Units	
Design Speed (km/h)	Maximum Radius (m)	Design Speed (mph)	Maximum Radius (ft)
20	24	15	114
30	54	20	203
40	95	25	317
50	148	30	456
60	213	35	620
70	290	40	810
80	379	45	1025
90	480	50	1265
100	592	55	1531
110	716	60	1822
120	852	65	2138
130	1000	70	2479
		75	2846
		80	3238

Reproduced with Permission from "A Policy on Geometric Design of Highways and Streets," 2001, by AASHTO, Washington, DC, USA.

11.16   It is intended to place a clock tower near an important horizontal curve of a two-lane roadway segment (Figure 11.23) with a posted speed of 20 mph (32.2 km/h) and a 0% grade. The radius of the horizontal curve is 1000 ft (304.8 m). Based on the stopping sight distance (SSD) on this curve, calculate the horizontal sightline offset where the clock tower can be placed without affecting the required SSD. Assume that the perception-reaction time for drivers is 2.5 sec and the superelevation rate is 0%.

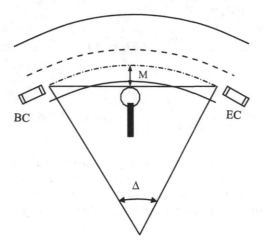

FIGURE 11.23   Determining the sightline offset for a simple horizontal curve for Problem 11.16.

Solution:

Based on the diagram above, the distance M from the centerline of the inside lane of the roadway at the curve represents the horizontal sightline offset at which the clock tower can be placed without affecting (reducing) the stopping sight distance.

The stopping sight distance as the vehicle moves from point BC to point EC on the curve is simply the length of the arc. Therefore:

$$SSD = R\Delta \tag{11.59}$$

Where:

    SSD = stopping sight distance
    R = radius of horizontal curve
    Δ = central angle (in radians)

$\Rightarrow$

$$SSD = R\frac{\Delta\pi}{180} \tag{11.60}$$

Where:

    Δ = central angle (in degrees)

Based on geometry:

$$\cos\frac{\Delta}{2} = \frac{R - M}{R} \tag{11.61}$$

Where:

    M = horizontal sightline offset

Or the following formula can be used to determine M:

$$M = R\left(1 - \cos\frac{\Delta}{2}\right)$$

The required stopping sight distance is also given by the following formula:

$$SSD = 1.47ut + \frac{u^2}{30\left(\dfrac{a}{g} \pm G\right)} \qquad (11.62)$$

Where:
  SSD = stopping sight distance (ft)
  u = design speed (mph)
  t = perception-reaction time of drivers (typical value = 2.5 sec)
  a = acceleration/deceleration rate (typical value = 11.2 ft/sec^2 = 3.41 m/sec^2)
  g = gravitational constant (32.18 ft/sec^2 = 9.81 m/sec^2)
  G = grade of tangent

Using a design speed of 30 mph in the above equation since the posted speed is 20 mph will provide the following value of SSD:

$\Rightarrow$

$$SSD = 1.47(30)(2.5) + \frac{(30)^2}{30\left(\dfrac{11.2}{32.18}\right)} = 196.4 \text{ ft} (59.9 \text{ m})$$

$$SSD = R\frac{\Delta\pi}{180}$$

$\Rightarrow$

$$196.4 = 1000\frac{\Delta\pi}{180}$$

$\Rightarrow$

$$\Delta = 11.25°$$

$$\cos\frac{\Delta}{2} = \frac{R - M}{R}$$

$\Rightarrow$

$$\cos\frac{11.25°}{2} = \frac{1000 - M}{1000}$$

Solving for M provides:

$$M = 4.82 \text{ ft} (1.47 \text{ m})$$

Therefore, the clock tower should be placed at a distance equal to 4.82 ft (1.47 m) from the centerline of the inner lane of the roadway curve so that it will not affect the required stopping sight distance.

The MS Excel worksheet used to perform the computations of this problem is shown in Figure 11.24.

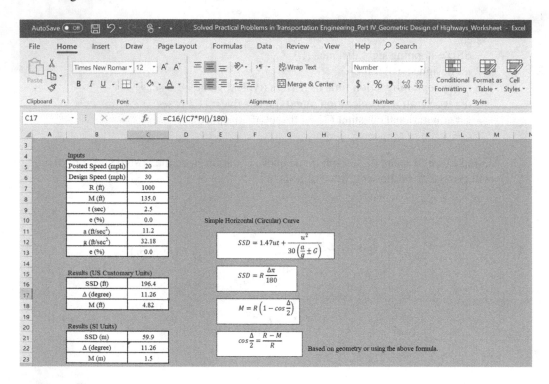

FIGURE 11.24   Image of MS Excel worksheet used for the computations of Problem 11.16.

12 Design of Other Facilities

Chapter 12 emphasizes the criteria needed to use a climbing lane for two-lane two-way highways and the design of on-street parking spaces and off-street parking lots. The design of on-street parking covers inclined design at five main inclination angles to the street curb: 0°, 30°, 45°, 60°, and 90°. The derivations of all formulas used in the design of on-street parking spaces and off-street parking lots are presented in detail. Additionally, the computations related to all geometric parameters of parking spaces and parking lots, the number of parking spaces, and the width of the aisles (corridors) of the parking lots are performed accurately for different types of inclination angles.

12.1 Determine if a climbing lane is needed on a 2000 ft (609.6 m) two-lane two-way highway segment with a traffic flow of 320 veh/h if the percentage of trucks on this segment is 10% and the grade of the segment is +4.0%.

Solution:

To check whether a climbing lane is needed, the following three criteria should be satisfied:

1. Traffic flow on the upgrade direction is higher than 200 veh/h.
2. Truck traffic flow on the upgrade direction is higher than 20 veh/h.
3. One of the following conditions should be present:
   - A speed reduction ≥ 15 km/h (10 mph) for a typical heavy truck.
   - A level of service (LOS) of E or F on the upgrade direction.
   - A reduction in LOS from the approach segment to the upgrade ≥ 2.

Criterion #1 is satisfied in this case since the traffic flow is equal to 320 veh/h > 200 veh/h.
Criterion #2 is also satisfied since the truck traffic is equal to 0.10 (320) = 32 truck/h > 20 veh/h.
Criterion #3: The speed reduction will be checked in this case since no information is available on the level of service of the highway. The speed reduction is determined using Figure 12.1.

Moving on the 4%-curve for a distance of 2000 ft results in a speed of 54 mph. It means that the reduction in speed is 70 − 54 = 16 mph > 10 mph. Therefore, criterion #3 is also satisfied. The critical length requirement can be checked instead of criterion #3 that determines the speed reduction. Figure 12.2 is used in this case to determine the critical length for a 4% grade and, using an upper limit for a speed reduction of 10 mph, the critical length in this case is equal to 1200 ft ($\cong$ 366 m). Therefore, the length of the highway segment (2000 ft) with an upgrade of 4% is not appropriate since it is higher than the critical length.
In summary, the three criteria are achieved and justify the need for a climbing lane on this highway segment.

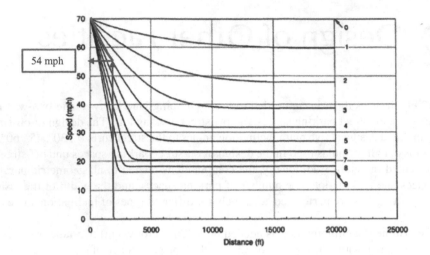

FIGURE 12.1 Speed-distance curves for a typical heavy truck of 120 kg/kW (200 lb/hp) for deceleration on upgrades. Used with permission from "A Policy on Geometric Design of Highways and Streets," 2001, by AASHTO, Washington, DC, USA.

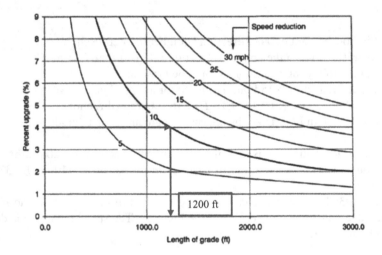

FIGURE 12.2 Critical lengths of grades for design, assumed typical heavy truck of 120 kg/kW (200 lb/hp), entering speed = 110 km/h (70 mph). Used with permission from "A Policy on Geometric Design of Highways and Streets," 2001, by AASHTO, Washington, DC, USA.

12.2 Determine the number of parking spaces for street parking inclined at 45° to the street curb along a distance of 608 ft (185.3 m).

Solution:

On-street parking facilities can be designed using three methods: (1) parking spaces (bays) parallel to the street curb (parallel to the direction of traffic), (2) parking spaces inclined to the street curb (inclined to the direction of traffic) at an inclination angle of 30°, 45°, or 60°, and (3) parking spaces (bays) perpendicular to the street curb (perpendicular to the direction of traffic); i.e., the inclination angle is equal to 90°.

The diagram below shows an illustration of on-street parking spaces inclined to the street curb by an inclination angle of 45°. The derivation of the formulas to compute the distances shown in Figure 12.3 is explained in detail.

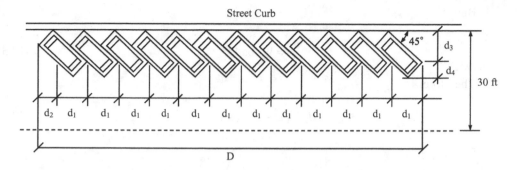

Street Curb

FIGURE 12.3   A street parking inclined at an angle of 45° for Problem 12.2.

The parking space has dimensions of 18 ft×8.5 ft (5.5 m×2.6 m) and is inclined to the street curb by an angle of 45°. The distance d_1 shown in Figure 12.4 is composed of two symmetric components as in the expression below:

$$d_1 = W \cos 45° + W \cos 45° = 2W \cos 45° \tag{12.1}$$

Where:
  d_1 = regular horizontal spacing (shown in the diagram)
  W = width of parking space
  45° = inclination angle with street curb

$\Rightarrow$

$$d_1 = 2(8.5) \cos 45° = 12.02 \text{ ft} \cong 12 \text{ ft} (3.66 \text{ m})$$

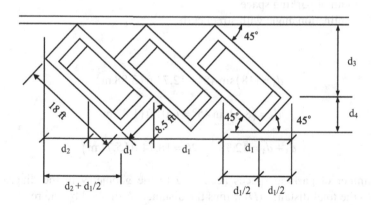

FIGURE 12.4   Schematic geometric diagram for the distances d_1, d_2, d_3, and d_4 for Problem 12.2.

The distance $(d_2 + d_1/2)$ in the diagram above is computed, as shown below:

$$d_2 + \frac{d_1}{2} = L \cos 45° \tag{12.2}$$

Where:
  d_1 = regular spacing (shown in the diagram)
  d_2 = remaining spacing of the total distance (D) for the last spacing (shown in the diagram)
  L = length of parking space
  45° = inclination angle with street curb

But:

$$\frac{d_1}{2} = W\cos 45° \tag{12.3}$$

Therefore:

$$d_2 = (L - W)\cos 45° \tag{12.4}$$

$\Rightarrow$

$$d_2 = (18 - 8.5)\cos 45° = 6.72 \text{ ft} (2.05 \text{ m})$$

The distances d_3 and d_4 in the diagram are determined using geometry and the following two formulas, respectively:

$$d_3 = L\sin 45° \tag{12.5}$$

$$d_4 = W\sin 45° \tag{12.6}$$

Where:
d_3 = perpendicular distance from street curb to right rear corner of parking space (shown in the diagram)
d_4 = vertical component of the width of the car (shown in the diagram)
L = length of parking space
W = width of parking space
$45°$ = inclination angle with street curb

$\Rightarrow$

$$d_3 = (18)\sin 45° = 12.73 \text{ ft} (3.88 \text{ m})$$

$$d_4 = (8.5)\sin 45° = 6.01 \text{ ft} (1.83 \text{ m})$$

$$d_3 + d_4 = 12.73 + 1.26 = 18.74 \text{ ft} (5.71 \text{ m})$$

The number of parking spaces according to the geometry in the diagram above is equal to the total distance (D) minus the distance d_2 divided by the regular spacing d_1. Consequently, the number of parking spaces (bays) for on-street parking when the parking spaces are inclined to the street curb at an inclination angle of $45°$ is computed using the following formula:

$$N = \frac{D - d_2}{d_1} \tag{12.7}$$

Or:

$$N = \frac{D - 6.72}{12.02} \tag{12.8}$$

Where:

    N = number of parking spaces

    D = distance along the street curb available for parking (ft)

In this case, the inclination angle is 45°; therefore, the formula derived above is used to compute the number of parking spaces as shown below:

$\Rightarrow$

$$N = \frac{608 - 6.72}{12.02} \cong 50 \text{ parking spaces}$$

12.3   Using the same methodology used in Problem 12.2 above, derive a formula to compute the number of parking spaces for on-street parking inclined at 30° to the street curb along a distance of 1210 ft (368.8 m).

Solution:

Figure 12.5 illustrates on-street parking spaces inclined to the street curb by an inclination angle of 30°. The derivation of the formulas to compute the distances shown in the diagram is explained in the procedure below.

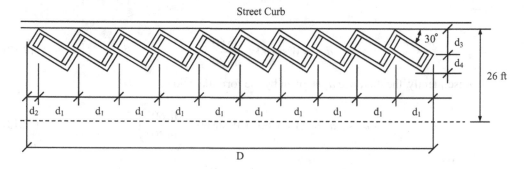

FIGURE 12.5 A street parking inclined at an angle of 30° for Problem 12.3.

The parking space has dimensions of 18 ft × 8.5 ft (5.5 m × 2.6 m) and is inclined to the street curb by an angle of 30°. Figure 12.6 illustrates the distances d_1, d_2, d_3, and d_4 clearly in an amplified illustration.

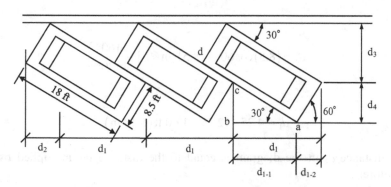

FIGURE 12.6 Schematic geometric diagram for the distances d_1, d_2, d_3, and d_4 for Problem 12.3.

The distance d_1 is composed of two components d_{1-1} and d_{1-2}. The two components are calculated, as shown below:

$$d_{1-2} = W \cos 60° \tag{12.9}$$

Using the sine law for the triangle abc, the following formula can be written:

$$\frac{\overline{ab}}{\sin \hat{c}} = \frac{\overline{bc}}{\sin \hat{a}} \tag{12.10}$$

But, $\overline{ab} = d_{1-1}$, angle $c = 60°$, and angle $a = 30°$, and

$$\overline{bc} = d_4 = W \sin 60° \tag{12.11}$$

Therefore:

$$\frac{d_{1-1}}{\sin 60°} = \frac{W \sin 60°}{\sin 30°}$$

$\Rightarrow$

$$d_{1-1} = W \sin 60° \left(\frac{\sin 60°}{\sin 30°} \right) \tag{12.12}$$

Consequently, the distance d_1 is given by the formula below:

$$d_1 = d_{1-2} + d_{1-1} = W \cos 60° + W \sin 60° \left(\frac{\sin 60°}{\sin 30°} \right) \tag{12.13}$$

Where:
   d_1 = regular horizontal spacing (shown in the diagram)
   W = width of parking space
   $30°$ = inclination angle with street curb
   $60°$ = angle c (in the diagram) = the angle that the width of the parking space makes with the line parallel to the street curb or the angle that the length of the parking space makes with the perpendicular to the street curb

$\Rightarrow$

$$d_1 = (8.5) \cos 60° + (8.5) \sin 60° \left(\frac{\sin 60°}{\sin 30°} \right)$$

$\Rightarrow$

$$d_1 = 4.25 + 12.75 = 17.0 \text{ ft} (5.18 \text{ m})$$

The distance d_2 in the diagram is equal to the distance $\overline{cd}$ multiplied by $\cos 30°$, therefore:

$$d_2 = \overline{cd} \cos 30° \tag{12.14}$$

But:

$$\overline{cd} = L - \overline{ac} \tag{12.15}$$

And using the sine law for the triangle abc and since angle $a=30°$, angle $b=90°$, and $\overline{bc} = d_4 = W\sin 60°$, it provides:

$$\frac{\overline{bc}}{\sin \hat{a}} = \frac{\overline{ac}}{\sin \hat{b}}$$

(12.16)

$\Rightarrow$

$$\frac{W\sin 60°}{\sin 30°} = \frac{\overline{ac}}{\sin 90°}$$

$\Rightarrow$

$$\overline{ac} = W\frac{\sin 60°}{\sin 30°}$$

(12.17)

Substituting Equation 12.17 into Equation 12.15 and then into Equation 12.14 provides:

$$d_2 = \left(L - W\frac{\sin 60°}{\sin 30°} \right)\cos 30°$$

(12.18)

Where:

d_2=remaining spacing of the total distance (D) for the last spacing (shown in the diagram)

L=length of parking space

W=width of parking space

30°=inclination angle with street curb

60° = angle c (in the diagram)=the angle that the width of the parking space makes with the line parallel to the street curb or the angle that the length of the parking space makes with the perpendicular to the street curb

$\Rightarrow$

$$d_2 = \left(18 - (8.5)\frac{\sin 60°}{\sin 30°} \right)\cos 30°$$

$\Rightarrow$

$$d_2 = 2.84 \text{ ft} (0.87 \text{ m})$$

The distance d_3 and d_4 are determined as shown in the following formulas, respectively:

$$d_3 = L\sin 30°$$

(12.19)

$$d_4 = W\sin 60°$$

(12.20)

Where:

d_3=perpendicular distance from street curb to right rear corner of parking space (shown in the diagram)

d_4=vertical component of the width of the car (shown in the diagram)

L=length of parking space

W=width of parking space

30°=inclination angle with street curb

$60° =$ angle c (in the diagram) = the angle that the width of the parking space makes with the line parallel to the street curb or the angle that the length of the parking space makes with the perpendicular to the street curb

$\Rightarrow$

$$d_3 = (18)\sin 30° = 9.00 \text{ ft}(2.74 \text{ m})$$

$$d_4 = (8.5)\sin 60° = 7.36 \text{ ft}(2.24 \text{ m})$$

$\Rightarrow$

$$d_3 + d_4 = 9.00 + 7.36 = 16.36 \text{ ft}(4.99 \text{ m})$$

The number of parking spaces according to the geometry in Figure 12.5 is equal to the total distance (D) minus the distance d_2 divided by the regular spacing d_1. Consequently, the number of parking spaces (bays) for the on-street parking when the parking spaces are inclined to the street curb at an inclination angle of 30° is computed using the formula in equation 12.7:

$$N = \frac{D - d_2}{d_1}$$

Or:

$$N = \frac{D - 2.84}{17.00} \tag{12.21}$$

Where:
$N =$ number of parking spaces
$D =$ distance along the street curb available for parking (ft)

$\Rightarrow$

$$N = \frac{1210 - 2.84}{17.00} \cong 71 \text{ parking spaces}$$

12.4   For an inclination angle of 60° with the street curb, derive the formulas used to calculate the number of parking spaces for on-street parking along a curb distance of 1210 ft (368.8 m).

Solution:

Figure 12.7 illustrates on-street parking spaces inclined to the street curb by an inclination angle of 60°. The derivation of the formulas to compute the distances shown in the diagram is explained in the procedure below.

Street Curb

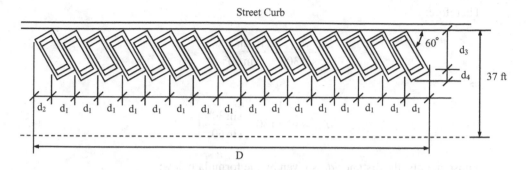

FIGURE 12.7 A street parking inclined at an angle of 60° for Problem 12.4.

The typical dimensions of the parking space are 18 ft×8.5 ft (5.5 m×2.6 m) as used in the previous two problems. The parking space is inclined to the street curb by an angle of 60°. Figure 12.8 is a magnified illustration for the distances d_1, d_2, d_3, and d_4.

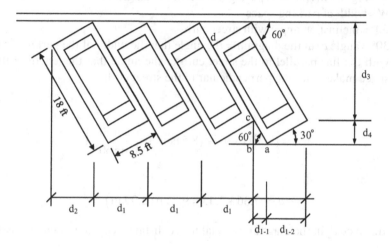

FIGURE 12.8 Schematic geometric diagram for the distances d_1, d_2, d_3, and d_4 for Problem 12.4.

The distance d_1 is composed of two components d_{1-1} and d_{1-2}. The two components are calculated, as shown below:

$$d_{1-2} = W \cos 30^\circ \qquad (12.22)$$

The triangle abc is considered and the sine law is used to write the following formula:

$$\frac{\overline{ab}}{\sin \hat{c}} = \frac{\overline{bc}}{\sin \hat{a}} \qquad (12.23)$$

But, $\overline{ab} = d_{1-1}$, angle $c = 30°$, and angle $a = 60°$, and

$$\overline{bc} - d_4 = W \sin 30^\circ \qquad (12.24)$$

Therefore:

$$\frac{d_{1-1}}{\sin 30°} = \frac{W \sin 30°}{\sin 60°}$$

$\Rightarrow$

$$d_{1-1} = W \sin 30° \left(\frac{\sin 30°}{\sin 60°} \right) \qquad (12.25)$$

Consequently, the distance d_1 is given by the formula below:

$$d_1 = d_{1-2} + d_{1-1} = W \cos 30° + W \sin 30° \left(\frac{\sin 30°}{\sin 60°} \right) \qquad (12.26)$$

Where:
    d_1 = regular horizontal spacing (shown in the diagram)
    W = width of parking space
    $60°$ = inclination angle with street curb
    $30°$ = angle c (in the diagram) = the angle that the width of the parking space makes
    with the line parallel to the street curb or the angle that the length of the parking
    space makes with the perpendicular to the street curb

$\Rightarrow$

$$d_1 = (8.5) \cos 30° + (8.5) \sin 30° \left(\frac{\sin 30°}{\sin 60°} \right)$$

$\Rightarrow$

$$d_1 = 7.36 + 2.45 = 9.81 \text{ ft} (2.99 \text{ m})$$

The distance d_2 in the diagram is equal to the distance $\overline{cd}$ multiplied by $\cos 60°$, therefore:

$$d_2 = \overline{cd} \cos 60° \qquad (12.27)$$

But:

$$\overline{cd} = L - \overline{ac} \qquad (12.28)$$

And using the sine law for the triangle abc and since angle $a = 60°$, angle $b = 90°$, and $\overline{bc} = d_4 = W \sin 30°$, it provides:

$$\frac{\overline{bc}}{\sin \hat{a}} = \frac{\overline{ac}}{\sin \hat{b}} \qquad (12.29)$$

$\Rightarrow$

$$\frac{W \sin 30°}{\sin 60°} = \frac{\overline{ac}}{\sin 90°}$$

$\Rightarrow$

$$\overline{ac} = W \frac{\sin 30°}{\sin 60°} \qquad (12.30)$$

Substituting Equation 12.30 into Equation 12.28 and then into Equation 12.27 provides:

$$d_2 = \left(L - W\frac{\sin 30°}{\sin 60°}\right)\cos 60° \tag{12.31}$$

Where:

d_2 = remaining spacing of the total distance (D) for the last spacing (shown in the diagram)

L = length of parking space

W = width of parking space

60° = inclination angle with street curb

30° = angle c (in the diagram) = the angle that the width of the parking space makes with the line parallel to the street curb or the angle that the length of the parking space makes with the perpendicular to the street curb

⇒

$$d_2 = \left(18 - (8.5)\frac{\sin 30°}{\sin 60°}\right)\cos 60°$$

⇒

$$d_2 = 6.55 \text{ ft} \left(2.00 \text{ m}\right)$$

Based on geometry, the distance d_3 and d_4 are determined as shown in the following formulas, respectively:

$$d_3 = L\sin 60° \tag{12.32}$$

$$d_4 = W\sin 30° \tag{12.33}$$

Where:

d_3 = perpendicular distance from street curb to right rear corner of parking space (shown in the diagram)

d_4 = vertical component of the width of the car (shown in the diagram)

L = length of parking space

W = width of parking space

60° = inclination angle with street curb

30° = angle c (in the diagram) = the angle that the width of the parking space makes with the line parallel to the street curb or the angle that the length of the parking space makes with the perpendicular to the street curb

⇒

$$d_3 = (18)\sin 60° = 15.59 \text{ ft} \left(4.75 \text{ m}\right)$$

$$d_4 = (8.5)\sin 30° = 4.25 \text{ ft} \left(1.30 \text{ m}\right)$$

⇒

$$d_3 + d_4 = 15.59 + 4.25 = 19.84 \text{ ft} \left(6.05 \text{ m}\right)$$

The number of parking spaces according to the geometry in the diagram above is equal to the total distance (D) minus the distance d_2 divided by the regular spacing d_1. Consequently, the number of parking spaces (bays) for the on-street parking when the parking spaces are inclined to the street curb at an inclination angle of 60° is computed using the formula in Equation 12.7:

$$N = \frac{D - d_2}{d_1}$$

Or:

$$N = \frac{D - 6.55}{9.81} \tag{12.34}$$

Where:
   N = number of parking spaces
   D = distance along the street curb available for parking (ft)

$\Rightarrow$

$$N = \frac{988 - 6.55}{9.81} \cong 100 \text{ parking spaces}$$

12.5   Use the same methodology in Problems 12.2, 12.3, and 12.4 to derive a formula for calculating the number of parking spaces for an on-street parking perpendicular to the street curb (the inclination angle is 90°) if the curb distance available for parking is 765 ft (233.2 m).

Solution:

Figure 12.9 illustrates on-street parking spaces that are perpendicular to the street curb (i.e., the parking spaces are inclined to the street curb by an inclination angle of 90°). The derivation of the formula in this case is simpler than the derivation of the formulas for parking spaces inclined to the street curb at 30°, 45°, or 60° (shown previously in earlier problems). The derivation of the formula is explained in detail in the following procedure.

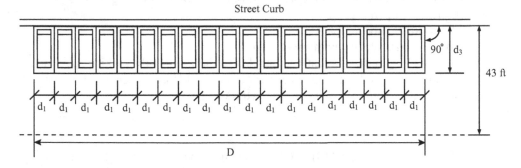

FIGURE 12.9   A street parking inclined at an angle of 90° for Problem 12.5.

The typical dimensions of the parking space are 18 ft×8.5 ft (5.5 m×2.6 m) as used in earlier problems. The parking space is perpendicular to the street curb; i.e., it is inclined to the street curb by an angle of 90°. Figure 12.10 is an enlarged illustration of the distances d_1 and d_3.

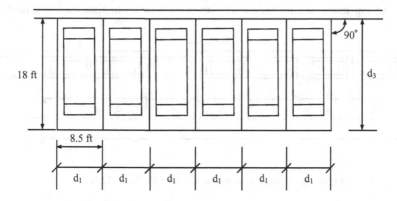

FIGURE 12.10 Schematic geometric diagram for the distances d_1, d_2, d_3, and d_4 for Problem 12.5.

The distance d_1 in this case is simply the width of the parking space (W). The distance d_2 does not exist (it is equal to zero in this case). Similarly, the distance d_4 does not exist (it is also equal to zero). The distance d_3 is equal to the length of the parking space (L) because it is perpendicular to the street curb. Therefore:

$$d_1 = W \tag{12.35}$$

$$d_3 = L \tag{12.36}$$

Consequently, the number of parking spaces according to the geometry in the diagram above is equal to the total distance (D) divided by the distance d_1, which is equal to the width of the parking space in this case (W). And hence, the number of parking spaces (bays) for the on-street parking when the parking spaces are perpendicular to the street curb is computed using the following formula:

$$N = \frac{D}{d_1} = \frac{D}{W} \tag{12.37}$$

Or:

$$N = \frac{D}{8.5} \tag{12.38}$$

Where:
  N = number of parking spaces
  d_1 = regular distance or spacing (in this case, it is equal to the width of the parking space)
  D = distance along the street curb available for parking (ft)
  W = width of parking space

$\Rightarrow$

$$N = \frac{765}{8.5} = 90 \text{ parking spaces}$$

12.6   For on-street parking that is parallel to the street curb, determine the number of parking
        spaces along a curb distance of 440 ft (134.1 m).

Solution:

Figure 12.11 illustrates on-street parking spaces that are parallel to the street curb (i.e.,
the parking spaces are inclined to the street curb by an inclination angle of 0°).

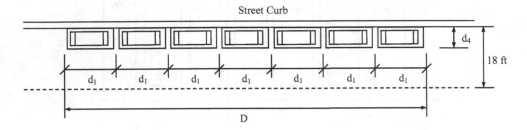

FIGURE 12.11   A street parking inclined at an angle of 0° for Problem 12.6.

In the case of on-street parallel parking, the typical dimensions of the parking space
are 18 ft×8 ft (5.5 m×2.4 m). The parking space is parallel to the street curb; i.e., it is
inclined to the street curb by an angle of 0°. Figure 12.12 is an enlarged illustration of
the distances d_1 and d_4.

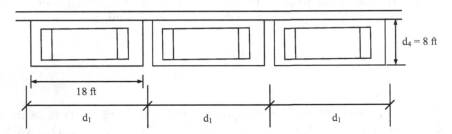

FIGURE 12.12   Schematic geometric diagram for the distances d_1, d_2, d_3, and d_4 for Problem 12.6.

The distance d_1 in this case is simply the length of the parking space (L) plus an additional spacing of 4 ft (2 ft from each of the two sides). The distance d_2 does not exist (it
is equal to zero in this case). In addition, the distance d_3 does not exist (it is also equal
to zero). The distance d_4 is equal to the width of the parking space (W) because it is
parallel to the street curb. Therefore:

$$d_1 = L + 4 \qquad\qquad\qquad (12.39)$$

$$d_4 = W \qquad\qquad\qquad (12.40)$$

Consequently, the number of parking spaces according to the geometry in the diagram
above is equal to the total distance (D) divided by the distance d_1. And hence, the
number of parking spaces (bays) for the on-street parking when the parking spaces are
parallel to the street curb is computed using the following formula:

$$N = \frac{D}{d_1} = \frac{D}{L + 4} \qquad\qquad\qquad (12.41)$$

Or:

$$N = \frac{D}{22} \qquad (12.42)$$

Where:

N = number of parking spaces
D = distance along the street curb available for parking (ft)
d_1 = regular distance or spacing (in this case, it is equal to the length of the parking space plus 4 ft)
L = length of parking space

$\Rightarrow$

$$N = \frac{440}{22} = 20 \text{ parking spaces}$$

It should be noted that the number of parking spaces decreases with the decrease in the inclination angle. Therefore, the highest number of parking spaces can be obtained when the inclination angle is 90°, i.e., when the parking space is perpendicular to the street curb. On the other hand, the lowest number of parking spaces can be obtained when the inclination angle is 0°, i.e., when the parking space is parallel to the street curb. However, in terms of safety, parallel parking is the safest style of parking compared to inclined parking due to the interference of parking with the traffic movement. Hence, the perpendicular parking (when the inclination angle is 90°) is considered the least safe on-street parking style, although it is the most efficient parking method.

12.7 A highway design engineer is required to design a parking lot for a new shopping center in the suburb of a city. If the available area for the parking lot is 600 ft × 700 ft (182.9 m × 213.4 m) as shown in Figure 12.13, determine the maximum number of parking spaces, taking into consideration that the parking lot should facilitate circulation of the traffic inside the parking area by providing a one-way traffic flow on each corridor (aisle) of the parking.

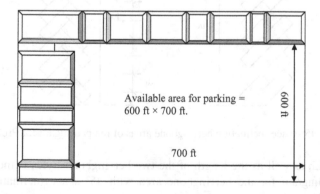

FIGURE 12.13   A parking lot for a new shopping center for Problem 12.7.

Solution:

A herringbone layout will be used with a 45°-space inclination and one-way traffic flow in parking aisles. The following procedure derives the formulas used to compute the number of parking spaces for a space inclination angle of 45°.

By considering a typical dimension for each parking space of 20 ft × 8 ft (6.1 m × 2.4 m) and the distances d_1 and d_2 shown in Figure 12.14, the number of parking spaces for each herringbone area is calculated according to the procedure described below. Figure 12.14 illustrates the general layout of a herringbone parking lot using any inclination angle and corresponding aisle (corridor) width.

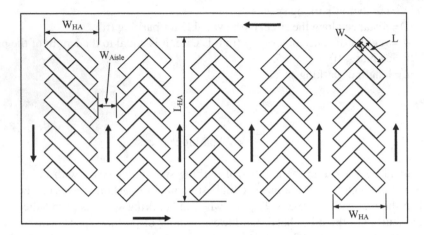

FIGURE 12.14   A general layout of a herringbone parking for Problem 12.7.

Two of the 45°-space inclination herringbone areas are plotted in Figure 12.15 to show all the distances involved in the derivation of the formula that will be used in the computation of the number of parking spaces in the entire parking lot.

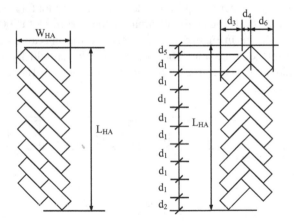

FIGURE 12.15   Two 45°-space inclination herringbone areas of the parking lot for Problem 12.7.

The width as well as the length of the two herringbone styles is the same. Another enlarged diagram for the herringbone area with 45° space inclination is shown in Figure 12.16.

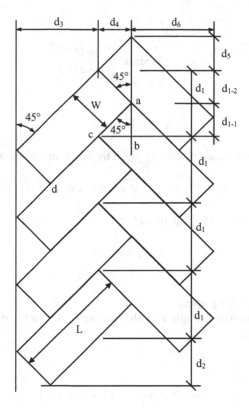

FIGURE 12.16 Enlarged geometric diagram for the herringbone area with 45° space inclination for Problem 12.7.

In this diagram (Figure 12.16), the length of the herringbone area (L_{HA}) is composed of several regular distances d_1, d_2, and d_5. The number of d_1 distances is dependent on the number of parking spaces in the row of the herringbone area. Whereas, the width of the herringbone area (W_{HA}) is composed of d_3, d_4, and d_6. The distances d_1, d_2, d_3, d_4, d_5, and d_6 are determined using geometry as described below:

The distance d_1 is composed of two components d_{1-1} and d_{1-2}. The two components are calculated, as shown below:

$$d_{1-2} = W \cos 45° \tag{12.43}$$

Considering the triangle abc in the diagram and using the sine law provides the following formula:

$$\frac{\overline{ab}}{\sin \hat{c}} = \frac{\overline{bc}}{\sin \hat{a}} \tag{12.44}$$

But, $\overline{ab} = d_{1-1}$, angle $c = 45°$, and angle $a = 45°$, and

$$\overline{bc} = d_4 = W \sin 45° \tag{12.45}$$

Therefore:

$$\frac{d_{1-1}}{\sin 45°} = \frac{W \sin 45°}{\sin 45°}$$

$\Rightarrow$

$$d_{1-1} = W \sin 45° \left(\frac{\sin 45°}{\sin 45°} \right)$$

Or simply:

$$d_{1-1} = W \sin 45° \qquad (12.46)$$

Consequently, the distance d_1 will be given by the formula below:

$$d_1 = d_{1-2} + d_{1-1} = W \cos 45° + W \sin 45° \qquad (12.47)$$

And since $\sin 45° = \cos 45°$, this provides:

$$d_1 = 2W \sin 45° \qquad (12.48)$$

Where:
   W = width of parking space
   45° = space inclination angle with the transverse line of the parking lot (vertical line in the diagram)

$\Rightarrow$

$$d_1 = 2(8)\sin 45° = 11.3 \text{ ft} (3.4 \text{ m})$$

The distance d_2 in the diagram is equal to the distance $\overline{cd}$ multiplied by $\cos 45°$, therefore:

$$d_2 = \overline{cd} \cos 45° \qquad (12.49)$$

But:

$$\overline{cd} = L - \overline{ac} \qquad (12.50)$$

And using the sine law for the triangle abc again and since angle $a = 45°$, angle $b = 90°$, and $\overline{bc} = d_4 = W \sin 45°$, it provides:

$$\frac{\overline{bc}}{\sin \hat{a}} = \frac{\overline{ac}}{\sin \hat{b}} \qquad (12.51)$$

$\Rightarrow$

$$\frac{W \sin 45°}{\sin 45°} = \frac{\overline{ac}}{\sin 90°}$$

$\Rightarrow$

$$\overline{ac} = W \frac{\sin 45°}{\sin 45°}$$

Or:

$$\overline{ac} = W \tag{12.52}$$

Substituting Equation 12.52 into Equation 12.50 and then into Equation 12.49 provides:

$$d_2 = (L - W)\cos 45° \tag{12.53}$$

Where:
  L = length of parking space
  W = width of parking space
  45° = space inclination angle with the transverse line of the parking lot (vertical line in the diagram)

$\Rightarrow$

$$d_2 = (20 - 8)\cos 45° = 8.5 \text{ ft} (2.6 \text{ m})$$

The distance d_3 and d_4 are determined as shown in the following formulas, respectively:

$$d_3 = L \sin 45° \tag{12.54}$$

$$d_4 = W \sin 45° \tag{12.55}$$

Where:
  L = length of parking space
  W = width of parking space
  45° = space inclination angle with the transverse line of the parking lot (vertical line in the diagram)

$\Rightarrow$

$$d_3 = (20)\sin 45° = 14.1 \text{ ft} (4.3 \text{ m})$$

$$d_4 = (8)\sin 45° = 5.7 \text{ ft} (1.7 \text{ m})$$

The distance d_5 is determined using geometry as in the expression below:

$$d_5 = W \cos 45° \tag{12.56}$$

$\Rightarrow$

$$d_5 = (8)\cos 45° = 5.7 \text{ ft} (1.7 \text{ m})$$

Similarly, the distance d_6 is determined using geometry as in the following expression:

$$d_6 = L \sin 45° \tag{12.57}$$

$\Rightarrow$

$$d_6 = (20)\sin 45° = 14.1 \text{ ft} (4.3 \text{ m})$$

Now all the distances d_1, d_2, d_3, d_4, d_5, and d_6 have been calculated, as shown above. The width of the herringbone area is determined as in the formula below:

$$W_{HA} = d_3 + d_4 + d_6 \tag{12.58}$$

Where:

W_{HA} = total width of one herringbone area as shown in Figure 12.16 above

$\Rightarrow$

$$W_{HA} = 14.1 + 5.7 + 14.1 = 33.9 \text{ ft} (10.3 \text{ m})$$

Based on the geometry in the above diagrams, the number of parking spaces in each row of one herringbone area can be determined as follows:

$$N_{1-1} = \frac{W_T - (d_2 + d_5) - 2W_{Aisle}}{d_1} \tag{12.59}$$

Where:

N_{1-1} = number of parking spaces in each row of one herringbone area
W_T = total width of the parking lot
W_{Aisle} = width of the aisle (corridor) used for traffic circulation (typical value of 11 ft is used in herringbone parking layout with 45°-space inclination angle)

$$N_{1-1} = \frac{600 - (8.5 + 5.7) - 2(11)}{11.3} = 49.9 \text{ parking spaces}$$

In this case, 49 parking spaces will be used for each row of the herringbone area. Therefore, the total number of parking spaces in the herringbone area will be:

$$N_1 = 2\left(\frac{W_T - (d_2 + d_5) - 2W_{Aisle}}{d_1}\right) \tag{12.60}$$

Where:

N_1 = number of parking spaces in one herringbone area (two rows)
W_T = total width of the parking lot
W_{Aisle} = width of the aisle (corridor) used for traffic circulation (typical value of 11 ft is used in herringbone parking layout with 45°-space inclination angle)

$\Rightarrow$

$$N_1 = 2(49) = 98 \text{ parking spaces}$$

The number of herringbone areas in the parking lot is also determined by geometry. The herringbone parking layout is composed of several herringbone areas; each one has one aisle (corridor) plus an additional aisle (corridor) at the end. Therefore, the width of one unit (one herringbone area plus one aisle) is equal to:

$$W_{HA+Aisle} = W_{HA} + W_{Aisle} \tag{12.61}$$

⇒

$$W_{\text{HA+Aisle}} = 33.9 + 11 = 44.9 \text{ ft} (13.7 \text{ m})$$

Therefore, the number of herringbone areas in the parking lot is equal to the total length of the parking lot minus the width of one aisle (the additional aisle) divided by the width of one unit (one herringbone area plus one aisle). In other words, the following expression can be used:

$$N_2 = \frac{L_T - W_{\text{Aisle}}}{W_{\text{HA+Aisle}}} \qquad (12.62)$$

Where:

N_2 = number of herringbone areas in the parking lot

L_T = total length of the parking lot

W_{Aisle} = width of the aisle (corridor) used for traffic circulation (typical value of 11 ft is used in herringbone parking layout with 45°-space inclination angle)

$W_{\text{HA+Aisle}}$ = width of one unit of one herringbone area with one aisle (corridor)

⇒

$$N_2 = \frac{700 - 11}{44.9} = 15.3 \text{ areas}$$

Fifteen herringbone areas will be used in this case. Consequently, the total number of parking spaces in the parking lot will be equal to the number of parking spaces in each herringbone area multiplied by the number of herringbone areas in the parking lot as shown in the following expression:

$$N = N_1 (N_2) \qquad (12.63)$$

Where:

N = total number of parking spaces in the parking lot

N_1 = number of parking spaces in each herringbone area

N_2 = number of herringbone areas

⇒

$$\cdot N = 98 (15) = 1470 \text{ parking spaces}$$

Then the parking area layout will look like the one shown in Figure 12.17.

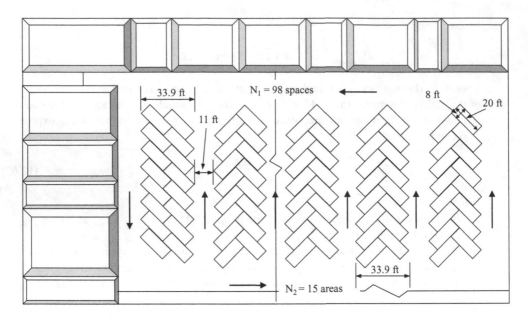

FIGURE 12.17 The final layout of the 45°-space inclination herringbone parking lot with the required dimensions for Problem 12.7.

12.8   Design a parking layout using 60° inclination angle and a 2.0-ft wide median between each successive parking areas with one-way traffic circulation on aisles for a 290 ft × 520 ft (88.4 m × 158.5 m) parking area of a small shopping center such that the entrance is from one side and the exit is from the other side as shown in Figure 12.18.

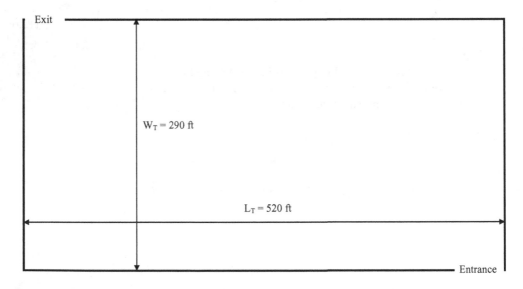

FIGURE 12.18   A 290 ft × 520 ft parking area for a small shopping center for Problem 12.8.

Solution:

A value of 18 ft for the width of aisles used for 60°-inclination parking is typically used. Typical dimensions of 20 ft×8 ft for the parking space will be used. A layout like this illustrated in Figure 12.19 will be used for this parking lot.

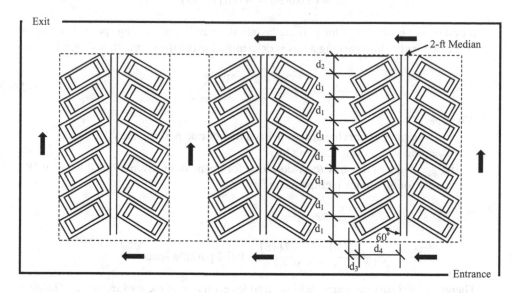

FIGURE 12.19   Layout of the 60°-space inclination herringbone parking lot for Problem 12.8.

The formulas derived earlier in previous problems to determine the distances d_1, d_2, d_3, and d_4 and used in Problem 12.4 for 60°-inclination parking will be used herein. Therefore:

$$d_1 = W\cos 30° + W\sin 30°\left(\frac{\sin 30°}{\sin 60°}\right)$$

$\Rightarrow$

$$d_1 = (8)\cos 30° + (8)\sin 30°\left(\frac{\sin 30°}{\sin 60°}\right) = 2.3 \text{ ft}(0.7 \text{ m})$$

$$d_2 = \left(L - W\frac{\sin 30°}{\sin 60°}\right)\cos 60°$$

$\Rightarrow$

$$d_2 = \left(20 - (8)\frac{\sin 30°}{\sin 60°}\right)\cos 60° = 7.7 \text{ ft}(2.3 \text{ m})$$

$$d_3 = L\sin 60°$$

$\Rightarrow$

$$d_3 = (20)\sin 60° = 17.3 \text{ ft}(5.3 \text{ m})$$

$$d_4 = W \sin 30°$$

$$\Rightarrow$$

$$d_4 = (8)\sin 30° = 4.0 \text{ ft} (1.2 \text{ m})$$

Based on the geometry of the parking layout, the number of parking spaces in each row of the 60°-inclination parking areas is determined using the expression shown below:

$$N_{1-1} = \frac{W_T - d_2 - 2W_{\text{Aisle}}}{d_1} \tag{12.64}$$

Where:
N_{1-1} = number of parking spaces in each row of one parking area
W_T = total width of the parking lot
W_{Aisle} = width of the aisle (corridor) used for traffic circulation (typical value of 18 ft is used for 60°-space inclination angle)

$$\Rightarrow$$

$$N_{1-1} = \frac{290 - 7.7 - 2(18)}{2.3} = 107.1 \text{ parking spaces}$$

Therefore, 107 parking spaces will be used for each row of the parking areas. The number of spaces per each parking area of two rows will be determined as follows:

$$N_1 = 2\left(\frac{W_T - d_2 - 2W_{\text{Aisle}}}{d_1}\right) \tag{12.65}$$

Where:
N_1 = number of parking spaces in one parking area (two rows)
W_T = total width of the parking lot
W_{Aisle} = width of the aisle (corridor) used for traffic circulation (typical value of 18 ft is used for 60°-space inclination angle)

$$\Rightarrow$$

$$N_1 = 2(107) = 214 \text{ parking spaces}$$

Using geometry, the number of parking areas in the parking lot will also be determined as in the expression shown below:

$$N_2 = \frac{L_T - W_{\text{Aisle}}}{W_{\text{PA+Aisle}}} \tag{12.66}$$

Where:
N_2 = number of parking areas in the parking lot
L_T = total length of the parking lot
W_{Aisle} = width of the aisle (corridor) used for traffic circulation (typical value of 18 ft is used for 60°-space inclination angle)
$W_{\text{PA + Aisle}}$ = width of one unit of one parking area with one aisle (corridor)

But:

$$W_{\text{PA+Aisle}} = W_{\text{PA}} + W_{\text{Aisle}} \qquad (12.67)$$

And:

$$W_{\text{PA}} = 2(d_3 + d_4) + W_{\text{median}} \qquad (12.68)$$

Where:
   W_{PA} = total width of one parking area as shown in the diagram above

$\Rightarrow$

$$W_{\text{PA}} = 2(17.3 + 4.0) + 2.0 = 44.6 \text{ ft} (13.6 \text{ m})$$

$\Rightarrow$

$$W_{\text{PA+Aisle}} = 44.6 + 18 = 62.6 \text{ ft} (19.1 \text{ m})$$

$\Rightarrow$

$$N_2 = \frac{520 - 18}{62.6} = 8.02 \text{ areas}$$

Eight parking areas will be used in this case. Therefore, the total number of parking spaces in the parking layout will be determined using the expression shown below:

$$N = N_1(N_2) \qquad (12.69)$$

Where:
   N = total number of parking spaces in the parking lot
   N_1 = number of parking spaces in each parking area
   N_2 = number of parking areas

$\Rightarrow$

$$N = 214(8) = 1712 \text{ parking spaces}$$

Multiple-Choice Questions and Answers for Chapter 8

1. The predicted hourly volume that is used in design which is normally a percentage of the annual daily traffic (ADT) is called:
   a. Peak hourly volume
   b. Peak hour factor (PHF)
   c. **Design hourly volume (DHV)**
   d. Annual average daily traffic (AADT)
   e. Maximum flow rate

2. The speed that is selected based on the functional classification of the highway, the topography of the area, and the land use of the adjacent area is called:
   a. Space mean speed
   b. **Design speed**
   c. Speed limit
   d. Mean free speed
   e. Time mean speed

3. The largest vehicle type that is most likely to use the highway with significant frequency and selected to represent all vehicles using the highway in the design process is called:
   a. Equivalent single-axle load (ESAL)
   b. Truck Factor
   c. Equivalent vehicle
   d. **Design vehicle**
   e. Frequent vehicle

4. One of the following elements is <u>not</u> among the main elements of a highway cross-section:
   a. Travel lanes
   b. Shoulders
   c. Medians
   d. **Adjacent mountains**
   e. None of the above

5. The width of travel lanes typically ranges from:
   a. **9 ft to 12 ft**
   b. 10 ft to 30 ft
   c. 4 ft to 80 ft
   d. 6 ft to 8 ft
   e. 2 ft to 12 ft

6. The width of roadway shoulders typically ranges from:
   a. 9 ft to 12 ft
   b. 10 ft to 30 ft
   c. 4 ft to 80 ft
   d. 6 ft to 8 ft
   e. **2 ft to 12 ft**

7. Marginal elements of a highway cross-section include:
   a. Curbs
   b. Gutters
   c. Guard rails
   d. Sidewalks
   e. **All of the above**
8. One of the following is <u>not</u> considered among the functions of a highway median:
   a. It serves as a stopping zone in emergency conditions
   b. It can be used for temporary lanes during maintenance jobs
   c. It can be used by pedestrians as a refuge
   d. It divides traffic in opposite directions
   e. **It can be used for motorcycles**
9. Flush, depressed, and raised are the three main types of:
   a. **Medians**
   b. Roadside barriers
   c. Curbs
   d. Gutters
   e. Side slopes
10. Median widths may vary in the range of:
    a. 9 ft to 12 ft
    b. 10 ft to 30 ft
    c. **4 ft to 80 ft**
    d. 6 ft to 8 ft
    e. 2 ft to 12 ft
11. Roadside barriers are used on highways to:
    a. Protect vehicles from obstacles and roadside slopes
    b. Protect pedestrians from traffic
    c. Shield property from traffic stream
    d. **a, b, and c**
    e. a and b
12. A structure in the longitudinal direction of (parallel to) traffic used to prevent an unruly vehicle from crossing the area separating the two opposite traffic directions in a divided highway is:
    a. Roadside barrier
    b. Curb
    c. **Median barrier**
    d. Gutter
    e. Side slope
13. Raised structures made of Portland cement concrete (PCC) or asphalt concrete that are used mainly in urban roadways to delimit road pavement edges and pedestrian sidewalks are called:
    a. Roadside barriers
    b. **Curbs**
    c. Median barriers
    d. Gutters
    e. Roadside slopes
14. Ditches located on the side of the roadway pavement near the curb and used for drainage purposes are called:
    a. Roadside barriers
    b. Curbs

    c. Median barriers

    d. **Gutters**

    e. Roadside slopes

15. The typical range for the height of roadway curbs is:

    a. 9 ft to 12 ft

    b. 10 ft to 30 ft

    c. 4 ft to 80 ft

    d. **6 ft to 8 ft**

    e. 2 ft to 12 ft

16. The main function of highway guard rails is to:

    a. Separate opposing traffic directions

    b. Protect pedestrians from traffic

    c. Provide a stopping area for vehicles in case of emergency

    d. **Prevent vehicles from leaving the highway**

    e. All of the above

17. Sloping and vertical are the two typical classes of the following roadway element:

    a. Roadside barrier

    b. **Curb**

    c. Median barrier

    d. Gutter

    e. Sidewalk

18. Cross (transverse) slopes for highway pavements, sloping from the middle downward to both sides, are used mainly to:

    a. **Remove surface water from the highway pavement**

    b. Enhance driving

    c. Reduce centrifugal force

    d. Improve the overall appearance of the highway

    e. Reduce speed and improve safety

19. The following roadway marginal element is constructed on fills to provide earthwork stability and serve as a rescue area for out-of-control vehicles:

    a. Roadside barriers

    b. Curbs

    c. **Side slopes**

    d. Median barriers

    e. Gutters

20. One of the following is *not* among the types of vertical curves:

    a. Crest curve

    b. Sag curve

    c. Spiral curve

    d. Compound curve

    e. **c and d**

21. The criterion used for the design of a crest vertical curve is:

    a. **Minimum stopping sight distance (SSD)**

    b. SSD provided by headlights

    c. Driving comfort

    d. Overall appearance of the curve

    e. Drainage at the lowest point of the curve

22. The design of the length of a sag vertical curve is affected by the following criterion:

    a. SSD provided by headlights

    b. Driving comfort

    c. Overall appearance of the curve

    d. Drainage at the lowest point of the curve

    e. **All of the above**

23. The coefficient K in the formula $L = KA$ which is used to design vertical curves is best defined as:

    a. The minimum length of the vertical curve

    b. The algebraic difference in grades

    c. **The length of the vertical curve per percent change in A**

    d. The minimum length based on comfort

    e. The length of the vertical curve per percent grade

24. The coefficient K used in the design of vertical curves is:

    a. A function of area type

    b. A function of roadway classification

    c. The length of the vertical curve per percent change in the tangent grade, G_1

    d. **A function of design speed**

    e. A function of design vehicle

25. The function that describes the vertical distance (coordinate) from the vertical curve tangent with respect to the horizontal distance from the beginning point of the vertical curve is:

    a. Exponential function

    b. **Parabolic function**

    c. Power function

    d. Cubical function

    e. Polynomial of the fourth degree

26. One of the following statements is <u>not</u> correct:

    a. **The length of the vertical curve is the length along the curve**

    b. The two tangents of the vertical curve are equal

    c. The vertical curve follows a parabolic function

    d. The rate of change of the vertical curve slope is constant

    e. The K factor is defined as the length of the vertical curve per percent change in A

27. One of the following is *not* among the types of horizontal curves:

    a. **Crest curve**

    b. Simple curve

    c. Spiral curve

    d. Compound curve

    e. Reversed curve

28. The minimum radius of a simple horizontal curve is directly proportional to one of the following factors:

    a. Superelevation

    b. **Design speed**

    c. Coefficient of side friction

    d. a and b

    e. b and c

29. The central angle that subtends at the center of an arc (or chord) length of 100 ft refers to:

    a. Deflection angle

    b. Intersection angle

    c. Turning angle

    d. **Degree of the curve**

    e. Curve angle

30. A simple horizontal curve with radius R and intersection angle (Δ) will have the highest middle ordinate (M) if Δ has the following value:
    a. 30°
    b. 45°
    c. 60°
    d. 75°
    e. **90°**

31. The angles between the tangent at the beginning point of the curve and the chords between successive stations on the curve and the beginning point are called:
    a. **Deflection angles**
    b. Intersection angles
    c. Turning angles
    d. Degree of the curve
    e. Curve angles

32. The sight distance of horizontal curves is measured from:
    a. The centerline of the outside lane
    b. The centerline of the roadway
    c. The chord of the curve
    d. The point of intersection of the two curve tangents
    e. **The centerline of the inside lane**

33. The following design vehicle requires the largest road widening:
    a. **WB-65**
    b. WB-50
    c. WB-40
    d. P
    e. SU

34. A compound curve consists of:
    a. A straight segment and a reversed curve
    b. Two or more spiral curves
    c. **Two or more simple curves**
    d. Two or more sag curves
    e. A sag curve and a crest curve

35. Reverse curves are rarely recommended to be used in highways due to:
    a. Driving comfort issues
    b. **Safety issues**
    c. Design issues
    d. Lower stopping sight distances available on these curves
    e. Lower radii used for these curves

36. Transition curves are introduced between two successive circular curves or between tangents and circular curves to:
    a. Increase speed
    b. Decrease speed
    c. Improve the passing sight distance
    d. **Gradually decrease or increase the radial force as the vehicles enters or leaves the circular curve**
    e. Increase the centrifugal force

37. One of the following is <u>not</u> among the advantages of transition curves:
    a. The appearance of the highway is enhanced
    b. A convenient arrangement for superelevation runoff is provided

    c. A smooth and easy-to-follow path for drivers is provided

    d. A transition in width for the widening of the traveled way section is facilitated

    e. **The driving speed is increased**

38. One of the following is <u>not</u> among the factors that control the maximum rate of superelevation used on highways:

    a. Climatic conditions

    b. **Highway geometric conditions**

    c. Type of terrain

    d. Type of area

    e. Frequency of slow-moving vehicles

39. One of the following statements is <u>not</u> true:

    a. **Superelevation rates above 8% should only be used for roads in areas with snow and ice**

    b. The highest superelevation rate commonly used on highways is 10%

    c. Superelevation may be ignored on low-speed urban roads

    d. A superelevation rate of 12% should not be exceeded

    e. A superelevation rate of 4 to 6% is appropriate for urban roads in areas with little or no constraints

40. The length of roadway needed to accomplish a change in the cross slope of the outside lane from zero to full superelevation or from full superelevation to zero is called:

    a. Transition length

    b. Circular length

    c. **Superelevation runoff**

    d. Tangent runout

    e. Tangent length

41. The length of roadway needed to accomplish a change in the cross slope of the outside lane from normal (2%) to zero or from zero to normal is called:

    a. Transition length

    b. Circular length

    c. Superelevation runoff

    d. **Tangent runout**

    e. Tangent length

42. When spiral curves are used in a transition design, it is recommended that the superelevation runoff is achieved over the length of the:

    a. Circular curve

    b. Tangent

    c. Tangent runout

    d. **Spiral curve**

    e. None of the above

43. When transition curves are not used, the common practice for attaining superelevation runoff is:

    a. Completely on a tangent

    b. Completely on a circular curve

    c. **At 67% on a tangent and 33% on a curve**

    d. At 33% on a tangent and 67% on a curve

    e. At 50% on a tangent and 50% on a curve

44. An extra lane in the upgrade direction of the roadway, used for slow-moving heavy vehicles whose speeds are significantly reduced by the grade, is called:

    a. Auxiliary Lane

    b. Storage lane

    c. Exclusive lane

    d. **Climbing lane**
    e. Emergency lane

45. A ramp that is provided on the downgrade of a highway for use by a truck that cannot slow down due to loss of control is called:
    a. Auxiliary ramp
    b. Safety ramp
    c. Exclusive ramp
    d. Downgrade ramp
    e. **Emergency escape ramp**

46. The geometric design of parking facilities considers the following aspect(s):
    a. Providing a safe arrangement of parking bays
    b. Providing safe, easy, and convenient access to parking areas
    c. No restrictions on the traffic flow of the adjacent road lanes
    d. a and b
    e. **a, b, and c**

47. The highest number of parking bays (spots) for an on-street parking facility can be obtained with an inclination angle of:
    a. **90° (perpendicular)**
    b. 60°
    c. 45°
    d. 30°
    e. 0° (parallel)

48. Interference with traffic movement in adjacent road lanes and accident rates are expected to be lower for an on-street parking facility with an inclination angle of:
    a. 90° (perpendicular)
    b. 60°
    c. 45°
    d. 30°
    e. **0° (parallel)**

49. The design of off-street parking facilities takes into account the following consideration(s):
    a. To obtain as many spaces as possible
    b. Parking a vehicle should only involve one distinct movement without reversing
    c. Traffic circulation should be facilitated in the aisles of the parking layout
    d. **All of the above**
    e. a and b

50. Access ramps connect the different levels of the parking garages; the required minimum lane widths for the curved ramps and the straight ramps, respectively, are:
    a. 12 ft for both
    b. **16 ft and 9 ft**
    c. 14 ft and 12 ft
    d. 14 ft and 10 ft
    e. 14 ft for both

References

American Association of State Highway and Transportation Officials (AASHTO), *A Policy on Geometric Design of Highways and Streets*, American Association of State Highway and Transportation Officials (AASHTO), Washington, DC, 2001.

Barry Kavanagh and Tom Mastin, *Surveying: Principles and Applications, International Edition*, 9th Edition, Pearson Education, Inc., New York, 2014.

Microsoft Excel, Office 365, 2016.

Microsoft Excel Solver Tool, Office 365, 2016.

Nicholas J. Garber and Lester A. Hoel, *Traffic and Highway Engineering*, 4th Edition, Cengage Learning, Inc., 2009.

Steven C. Chapra and Raymond P. Canale, *Numerical Methods for Engineers*, McGraw Hill Education, New York, 2015.

The Oahu Metropolitan Planning Organization. https://www.oahumpo.org/resources/publications-and-reports/.

William Martin, Nancy A. McGuckin, and Barton-Aschman, National Highway Cooperative Research Program (NCHRP) Report 365, Travel Estimation Techniques for Urban Planning, Transportation Research Board, National Research Council, National Academy Press, Washington, DC, 1998.

Index

Printed in the United States
by Baker & Taylor Publisher Service

Printed in the United States
by Baker & Taylor Publisher Services